HUMAN ANATOMY AND PHYSIOLOGY SERIES

BONE REGENERATION: GROWTH FACTORS, AUGMENTATION PROCEDURES AND TISSUE ENGINEERING APPLICATIONS

HUMAN ANATOMY AND PHYSIOLOGY SERIES

Handbook of Granulocytes: Classification, Toxic Materials Produced and Pathology
Reuben Hägg and Soren Kohlund (Editors)
2009. ISBN: 978-1-60741-582-4

Skin Anatomy and Physiology Research Developments
Leon F. Bukowskiy (Editor)
2009. ISBN: 978-1-60741-469-8

Bone Regeneration: Growth Factors, Augmentation Procedures and Tissue Engineering Applications
Vincent Legard and Rémi Schluter (Editors)
2010. ISBN: 978-1-60876-621-5

HUMAN ANATOMY AND PHYSIOLOGY SERIES

BONE REGENERATION: GROWTH FACTORS, AUGMENTATION PROCEDURES AND TISSUE ENGINEERING APPLICATIONS

VINCENT LEGARD
AND
RÉMI SCHLUTER
EDITORS

Nova Science Publishers, Inc.
New York

For permission to use material from this book please contact us:
Telephone 631-231-7269; Fax 631-231-8175
Web Site: http://www.novapublishers.com

Library of Congress Cataloging-in-Publication Data
Available upon request.
ISBN: 978-1-60876-621-5

Published by Nova Science Publishers, Inc. ✢ New York

Contents

Preface

Bone tissue plays a vital role in the human system for its biological and mechanical activities, resulting from its regenerative capacity and permanent remodeling process. Bone damage, either due to pathologies or traumas, is a very common occurrence and represents a major problem in orthopedics. This book briefly summarizes recent literature concerning modification and applications of osteoinductive materials for reconstructive surgery. Several recent developments of biocomposites containing silica nanoparticles or calcium sulphate intended for bone regeneration are also reported. In addition, new therapies are identified for bone regeneration in an injured or diseased state, as well as for bone loss in aging. Also discussed in this book is normal bone development, mesenchymal stem cells in osteogenesis, and essential elements that regulate osteogenesis at the extracellular, intracellular and nuclear levels.

Chapter I - Osteoinductive materials are keenly sought in reconstructive surgery. Polymer-based composites are ideal candidates, as they couple the ease of processing of polymers (by melt or by injection and hardening) with the mechanical reinforcement of fillers, once interfacial adhesion is assured. Moreover, an appropriate choice of the filler may enhance the bone regeneration activity, through cell adhesion or osteoinduction. In this respect, hydroxyapatite and calcium sulphate are good candidates. As organic binder, a wide choice of different polymers is possible: biocompatible and biodegradable natural polymers such as polysaccharides and, in more recent years, biocompatible and biodegradable synthetic polyesters are normally employed, with processing methodologies which depend on the intrinsic properties of the polymers (paste injection or injection molding).

In the present chapter, after briefly summarizing recent literature concerning modification and applications of these materials, several recent developments of biocomposites containing silica nanoparticles or calcium sulphate intended for bone regeneration are reported. The composites are characterized with respect to their chemical-physical and mechanical properties. Their biocompatibility and capacity to induce the osteoblastic phenotype in human bone marrow mesenchymal stem cells have been assessed. The authors will focus on two particular systems based on either natural or synthetic biopolymers with different biofillers: alginate/chitosan blends with calcium sulphate and poly(ε-caprolactone) with silica nanoparticles.

Chapter II - Bone formation is a complex and coordinated series of events resulting in a macro-structure that has the ability to continually repair and remodel throughout the life of a human. The human skeleton not only provides structural support but is also a complex network involving multiple organ systems such as the hematopoietic and endocrine systems. Integral in the steps of bone formation is the highly regulated process of osteoblastic differentiation. This process is critical for the normal development of bone, physiologic remodeling, and repair in an injured state. Over the past decade, the authors have begun to dissect the important regulatory elements of this cascade of events for osteogenesis. These key elements include morphogens, signal transduction proteins, and transcription factors. Aberrations in key regulatory genes result in a number of human pathologic musculoskeletal conditions. This chapter will discuss normal bone development, mesenchymal stem cells in osteogenesis, and essential elements that regulate osteogenesis at the extracellular, intracellular, and nuclear levels. Focus will be directed to Bone Morphogenetic Protein (BMP) signaling in osteogenesis. Like TGFβ, BMP signaling is through the Smad pathway and regulates a number of downstream effecter genes that leads to bone formation. In the past five to ten years, the authors have begun to use recombinant human BMPs in the clinical setting to augment bone healing in craniofacial surgeries, spinal fusions, and fracture care. As the authors further understand the normal processes of osteogenesis and bone formation, the authors can potentially identify new therapies for bone regeneration in an injured or diseased state, as well as for bone loss in aging.

Chapter III - Bone tissue plays a vital role in human system for its biological and mechanical activities, resulting from its regenerative capacity and permanent remodeling process (Gautier et al., 2005). Bone damage, either due to pathologies or traumas, is a very common occurrence and represents a major problem in orthopedics (Wiesmann et al, 2004). Most bone lesions, such as fractures or small size defects, heal well with conventional therapy due to the high regeneration potential of bone (Wiesmann et al, 2004). However, in large traumatic or post-surgical bone defects, fractures result in non-unions, others exhibit delayed healing, indicating that, in some cases, bone cannot heal through its normal wound healing response (Zimmermann & Wentzensen, 2007). In this context, there is a critical need for technologies that are able to enhance bone healing (Gautier et al., 2005). The development of effective therapies for bone regeneration is one of the most clinically important long-term goals of research in the mineralized tissue field. Gene therapy, tissue engineering and biomaterials have emerged as promising approaches for bone repair and reconstruction (Humatcher et al, 2000).

Chapter IV - There are different approaches for treatment of bone fractures; one of them is bone tissue engineering. There are great varieties of clinical cases when the bone grafts are needed. All of them implies to fill a relevant big gap of bone with bone tissue or to supply the bone with a good bioconductive material. For this purpose synthetic materials are used. These materials are bioactive — they produce a specific reaction on the boundary of the bone and material — they also provides osteoconductive and osteoinductive properties of the transplant. These material serve not only for mechanical stability of bone, but also as cells and/or factors delivery system for bone growth. There are different approaches in creation of bone tissue transplant. All of them means to use different scaffolds (bioceramics, bioglass), with different properties, different cells (differentiated or stem cells), and different methods of stem cell placement and retention in those materials — it implies different number of cells, different methods of loading those cells, usage of some factors that helps for bone formation and blood vessels invasion. Here the authors would like to make a review of different approaches in bone tissue engineering and to tell what was done in the author's laboratory in this area.

Chapter V - Autologous activated platelet rich plasma (PRP) is used to accelerate bone repair as a possible alternative to recombinant growth factors (GF), through the osteogenic GF released from α-granules. The advantages of PRP, which is usually prepared by adding thrombin and calcium to a platelet concentrate, lie in its mimicking the GF effects of the physiological bone healing and regenerative processes, in addition to a relatively simple and low cost technique. Moreover, if autologous PRP is used, immunological reactions are avoided. However, the efficacy of PRP in orthopaedic applications is still controversial because of the unpredictability of GF concentration in different platelet preparations, due both to individual factors and to different techniques.

Moreover, the effects of PRP are not dependant on a single GF, but on the combination of the effects of some molecules released by platelets, each of them acting on the different cells of bone microenvironment. The cells, in consequence of the stimulation with GF, in their turn influence other cells with various effects. In *in vitro* systems, PRP stimulated osteogenic differentiation of bone marrow stromal cells, while it inhibited osteoclastogenesis from the peripheral blood precursors, particularly complete osteoclast differentiation and activation. Moreover, PRP favored endothelial cell proliferation, tubulogenesis and expression of pro-osteogenic functions. PRP induced also the expression of ICAM-1, which is an adhesion molecule for monocytes. In experimental animals and in clinical application, the PRP efficiency to improve bone regeneration was increased by the combination with bone allografts, acting as scaffolds, and by the addition of bone marrow stromal cells.

Chapter VI - Octacalcium phosphate ($Ca_8H_2(PO_4)_6{\cdot}5H_2O$; OCP) is a calcium phosphate and has been suggested to be a precursor of biological apatite crystals in bones and teeth. The importance of synthetic OCP as a bone substitute material was recently recognized in the field of tissue engineering for bone regeneration. The author's previous studies elucidated the below-mentioned phenomena by using a rat calvarial defect model. OCP irreversibly converts to the apatitic phase, and the process of OCP-apatite conversion is involved in bone regeneration. Implanted OCP enhances bone regeneration more than hydroxyapatite (HA) and β-tricalcium phosphate (β-TCP), which have already been used clinically as a scaffold for bone regeneration. OCP can be used as an effective scaffold for growth factors, such as bone morphogenetic protein-2 (BMP-2) and transforming growth factor-β_1 (TGF-β_1), to enhance bone regeneration. Although OCP possesses many desirable properties, OCP cannot be molded by a sintering process, unlike bioceramics, such as HA or β-TCP. To improve the performance of bone regeneration and handling performance, a composite synthetic bone substitute constructed of synthetic OCP and porcine atelocollagen sponge (OCP/Col) has been developed. The synergistic effects of OCP/Col on bone regeneration are the result of bone nucleation by OCP and cell infiltration by collagen, and OCP/Col shows stable bone regeneration. OCP/Col significantly enhances bone regeneration more than OCP *per se*, β-TCP collagen composite, and HA collagen composite if implanted into a critical-sized defect model in a rat calvarium. During optimization of OCP/Col, it was confirmed that bone regeneration by OCP/Col was influenced by dehydrothermal treatment in the fabrication of OCP/Col or the composition of OCP and collagen. As OCP/Col is applicable without cell transplantation and exogenous osteogenic cytokines, cost-effective bone regenerative therapy is expected in the future.

Chapter VII - Bone tissue engineering requires a suitable osteoconductive and –inductive matrix and, dependent on the severity of the defect, additional sources of osteogenic cells. Traditionally, the osteoconductive matrix is a prefabricated three-dimensional scaffold (3D) composed of degradable polymers and/or ceramics. However, the increasing popularity of arthroscopic procedures and the requirement to bridge irregular bone defects resulted in great interest in fixation materials that are injectable, *in situ* forming and biodegradable.

To restore damaged bone defects additional bone-forming cells are needed. *In situ* forming materials (e.g. moldable calcium phosphate cements, injectable polymers hardening by redox- or photocrosslinking) capable of filling irregular bone defects are mostly inconvenient to mix directly with cells. The use of cell delivery systems which can be colonized by cells in which they are protected during the polymerization process of the biomaterial could be advantageous. The use of macroporous gelatin-based microcarriers in tissue engineering offers several advantages: (1) cells can be expanded in vitro, (2) biodegradable, (3) cells are in a "natural" 3D-environment and (4) cells within the interior of the microcarrier are in a protected environment.

In the author's study, bone marrow cells were cultured on *in situ* photopolymerizable scaffolds based on D, L-lactide, ε-caprolactone and trimethylene carbonate. Bone formation was demonstrated at the polymer interface. Cellular infiltration was possible but was limited to the edge of the polymer scaffold.

In the second part of this study, bone marrow cells were seeded on macroporous gelatin microcarriers (CultiSpher S) (diameter 130-380 μm). The cells proliferated and differentiated well on the microcarriers and were completely colonized after 14 days in culture. The polymer paste allowed to mix the cell seeded microcarriers and the porogen (e.g. gelatin) properly. After the photopolymerization process, microcarriers and gelatin were evenly distributed throughout the scaffold. After leaching out the gelatin, a porous scaffold was obtained. The microcarriers remained in the scaffold even after 7 days which demonstrates that they were well entrapped in the polymer. Dependent on the composite type, the encapsulated cells survived the handling of the viscous paste and the photopolymerization proces.

These cell seeded microcarriers combined with different combinations of polymer and alfa-tricalcium phosphate were tested in a unicortical tibial defect model in goats. The results demonstrated cell survival and proliferation in the polymer-substituted bone defects. Hence cell seeded microcarriers combined with *in situ* photopolymerizable materials are promising for bone regeneration.

Chapter VIII - Bone regeneration in pre-clinical and clinical contexts requires three key components: an osteogenic soluble signal; an insoluble signal or substratum which delivers the signal and acts as a scaffold for the initiation of bone formation; and host cells capable of differentiation into bone cells in response to the osteogenic molecular signal. The '*bone induction principle*' has identified important rules for tissue engineering and regenerative medicine at large, i.e. soluble osteogenic molecular signals, insoluble signals or substrata including biomimetic bioactive matrices and responding stem cells. The osteogenic soluble molecular signals of the transforming growth factor-β (TGF-β) superfamily, i.e. the bone morphogenetic/ osteogenic proteins (BMPs/OPs), induce endochondral bone formation in non-human and human primates. Tissue induction in the non-human primate *Papio ursinus* develops as a mosaic structure in which the osteogenic proteins of the TGF-β superfamily singly, synergistically and synchronously initiate and maintain tissue induction and morphogenesis. Tissue engineering starts by erecting scaffolds of biomimetic biomaterial matrices that mimic the supramolecular assembly of the extracellular matrix of bone. Naturally-derived BMPs/OPs and gamma-irradiated recombinant human osteogenic protein-1 (hOP-1) delivered by allogeneic and xenogeneic insoluble collagenous bone matrices initiate the induction of bone formation in heterotopic extraskeletal and orthotopic skeletal sites of the non-human primate *Papio ursinus*. This process culminates in complete calvarial regeneration by day 90 and maintains the regenerated constructs by day 365. Uniquely in the non-human primate *Papio ursinus* and thus by extension to the primate *Homo sapiens*, the three mammalian TGF-β isoforms induce rapid and substantial endochondral bone formation in heterotopic intramuscular sites. In marked contrast, the TGF-β proteins induce only fibro-angiogenesis with potential scarring when implanted in rodents, lagomorphs and canine models. It is proposed that the translation in clinical contexts of the phenomenon of '*Bone: Formation by autoinduction*' may be achieved in humans by the synergistic binary application of relatively low doses of TGF-β proteins with a recombinant human osteogenic protein.

Chapter IX - Growth plate cartilage is responsible for lengthening of long bones in children. However, its trauma injuries are common and are often repaired undesirably by bony tissue which results in bone growth defects in children. This chapter reviews recent studies using a tibial growth plate injury model in young rats, which have characterised the injury responses and bony repair mechanisms. In this model, bony tissue formation at injured growth plate occurs within 2 weeks and sequentially through inflammatory, fibrogenic, osteogenic, and remodelling responses. The initial inflammatory response is rapid and acute at the injured growth plate, with neutrophil-dominant inflammatory infiltration and up-regulation of inflammatory cytokines and growth factors (TNF-α, IL-1β, TGF-β1, PDGF-B, BMP-3 and -4) and inflammatory mediators (Cox-2 and iNOS) peaking on day 1 and subsiding on day 3. After the initial inflammatory response, infiltrated mesenchymal cells undergo osteogenic or chondrogenic events, resulting in both intramembranous and endochondral bone formation and bony repair. Taken together, bony repair at the injured growth plate is preceded sequentially by inflammatory, fibrogenic, and osteogenic responses, and is through both intramembranous and endochondral ossification mechanisms.

Chapter X - There has been a great deal of progress in regenerative medicine over the last several years, but scaffolds are still required for tissue construction. Various biomaterials have been studied as scaffold candidates, but no functional scaffolds capable of acting synergistically with the tissue regeneration capacity of the cells and cytokines have been developed. Regenerative medicine for bone tissue is currently in the clinical application phase, but without an optimal scaffold, sufficient results have yet to be achieved. The authors found that carbon nanotubes (CNTs) promote osteogenic capability, and are currently investigating its mechanism. CNTs may be useful as a functional scaffold to markedly increase the bone regeneration capacity of the cells and cytokines.

Chapter XI - Proteases play essential roles in a variety of biological processes such as protein activation and degradation, cell death, development, and maintenance of morphology. Proteolytic activities in bone regeneration are divided into two functions, degradation and proteolysis. Proteolysis is distinct from degradation in that it represents highly specific and limited substrate cleavage that results in a specific change of protein function. In contrast, degradation such as collagen digestion is the breakdown of structural building blocks or the complete destruction of proteins to be recycled or discarded. Bone is a dynamic tissue that undergoes repeated resorption and reformation over its lifetime. Once bone regeneration is initiated, osteoclasts play a role in resorption, and osteoblasts are responsible for bone tissue formation. Matrix metalloproteinases are believed to be specifically implicated in osteoclastic bone resorption. MMP-2, -9, -13, and -14 are involved in endochondral bone growth. MMPs are also associated with estrogen deprivation, which often results in bone disruption. Cathepsin K is a cysteine protease that can degrade key bone matrix proteins such as collagen. While both MMPs and cathepsin K directly degrade skeletal structural proteins, site-1 protease (S1P) functions in the conversion of transcription factors involved in endochondral ossification to their active form. In addition, receptor activator of NF-κB ligand (RANKL), a key regulator of osteoclastogenesis, is converted to a soluble form by proteolysis. Thus, proteases perform essential functions in bone regeneration by either direct degradation of structural proteins or by proteolytic control of regulatory proteins. Knowledge of proteolytic activities in bone regeneration and elucidation of methods to control their activities using inhibitors will facilitate the development of new therapeutic targets and agents for bone diseases.

Chapter XII - In therapeutic bone repairs, autologous bone grafts conventional or vascularized, allografts, and biocompatible artificial bone substitutes all have their shortcomings. The bone formed from peptides (rhBMPs), demineralized bone powder or combinations of both is small in size. The tissue engineering may be an alternative for cranial bone repair.

The authors develop animal models to test the hypothesis that *ex vivo* replication defective Adenovirus mediated human BMP-2 gene transfer to bone marrow mesenchymal stem cells (MSCs) enhances the autologous bone formation for repairing a critical size craniomaxillofacial defect.

The MSC was implanted by using *ex vivo* adenovirus BMP-2 medicated gene transfer to the expanded MSC 7 days before implantation with a concentration was $5x10^7$/ml of collagen type I, and the control site using adenovirus β-galactosidase under the same condition.

The BMP-2 cell/construct site showed complete whitish solid bone formation after 3 months and the HE and Von Kossa stains demonstrated that mature woven bone with well mineralization. 3D CT image showed a wholesome repair in both cranium and maxilla. Quantitative analysis demonstrated that there was significant different ($p<0.001$) in bone formation. The biomechanic test did reveal no significant statistical difference in the stiffness of new bone formed by BMP-2 cell construct and the normal maxilla bone. The data evidenced *de novo* bone formation with quality as good as normal craniofacial bone, and thus capable of sustaining gravity load. The results demonstrated that *ex vivo* gene therapy by adenovirus BMP-2 transduction will induce and enhance craniomaxillofacial bone regeneration.

In: Bone Regeneration: Growth Factors... ISBN: 978-1-60876-621-5
Editors: V. Legard et al., pp. 1-57

Chapter I

Biodegradable Composites for Bone Regeneration

Luigi Calandrelli*[1]*, Fulvio Della Ragione*[2]*,
Giovanna Gomez d'Ayala*[1]*, Paola Laurienzo*[1]*,
***Mario Malinconico*[1] *and Adriana Oliva*[2]**
[1]Institute of Polymers Chemistry and Technology, CNR,
Via Campi Flegrei 34, 80078 Pozzuoli, (Naples), Italy
[2]Department of Biochemistry and Biophysics, Faculty of Medicine, Second University of Naples, Via L. De Crecchio 7, Naples, Italy

Abstract

Osteoinductive materials are keenly sought in reconstructive surgery. Polymer-based composites are ideal candidates, as they couple the ease of processing of polymers (by melt or by injection and hardening) with the mechanical reinforcement of fillers, once interfacial adhesion is assured. Moreover, an appropriate choice of the filler may enhance the bone regeneration activity, through cell adhesion or osteoinduction. In this respect, hydroxyapatite and calcium sulphate are good candidates. As organic binder, a wide choice of different polymers is possible: biocompatible and biodegradable natural polymers such as polysaccharides and, in more recent years, biocompatible and biodegradable synthetic polyesters are normally employed, with processing methodologies which depend on the intrinsic properties of the polymers (paste injection or injection molding).

In the present chapter, after briefly summarizing recent literature concerning modification and applications of these materials, several recent developments of biocomposites containing silica nanoparticles or calcium sulphate intended for bone regeneration are reported. The composites are characterized with respect to their chemical-physical and mechanical properties. Their biocompatibility and capacity to induce the osteoblastic phenotype in human bone marrow mesenchymal stem cells have been assessed. We will focus on two particular systems based on either natural or synthetic biopolymers with different biofillers: alginate/chitosan blends with calcium sulphate and poly(ε-caprolactone) with silica nanoparticles.

Introduction

Principles of Tissue Engineering

The term "tissue engineering" was defined by the pioneers Langer and Vacanti as "an interdisciplinary field that applies the principles of engineering and life sciences toward the development of biological substitutes that restore, maintain, or improve tissue or organ function" [1]. More recently, MacArthur and Oreffo defined tissue engineering as "understanding the principles of tissue growth, and applying this to produce functional replacement tissue for clinical use" [2].

Tissue engineering can be viewed as the combination of cells, an appropriate scaffold, biologically active molecules and/or suitable mechanical factors, aimed at the replacement, repair, maintenance, and/or improvement of tissue function, thus allowing the advancement of medicine. In effect, "regenerative medicine" is often used synonymously with "tissue engineering", although the former emphasizes the use of stem cells, rather than already differentiated cells, to produce tissues. Among the numerous regenerative medicine fields, bone tissue engineering is at the forefront of research and clinical applications, representing a great challenge to repair skeletal defects resulting from traumatic injuries, infections, biochemical disorders, tumors and other pathologies.

As previously mentioned, the main components of a tissue engineered construct are: 1) cells, 2) a scaffold, 3) bioactive factors and/or physical signals.

As regards the cellular component, it can consist of either differentiated cells from the tissue of interest or stem cells/precursor cells, but bone tissue engineering using bone marrow stem cells appears the most promising technique for reconstructing bone defects.

Bone marrow is a complex cellular array of stem cells of both hematopoietic and nonhematopoietic lineages. In 1987 Friedenstein et al. showed that a specific set of cells (colony forming unit fibroblasts, CFU-F) present in bone marrow can differentiate into different cell types, including osteoblasts [3]. These cells, referred to as mesenchymal stem cells or marrow stromal cells (MSCs), have a great capacity of self-renewal and multipotency. These cells, indeed, can give rise to precursors for bone, cartilage, adipose tissue, tendon, muscle, and marrow stroma [4,5]. In addition, MSCs have been demonstrated to be capable to differentiate into cardiomyocytes, astrocytes and neurons [6-8]. A number of studies on MSCs have demonstrated the great plasticity of these cells and that their fate strictly depends on various environmental signals [9,10].

An ideal bone tissue engineering strategy implies the use of autologous bone marrow MSCs, even if their demonstrated immunoregulatory activity might allow also the use of allogeneic cells [11].

The typical bone tissue regeneration strategy involves either the isolation and *ex vivo* expansion of cell population within the engineered construct in a bioreactor, followed by implantation, or in situ recruitment of endogenous cells to the implantation site through the use of osteoconductive scaffolds possibly delivering bioactive molecules, such as osteoinductive growth factors ("smart" scaffolds).

A serious drawback to the former approach is represented by the low number of MSCs in bone marrow. In this respect, the utilization of a person's own platelet-rich plasma, that is a rich source of a number of growth factors essential in the modulation of the wound-healing process, can be a method for the rapid *ex vivo* expansion of MSCs before their clinical application, avoiding the use of heterologous human- or animal-derived growth supplements as well as expensive recombinant growth factors [12].

While in a great number of animal studies the feasibility of bone formation has been reported, in human bone tissue engineering it has not always been a clinical success. Among the factors preventing the reaching of a successful outcome, lack of sufficient vascular supply is thought to be the main cause of failure since the vascularization is a key factor in bone tissue engineering [13].

Even if MSCs themselves secrete a large number of bioactive factors capable of creating a regenerative microenvironment at the injury site, which Caplan defines as "trophic activity" [14], and are able to promote the neovascularization through the release of angiogenic factors, it could be necessary, particularly for large bone defects, to stimulate vessel growth and thus improve the oxygen and nutrients supply by adding angiogenetic growth factors, or even endothelial cells, to the tissue engineered construct [13,15].

Biocompatibility Testing of Medical Devices: "Tests for in vitro Cytotoxicity (EN ISO 10993-5)"

Upstream to clinical applications in humans, all materials designed for biomedical devices must be submitted to *in vitro* and *in vivo* tests to verify biocompatibility, response and behavior of the interacting cells according to the rules of EN ISO 10993. The standard EN ISO 10993-5, that has been just revised substituting the previous edition 10-years old, contains methods designed to assess the in vitro cytotoxicity of medical devices, namely the biological response of mammalian cells using appropriate biological parameters.

Indeed, ISO 10993-5:2009 describes test methods that specify the incubation of cultured cells in contact with a device and/or extracts of a device either directly or through diffusion. One of these methods, the so-called indirect test (or elution test), represents the first step of the biocompatibility analysis and is widely used and applicable to every type of medical devices. In this method, culture medium containing extractables derived from the tested material, obtained under standard conditions, is applied to a cell layer. The cultures are then returned to the 37°C incubator and periodically examined by microscope for as long as three days. During incubation, any leachable chemicals from the test material can directly interact with the cell layer and induce or not signs of toxicity.

Once proved the absence of cytotoxicity by means of the elution method, materials projected for bone tissue regeneration devices should be subjected to the another *in vitro* approach, that is a direct contact method. In this approach, cell populations typical of the implant site, as well as their progenitors, are cultured directly onto the material under investigation and observed, for a period of days or weeks, in terms of morphological and functional features.

Principles of Synthetic Bone Substitutes

Bone is a complex and a highly specialized form of connective tissue pertaining to the formation of the skeleton of the body. Bone not only provides mechanical support but also serves as a reservoir for minerals, particularly calcium and phosphate. It is a good example of a dynamic tissue, since it has a unique capability of self-regenerating or self-remodeling to a certain extent throughout the life without leaving a scar [16]. However, many circumstances call for bone grafting owing to bone defects either from traumatic or from non-traumatic destruction. There are multiple methods available for the treatment of bone defects, which includes the traditional methods of autografting and allografting

Although autografting and allografting are clinically considered as good therapies, they have limitations. For example, supply of autograft is limited and there is a possibility of pathogen transfer from allograft. Accordingly, there is a great need for the use of synthetic bone grafts. Nowadays, numerous synthetic bone graft materials, both single- and multi-phases, are available which are capable of alleviating some of the practical complications associated with the autogenous or allogeneic bones.

Several classes of materials such as ceramics, metals and polymers have been used as artificial bone to fill bone cavities and to refill bony structures [17,18]. In most of the cases, metals and ceramics are used in hard tissue applications, whilst polymers in soft tissue applications due to their mechanical properties. Although there is good progress in bone grafting using synthetic bone grafts, the way in which they execute their functions *in vivo* is quite different and most of them differ from natural bone either compositionally or structurally. Further, a single-phase material (also called monolithic) does not always provide all the essential features required for bone growth, which leads to incessant investigation in search of an ideal bone graft. There is, therefore, a great need for engineered multi-phase materials (also called composites) with structure and composition similar to natural bone [19].Composites are widely used in both soft and hard tissue applications.

Being the mineral phase of bone mainly composed by natural hydroxyapatite (HA), cements based on synthetic HA are among the most investigated materials for dental and orthopedic applications in reconstructive surgery. HA is employed as substitute for defective bones because of its biocompatibility, bone bonding ability, osteoconductivity, non-toxic and non-inflammatory effects [20-26]. Nevertheless, HA cements are hard (not injectable and mouldable) [27,28], brittle, and not suitable for handling and processing [28-31]. Consequently, several composite and nanocomposite materials based on natural or synthetic biodegradable polymers with HA or its precursors (i.e., calcium phosphates) have been proposed and tested [32-38]. Biocomposite implantation materials based on calcium phosphates substantially expanded the possibility of restorative and substitutive osteoplastic surgery, mainly in dentistry, maxillofacial surgery, and neurosurgery.

The term nanocomposite can be defined as a heterogeneous combination of two or more materials in which at least one of those materials is on a nanometer-scale. Natural bone itself is a true nanocomposite composed mainly of HA nanocrystallites in the collagen-rich organic matrix [39,40]; thereby choosing a HA/collagen nanocomposite as a bone graft material is an added advantage [41], as it mimics the bone components [42]. Nanocrystalline HA promotes osteoblast cells adhesion, differentiation and proliferation, osteointegration and deposition of calcium containing minerals on its surface better than microcrystalline HA, thus enhancing the formation of new bone tissue within a short period [43-45]. Nanocomposites have been largely studied mainly for their superior mechanical characteristics; nevertheless, more recent studies are aimed at giving and using bioactivity and osteoconductivity. The main role of osteoconductive grafts is to serve as a structural framework through which the host bone cells infiltrate and regenerate a new bone tissue. Many researchers have attempted to give osteoconductivity to bone cement based on natural or synthetic biopolymers by introducing different bioactive fillers. Biological and clinical testing made it possible to establish that the biological activity of materials and their compatibility with an organism are primarily related to the presence in their composition of HA and tricalcium phosphate (β-TCP); more recently, ceramic fillers as silica nanoparticles [46-48] or different salts, as calcium sulphate [49-50], have been proposed. However, the interfacial strength between bioactive filler and the organic polymer is of concern because of the lack of adhesion between the two phases, resulting in an early failure at the interface. In addition, the distribution of the nanometric filler in the polymeric matrix must be homogeneous, in order to increase its surface exposure. These parameters are fundamental to induce the osteoconducting ability of the biocomposite.

Nanocomposites can be made, conventionally, by blending a heterogeneous combination of two or more materials, differing in morphology or composition. It is well known that blending of multiple materials with different characteristics leads to composites with tailor-made properties, but it is quite difficult to control homogeneity and uniformity of the reinforcing phases. The crystallites are not uniform in size, and are often aggregated and randomly distributed into the matrix; therefore there is no structural uniformity observed that is close to natural bone. Further, there is no sign of chemical interaction between mineral and organic phases, which is probably due to the lack of suitable interfacial-bonding. So, it is clear that the first research effort must consist in optimizing nanocomposite processing conditions and interfacial characteristics. Therefore, if nanometer-sized bioactive ceramics can be dispersed homogeneously and bonded covalently to the polymer phase at the molecular level, all problems are likely to be solved. The improvement of the adhesion between the ceramic fillers and the biopolymer could be achieved using a coupling agent; surface modification of the inorganic particles by organic molecules also is an effective means to manipulate the inter-surface properties, and develop optimized biomaterials [51]. A brief overview of the most promising polymers and inorganic fillers used to realize biocomposites for bone tissue engineering, together with the chemical strategies developed for the optimization of their properties, is hereafter reported.

Polymers for Bone Tissue Engineering

Biodegradable natural and synthetic polymers as well as some non-biodegradable polymers, which are currently used for cartilage tissue engineering, have been described as organic components in materials and composites for bone regeneration [52,53]. Polymers have distinct advantages over the ceramic materials. Bioactive ceramics have chemical composition resembling that of natural bone but they are inherently brittle and have low biodegradation rates, which limit their clinical uses [54]. On the other hand, the mechanical properties and biodegradation rate of polymers can be tailored to a certain extent for specific applications. Re-absorbable or degradable biomaterials containing the high MW hyaluronic acid have been extensively explored for tissue engineering applications. Biomaterials based on hyaluronic acid, with excellent histocompatibility and biodegradability, were prepared for the synthetic bio-skin and biosynthetic osteo-cartilage used for the reconstruction of tissues. The addition of growth factors and cytokines promoted cell migration, proliferation and formation of a new tissue [55]. Also natural polyesters based on hydroxyvalerate were tested for repair of the musculoskeletal tissues. *In vivo* experiments indicated that, for some compositions, they may perhaps be the most suitable polymeric material for this application [56].

Advanced tissue engineering involves the use of 3D polymeric scaffolds implanted at the defective site. They provide frameworks for cells to attach, proliferate, and form extracellular matrix (ECM). Biodegradable scaffolds should be bioabsorbed at a predetermined rate and the space initially occupied by them should be fully replaced by the regenerated tissue [57]. Several methods are used to produce scaffolds from both natural and synthetic polymers; among them, the electrospinning technique allows the preparation of scaffolds from uniform nanofibers with mean diameter of 260 nm, that closely mimic the native ECM found in the human body and are expected to support active tissue regeneration. Scaffolds suitable for bone tissue engineering were prepared by combining ceramic materials with biocompatible, water soluble polymers based on N-vinylpyrrolidone and acrolein diethyl acetal [58]. Synthetic biodegradable polyesters, as poly(L-lactide) (PLLA), poly(D,L-lactide-co-glycolide) (PLGA), polycaprolactone (PCL) and their copolymers or combinations of them have been often used for the preparation of porous scaffolds. However, the hydrophobicity, the acidity of their decomposition products, and the self acceleration of their degradation, have constituted serious drawbacks. Hybrid materials based on synthetic polymers and collagen, the main organic constituent of the human bone [59], have been proposed. *In vivo* studies showed that a hybrid scaffold has a high degree of tissue compatibility. Calcium deposition appeared 4 weeks after insertion [58]. PLLA/collagen scaffolds seemed to be preferable to pure collagen sponge as maintained the original shape throughout the implantation period of 8 weeks, while the collagen sponge collapsed [60]. Development of bone analogues on a biomimetic PCL/ceramic polymer was explored. Human marrow stem cells were able to adhere, migrate, and differentiate on these biodegradable scaffolds, which degraded much faster than PCL alone [61].

The usually low manufacture costs related to their large agricultural availability and renewability are additional advantages of natural biopolymers. Furthermore, their versatility of chemical structures and their well-known chemistry allow the development of advanced functionalized materials that can match several varied requirements [62-64]. Among the biopolymers, polysaccharides have the widest medical applications due to their nontoxicity, water solubility, or high swelling ability induced by simple modification, stability to pH variation, and a broad variety of chemical structures [65,66]. In particular, natural polysaccharides are widely employed in bone regeneration techniques, in order to overcome drawbacks of mineral bone graft substitutes, such as hydroxyapatite and calcium sulphate.

Special attention has been given recently to chitosan and its derivatives, in combination with HA and various natural and synthetic polymers. Grafted chitosans are extensively investigated for applications in cell transplantation and for tissue regeneration [67-69]. Chitosan has been shown to improve wound healing [70-72] and enhance bone formation [71-74]. In addition, it is non-toxic, bioresorbable and non-immunogenic [75-77]. For these reasons, it has been suggested to possess biological properties suitable for clinical applications.

Chitosan scaffolds for bone tissue engineering have been widely investigated [78-80] and showed to enhance bone formation both *in vitro* and *in vivo* [81]. In addition, chitosan has excellent filmogenic properties, and its biodegradation kinetics is slower than that showed by other biocompatible polymers (i.e., collagen) [82]. Feng Zhao *et al.* [36] prepared a biodegradable HA/chitosan-gelatine network composite as a three dimensional biomimetic scaffold for bone tissue engineering. Besides, an interesting case was reported by Murugan et al. [83] who prepared a bioresorbable composite bone paste based on HA in conjugation with chitosan for bone repair and regeneration. They used a wet chemical method at low temperature and obtained a HA composite with a decreased bioresorption and particle migration rate with respect to HA bioceramics.

Biocomposite nano-scaffolds of chitosan/poly(vinylalcohol)/HA prepared by electrospinning showed a well interconnected pore network structure with nanofibrous morphology and were considered as a promising material for bone tissue regeneration [84]. However, several efforts have been devoted to functionalize chitosan, because of its poor workability and scarce water solubility at neutral pH [85,86].

Great interest in biomedical field has been given also to alginate. Alginate is a safe, biocompatible natural polysaccharide obtained from marine algae used in food and biomedical applications, mainly for its capacity to crosslink in presence of bivalent ions under very mild conditions [87-89]. Alginate/HA and alginate/collagen/HA composites have been proposed as scaffold materials for bone tissue engineering [90]. Parhi et al. developed an alginate-hydroxyapatite biocomposite, in which the mineral phase was nucleated on alginate chains by precipitation method to obtain a biomimetic artificial bonelike composite [91]. A polymeric composite made up of nanohydroxyapatite and alginate was formed as a result of in situ nucleation of hydroxyapatite on alginate polymeric chain [92].

Silica-Based Materials and Nanoparticles for Bone Regeneration

Silica, the mineral of which sand is made, is generally inert in the body and can be modified easily using a variety of well-established chemical reactions. As such, researchers have considered silica an ideal candidate material from which to create multifunctional nanoparticles. Silica nanoparticles (SNP) have been widely used for biosensing and catalytic applications due to their large surface area to volume ratio, straightforward manufacture and the possibility to be doped and/or functionalized with fluorescent molecules [93,94], magnetic nanoparticles [95] or semiconducting nanocrystals [96]. Porous silica nanoparticles hold promise as drug delivery vehicles, imaging agents, and even nanoscale collection devices for cancer markers [97]. SNP can be prepared using several techniques, including the widely used Stobër technique and the water-in-oil nanoemulsion system. The former is based on the simple hydrolysis of a silica precursor in alcoholic medium in the presence of ammonia; the latter uses water droplets inside of reverse micelles as nanoreactors. The size of the final nanospheres is mainly regulated by the dimension of the water droplets and, therefore, by the molar ratio of water to surfactant and by the molar ratio of water to the precursor. Other relevant parameters are the molar ratio of the precursor to the catalyst, the reactivity of the precursor, the total time and temperature of the reaction. As previously said, silica based nanoparticles find their use particularly in bioanalysis if these are conjugated to recognize biocomponents. In the conjugation process, special elements that interact with the analyses are carefully selected. Using various methods these elements can suitably attached to the silica nanoparticle surface. There are several biochemical binding approaches for modifying the silica nanoparticle surfaces to contain functional groups. These functional groups are useful for further modification in surface and biomolecule immobilization. So far the research has been successfully completed on biochemical groups including amines, carboxyls, thiols, however various other molecules including oligonucleotides, enzymes and other proteins are being used for immobilizing onto the nanoparticles. Application of such organically modified silica nanoparticles as a nonviral vector for efficient *in vivo* gene delivery has been reported, where highly monodispersed, stable aqueous suspension of nanoparticles were surface-functionalized with amino groups for binding of DNA [98].

For what concerns applications of silica nanoparticles in bone tissue regeneration, several examples of nanocomposites based on acrylic matrices containing silica nanoparticles as bioactive filler for bone cement [46,99] and resin for dental applications [100,101] have been reported. The authors always observed high mechanical properties in addition to crystalline apatite formation on implants in simulated body fluid (SBF), proving that silica nanoparticles act as a nucleation site for several salts, as calcium phosphates, that promote formation of HA in *in vivo* simulating conditions. Formation of an apatite layer and favourable mechanical properties were also observed on PCL/silica nanocomposites [47,102,103].

To improve the interface with the organic matrix, silica based nanoparticles can be chemically or physically modified [104]. Different non-covalent mechanisms, as hydrophobic interactions [105-107], ionic interactions [108,109], and π-π interactions [110-112], are involved in the physical adsorption of functional molecules onto the surface of nanoparticles. The silanization method is widely used for chemical modification; as a conventional silanization reagent, trimethoxysilylpropyl–diethylenetriamine (DETA) has been reported to introduce amine groups onto the surfaces of silica nanoparticles [113]. Similarly, trimethoxysilane–PEG and trifluoroethylester PEG silane have been shown to link to silica based nanospheres and introduce PEG molecules onto the nanosphere surface [104,114-117]. Vinyl modified silica nanoparticles have been photopolymerized with acrylic monomers to develop polymethacrylate-silica hybrid/nanocomposite dental materials characterized by low volume shrinkage [118].

Nowadays, along with the use of nanoparticles, several examples can be found in literature concerning multi-phase materials intended for bone regeneration based on silica. Non-woven silica gel fabric prepared by the electrospinning method provided a suitable morphological architecture for the differentiation of pre-osteoblastic cells and osteoconductivity [119]. Nano-silica-fused whiskers combined with calcium phosphate cements (CPC) were used as fillers in a resin matrix-based composite. The role of nano-silica was fundamentally to reinforce the CPC-based composites. The mechanical properties of the CPC/silica-whisker composites nearly matched those of cortical bone and trabecular bone [120]. As the general drawback of biocomposite based on HA and β-TCP consists in a substantial decrease in the activity of calcium phosphates in a physiological medium, since their synthesis requires a high temperature (above 1200°C), biocomposite synthesized by sintering HA and β-TCP powders and an amorphous silicate matrix at temperatures below 800°C makes possible to preserve the defects of the HA structure in the material and, accordingly, its relatively high resorption capacity. Further, the main advantages of such implant materials are their mechanical strength, which is comparable to that of bone tissue, and their suitability for the production of structural implants. At the same time, the related porous biocomposites in their physicomechanical parameters are at the same level as spongy bone tissues [121].

The Use of Calcium Sulphate in Bone Regeneration

The hemihydrate form of calcium sulphate ($CaSO_4 \cdot ½ H_2O$, CHS), better known as Plaster of Paris, is one of the calcium salts most investigated as a substitute for bone graft [49,50]. It has been demonstrated that it can prevent the growth of soft connective tissue in the cavity and that it also has osteoconductive properties [122]. It is in fact reported that gypsum, in contact with body fluids, forms calcium phosphate deposits which are finally responsible for conducting and accelerating bone formation [123-125]. Addition of water to CHS powder elicits an exothermic reaction with the end product being the dihydrate form of calcium sulphate ($CaSO_4 \cdot 2H_2O$, gypsum). The as obtained product is a paste with good handling properties and mouldability which in several minutes gets a hard cement. Despite its efficiency, the potential applications are limited by its rapid *in vivo* resorption and its brittleness [126]. Novel calcium sulphate/polymer systems in which CHS can be encapsulated in a polymeric biodegradable and biocompatible matrix, in order to retain the structural integrity and decrease the bioresorption rate, have been recently developed [127,128]. In one case, poly(ε-caprolactone) (PCL), a biodegradable and non-cytotoxic polyester [30,32,129], currently regarded as a soft and hard-tissue compatible material including resorbable sutures, drug delivery system and, recently, bone graft substitutes [30,129,130], has been used to realize these networks. In the other case, encapsulation of calcium sulphate in hydrogels based on blends of natural, biocompatible polysaccharides got a hard cement, characterized by good mechanical properties and slower resorption rate.

In this chapter, some novel composite materials for bone tissue regeneration based on biodegradable polymers are described. We pointed on two different systems.

The first system we dealt with is a PCL-based composite with silica nanoparticles. The nanoparticles have been chemically modified on their surface with PCL of reduced molecular weight to improve interface adhesion with the polymer matrix; successive introduction of vinyl groups allowed grafting onto PCL. The resulting nanocomposite is characterized by improved mechanical and osteoconductive properties.

The second one consists of a blend of natural polysaccharides, chitosan and alginate, entrapping calcium sulphate as osteoinducting phase. We showed how it was possible to obtain a material of superior characteristics, due to a synergistic effect between the two polymers at a given composition of the blend.

Realization and Performance Analysis of PCL/ Silica Nanocomposites for Bone Regeneration

Nanocomposites of PCL and silica, in which the silica nanoparticles have been modified to provide a chemical graft onto PCL and so improve interfacial adhesion, are hereafter described.

The nanoparticles were prepared following the Stöber method [131], as described in the Experimental (see below), and then characterized by Transmission Electron Microscopy (TEM) (**Figure 1**). From the morphological analysis we deduce that the so obtained nanoparticles are spherical in shape with dimensions comprised between 100 and 200 nm.

Mechanical Alloys (Chemical Physical and Mechanical Properties)

Before the realization of the functionalised nanocomposites, mechanical alloys, i.e., nanocomposites obtained simply by mixing unmodified PCL and unmodified silica nanoparticles, A90 (see Experimental), in matrix PCL with a percentage of silica of 1, 3 and 5% have been prepared and characterized. The scope of this preliminary study was to verify if in the absence of specific interfacial interactions, but simply with the addition of a nanometric filler to the polymer matrix, the physical and mechanical chemical properties of the PCL could improve. As it is shown from **Table I**, the calorimetric analysis (DSC) of the PCL evidences the temperatures of fusion (T_f) and of onset and endset of fusion (Tonset and Tendset), the degree of crystallinity in heating (Xc^1) and cooling (Xc^2) and the temperatures of crystallization (Tc), glass transition (Tg) and degradation (Td). The presence of the nanosilica has not caused any variation of the transition temperatures (T of fusion, T of crystallization and T of glass transition). Also the thermal stability of composites did not seem to be influenced from the presence of the silica.

The tensile tests carried out in order to estimate parameters such as stress at break, Young's modulus and extensibility, indicate that increasing the percentage of silica leads to a corresponding increase of the modulus and a diminution of the extensibility (**Table II**). This to demonstrate that it is possible to increase the structural properties of the PCL without depressing, in considerable measure, the plastic properties of the material.

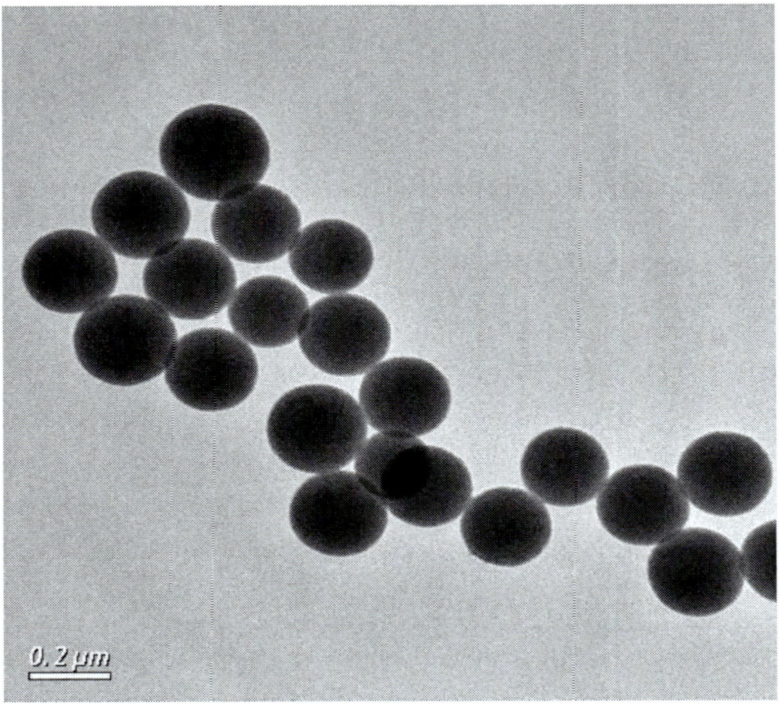

Figure 1. TEM micrograph of Aerosil® 90 silica nanoparticles.

Table I. Thermal parameters of PCL

PCL	T_f (°C)*	Xc[1]	Tc (°C)	Xc[2]	Tg (°C)	Td (°C)
	58 (52-65)	42	32	38	-60	383

*in parentheses, the T of onset and end of fusion are reported

Table II. Mechanical parameters of plain PCL and of PCL/A90 silica nanocomposites

PCL/A90	Stress at break (MPa) (± 15%)	Young's Modulus (MPa) (± 4%)	Elongation at break (%) (± 15%)
100/0	44	380	1500
99/1 (1%)	40	430	1366
97/3 (3%)	37	500	1400
95/5 (5%)	37	520	1300

Chemical Modification and Characterization of Nanocomposites

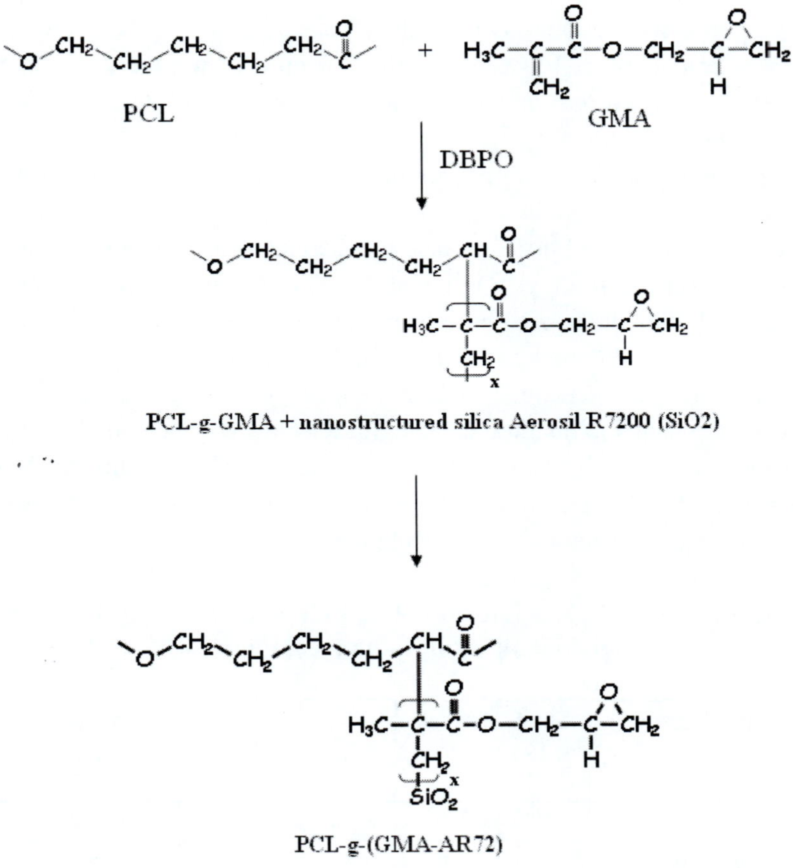

Figure 2. The sequence of reactions which lead to the formation of functionalised nanocomposites.

It is found in literature that in presence of specific interfacial interactions between matrix and nano-reinforcement, it is possible to exalt the performances of the nanocomposites. In order to improve the interfacial adhesion between the matrix and the nanoparticles, to promote the formation of a reactive interphase and to increase the hydrophylicity, opportune chemical modifications have been carried out either on the polyester chain or on the surface of the nanoparticles. In particular, the PCL has been functionalised introducing in chain approximately 10% by weight of glycidilmethacrylate (GMA) with the help of dibenzoylperoxide (DBPO). Nanoparticles of silica, characterized from the presence on the surface of vinyl groups (Aerosil® R7200), hence in a position to link them to the GMA through the creation of a covalent bond during the GMA radical homopolymerization, has been therefore added in the Rheocord and the mixing was continued to 100°C for further 10 min. In **Figure 2** we show the sequence of the reactions that leads to the formation of the functionalised nanocomposites.

The functionalised nanocomposites (containing 1, 3 or 5% of nanoparticles, as well as previously described mechanical alloys) do not have demonstrated substantial and meaningful differences to the thermal analysis compared to results obtained for mechanical alloys. On the systems so prepared, mechanical tests at low and high deformation speed have been carried out. In **Table III** the results of the tensile tests are reported. As it is possible to observe, the presence of the nanoparticles of silica generates an increment of the Young's modulus. In turn, such increment results to be more meaningful once the chemical modifications so far illustrated are introduced (see Table II for a comparison). In this last case, the variations recorded are in accordance with the content of nanoreinforcement that acts hence as stress concentrator, hardening the system.

Table III. Mechanical parameters of plain PCL and of PCL-g-(GMA-AR72) functionalised silica nanocomposites.

Sample (silica %)	Stress at break (MPa) (±15%)	Young's Modulus (MPa) (± 4%)	Elongation at break (%) (±15%)
PCL	44	380	1500
PCL-g-(GMA-AR72) 1%	24	509	563
PCL-g-(GMA-AR72) 3%	19	566	443
PCL-g-(GMA-AR72) 5%	20	563	510
PCL-g-GMA/A90 1%	22	517	577
PCL/A90 1%	40	430	1366

Moreover, we notice from Table III that the increment of the Young's modulus, although significant when increasing the percentage of nanosilica, turns out to be nearly equal in the percentage of the 3 and 5% thus demonstrating that the increase of the total surface of the nanoparticles determines the impossibility for the polymer matrix to coat it all. As a consequence, high percentages of nanosilica caused the birth of structural defects with worsening of the mechanical properties or at least an invariance of properties. Moreover, tensile tests have been carried out also on the nanocomposite in which the functionalisation regards only PCL (PCL-g-GMA/A90 1%) and on the corresponding mechanical alloy (PCL/A90 1%). In the direct comparison we notice that the single chemical modification of the PCL phase itself determines an increment of the mechanical properties that approach those of the functionalised nanocomposite, while the mechanical alloy has slightly better properties compared to neat PCL but clearly inferior to those of the PCL-g-GMA/A90 1% composite.

Morphologic Analysis (SEM)

Fracture surfaces of mechanical as well as functionalised systems have been observed to the Scanning Electron Microscope (SEM). Nanocomposites between not functionalised PCL and not functionalised silica (mechanical alloys PCL/A90 1%, PCL/A90 3%, PCL/A90 5%, (**Figure 3**), show, at least till 3%, a good dispersion of the nanofillers in the polymer matrix, but an insufficient interfacial adhesion is found. At 5% begins to evidence aggregations of the nanofillers. The alloy prepared with the functionalised PCL and the not functionalised silica (PCL-g-GMA/A90 1%) shows a morphology significantly different from the analogous mechanical alloy (PCL/A90 1%) (**Figure 4**) characterized by an elevated interconnection between the phases evidenced from the extensive covering of the nanofillers with the polymer matrix. In **Figure 4b** the interconnection is better appreciated. These phenomena boast in the nanocomposites where both the components are functionalised (PCL-g-(GMA-AR72) 1%, PCL-g-(GMA-AR72) 3%, PCL-g-(GMA-AR72) 5%; **Figure 5**). In fact in the composite at 1% a greater adhesion of the nanofiller to the polymer matrix is observed. The phenomenon then becomes evident in the composite with 3% of functionalised silica and seems to become stabilized, without great changes, in the composite with 5% of functionalised silica. It is interesting to emphasize that, as a result of this covering of the nanofillers with the matrix, for the same composition the surface of the functionalised system seems much more covered from nanofillers than the analogous alloy not made compatible.

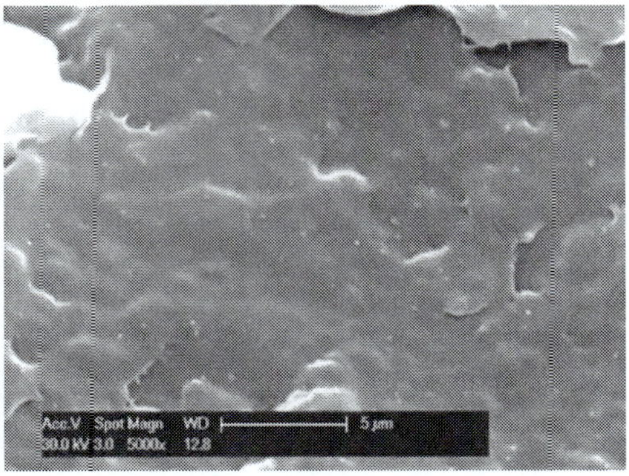

(a)

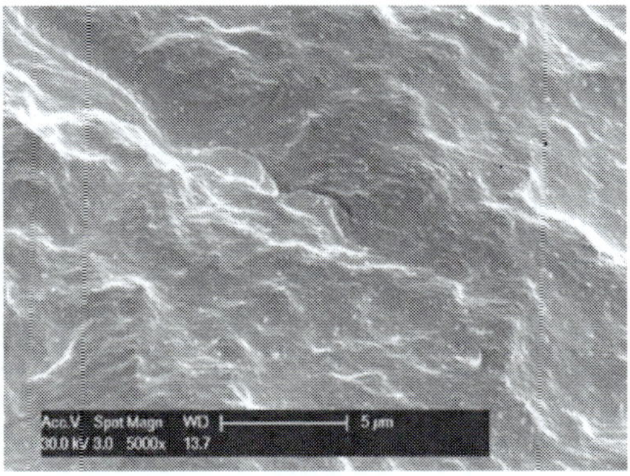

(b)

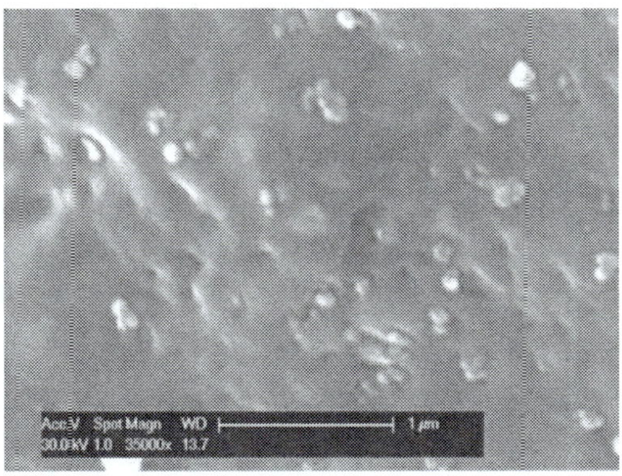

(c)

Figure 3. SEM micrographs of: (a) PCL/A90 1%, (b) PCL/A90 3%, (c) PCL/A90 5% nanocomposites.

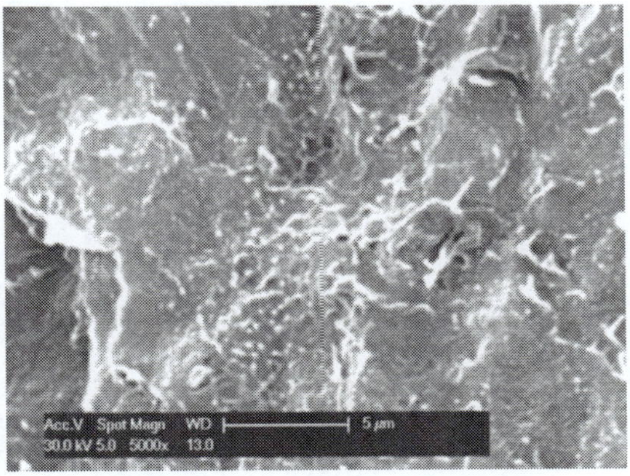

(a)

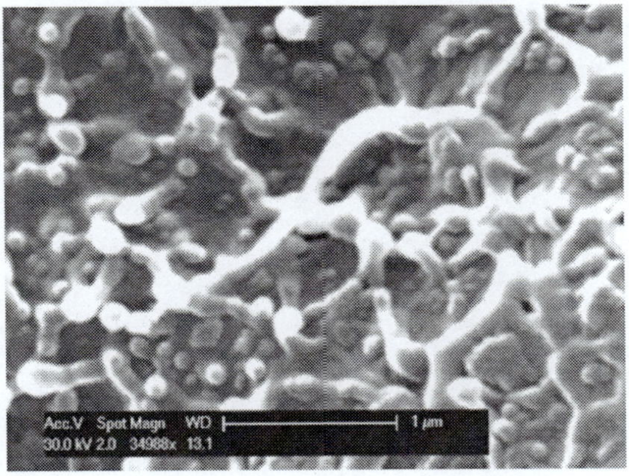

(b)

Figure 4. SEM micrographs of PCL-g-GMA/A90 1% at different magnifications.

As above mentioned, the functionalisation of the PCL with glicydilmethacrylate (GMA) was required to improve the hydrophilicity of the employed PCL. On the basis of preliminary biological results (see next paragraph), it appeared that the epoxy function could be responsible for the decreased adhesion of cells compared to PCL. GMA has been therefore replaced with butylmethacrylate (BMA). BMA is to be considered a monofunctional monomer as the only reactive group is the double bond, while GMA is a bifunctional acrylic with two reactive groups, namely the double bond and the epoxy group. In **Figure 6** we show the reactions that lead to the functionalisation of the PCL with the BMA and the nanostructured silica. The preparation of the nanocomposites is the same as previously shown.

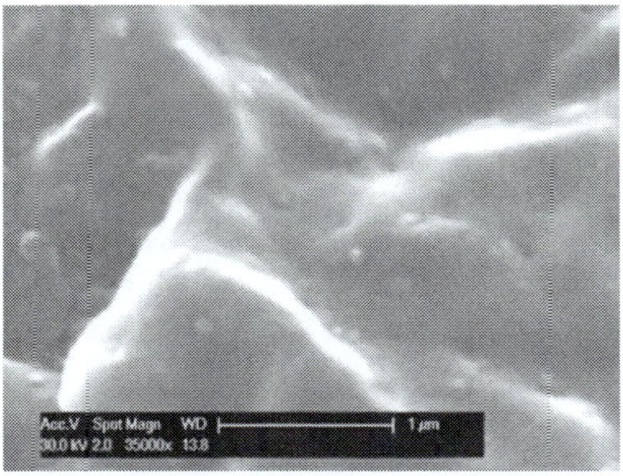

(a)

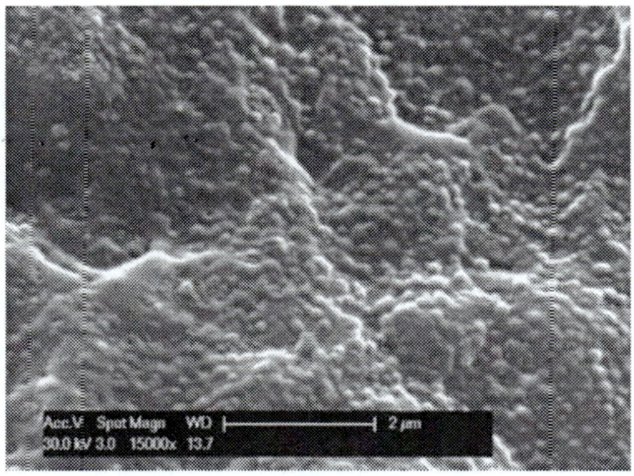

(b)

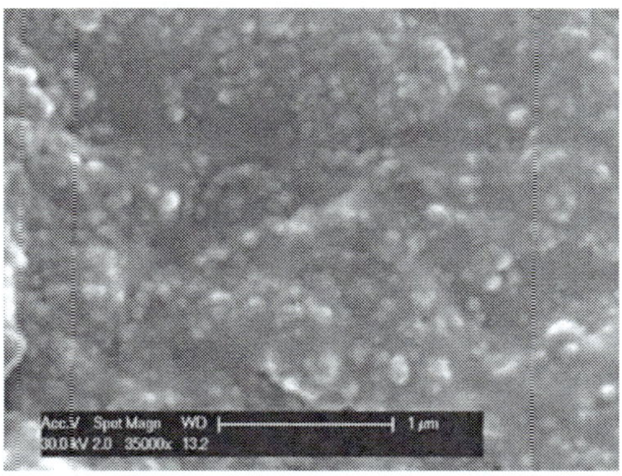

(c)

Figure 5. SEM micrographs of : (a) (PCL-g-(GMA-AR72) 1%, (b) PCL-g-(GMA-AR72) 3%, (c) PCL-g-(GMA-AR72) 5%.

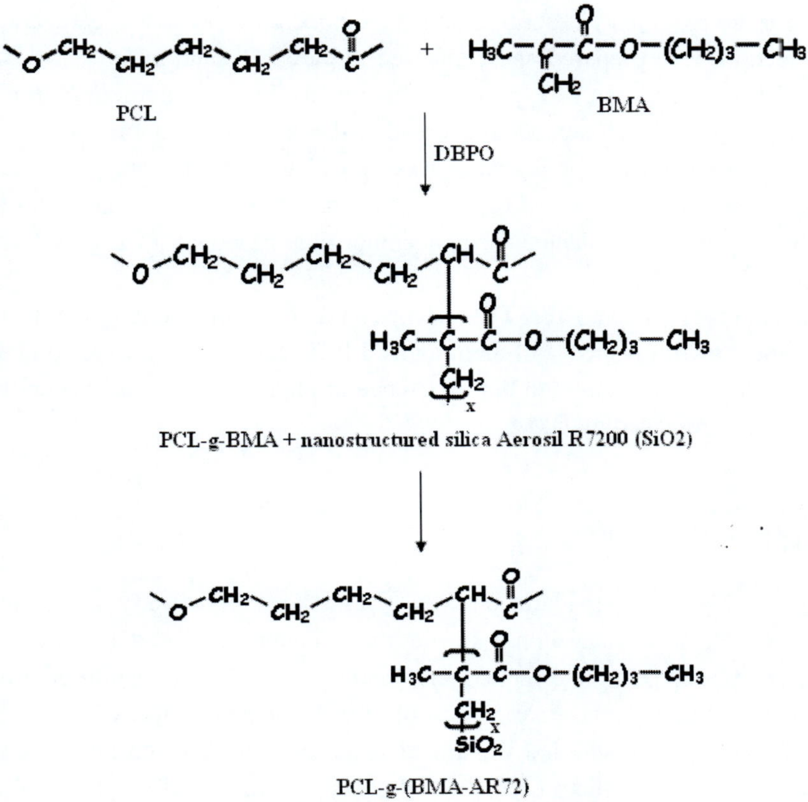

Figura 6. Preparation of PCL-g-(BMA-AR72) nanocomposite.

Table IV. Comparison of mechanical parameters of PCL-g-GMA, PCL-g-BMA, and their functionalized silica nanocomposites.

Sample	Stress at break (MPa) (± 15%)	Young's Modulus (MPa) (± 4%)	Elongation at break (%) (± 15%)
PCL-g-GMA	23	510	567
PCL-g-(GMA-AR72) 3%	19	566	443
PCL-g-BMA	21	520	559
PCL-g-(BMA-AR72) 3%	22	560	479

As for the previously described system, nanocomposites with a percentage of 3% by weight were prepared and compared to pure PCL and pure PCL-g-BMA. As for PCL-g-GMA, also PCL-g-BMA was purified by dissolution in chloroform and precipitation in cyclohexane to get read of the unreacted monomers.

On the prepared nanocomposites mechanical tensile tests are performed as previously described, in view of getting a comprehensive comparison of the properties with those collected on nanocomposites based on PCL-g-GMA, i.e., PCL-g-(GMA-AR72).

The results shown in **Table IV** evidence an increase in Young modulus of the nanocomposite with respect to functionalized PCL, but the values are so close to each other that we can state no big difference in physical properties occurs when GMA group is replaced by BMA.

Biological Evaluation

ISO 10993-5:2009 standards describe test methods that specify the incubation of cultured cells in contact with a device (direct contact test) and/or extracts of a device (elution test). In the direct contact method, the cells were cultured directly on the material under investigation and observed, for a period of days or weeks, in terms of morphological and functional features. Although the limits of all in vitro studies, including the present, are related to the impossibility of reproducing the complex of the biological events that happen in vivo, nevertheless this type of approach can allow to investigate in detail the interaction between cells and implant surface. This initial phase is crucial, being the contact of cells with the biomaterial the basis for all the following events, including the expression of biochemical specific markers, as well as the deposition of an organized extracellular matrix and its mineralization.

We investigated *in vitro* the biocompatibility of the novel biocomposites as well as their capacity to induce the osteogenic differentiation of human bone marrow stromal cells. Cell proliferation and osteogenic differentiation were evaluated by MTT assay, alkaline phosphatase (AP) activity, osteocalcin (OC) content measurement.

Upstream to the *in vitro* studies with human cells, we analyzed the biocompatibility of the unmodified nanocomposite (PCL/A90) with different amount of silica nanoparticles (1, 3, and 5%) performing a set of elution tests in which murine cells (NIH-3T3) were subjected for 24 and 48 h to both 100% and 50% extracts of these materials (**Figures 7A** and **7B**, respectively). A certain toxicity at increasing percentages of silica was evident mostly with 100% extracts.

On the basis of this result and of the mechanical features of the PCL/A90 3% nanocomposite, which resembled the characteristics of bone tissue more than the others, we selected this composition as the starting material for the functionalised nanocomposites.

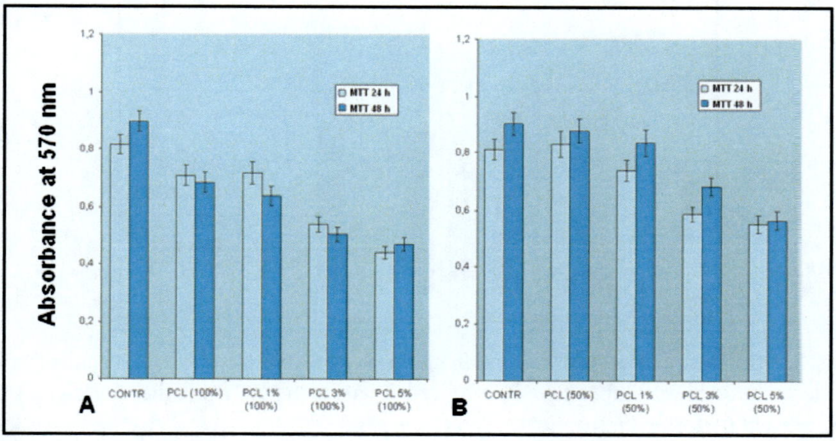

Figure 7. MTT test on 3T3 fibroblasts incubated for 24 and 48 h in presence of either 100% (A) or 50% (B) mechanical alloys extracts.

The following list reports all the materials tested on bone marrow stromal cells in a direct contact method:

- PCL
- PCL/A90 (not functionalized nanocomposite)
- PCL-g-GMA (functionalized PCL)
- PCL-g-GMA/A90 (functionalized PCL + nanosilica)
- PCL-g-(GMA-AR72) (functionalized PCL + functionalized nanosilica)
- PCL-g-BMA (functionalized PCL)
- PCL-g-(BMA-AR72) (functionalized PCL + functionalized nanosilica)

The composite samples (21 mm diameter-0.1 mm thickness) were cut in semilunar form with a surface of 3.6 cm^2, attached with bioinert silicon on the bottom of polystyrene 12-multiwell plates (surface of each well = 4 cm^2) and sterilized 24 h as indicated in the Experimental section.

The 0.4-cm^2 area left uncovered from the sample was intended to allow the observation of the cells growing in the immediate proximity of the materials. In fact, they were opaque to microscopic observation due to their semicrystalline nature. Each material was tested at least in triplicate.

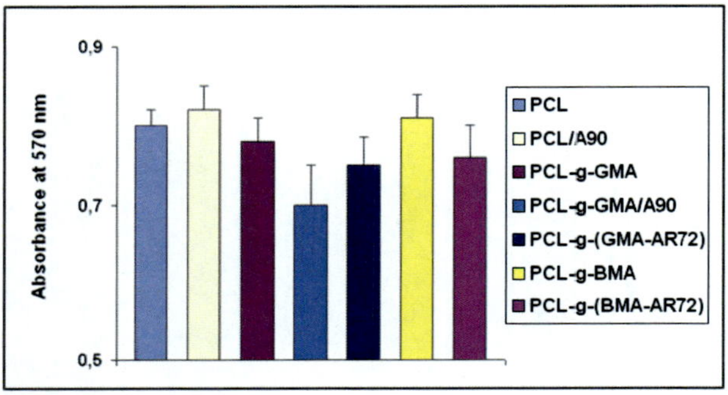

Figure 8. MTT test carried out 24 h after plating 60,000 MSCs ontoPCL (control) and its new derivatives and composites. The reported values result from the average of three analyses,each performed in triplicate.

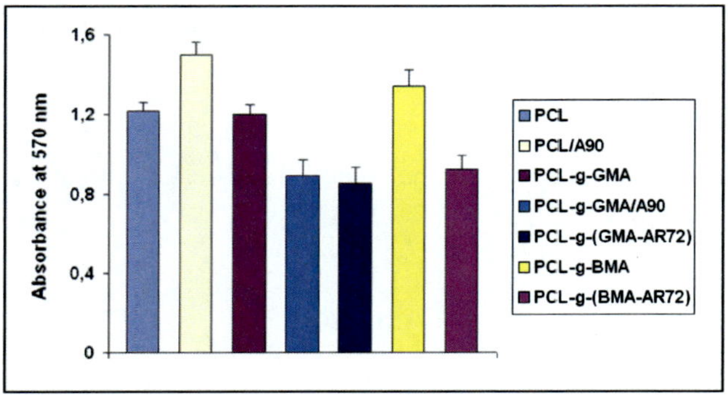

Figure 9. MTT test carried out 7 days after plating 60,000 MSCs onto PCL (control) and its new derivatives and composites. The reported values result from the average of three experiments, each performed in triplicate.

MSCs were seeded on discs at a density of 15,000 cells/cm^2 and incubated at 37°C in a 5% CO_2 humidified atmosphere. Cell vitality at 24 hours were assessed by MTT test (**Figure 8**).

The reported values were directly related to the affinity of cells towards the materials. It was apparent that the adhesion varied in a narrow range from a minimum of 0.70 for PCL-g-GMA/A90 to a maximum of 0.82 for PCL/A90 with a value of 0.80 for the control PCL.

After7-days of culture, MTT test was carried out again in order to assess cell growth on the various materials (**Figure 9**). The proliferation pattern reflected and overlapped that relative to adhesion of cells to the different discs, being PCL/A90 and PCL-g-BMA the composites that, more than others, had favoured the adhesion and then the cell growth.

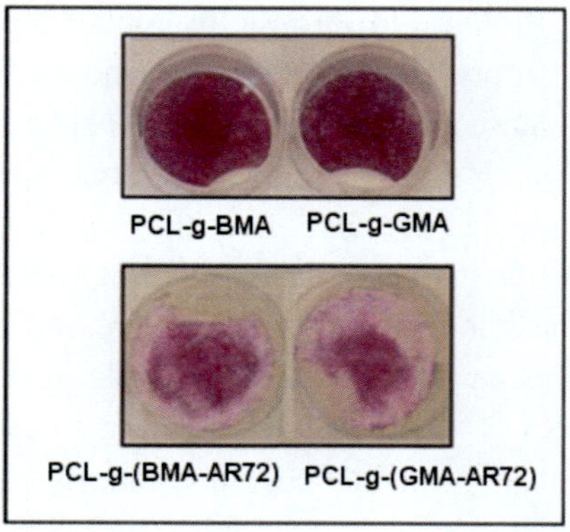

Figure 10. The macroscopic appearance of some PCL composites after MTT test carried out 7 days after cell seeding.

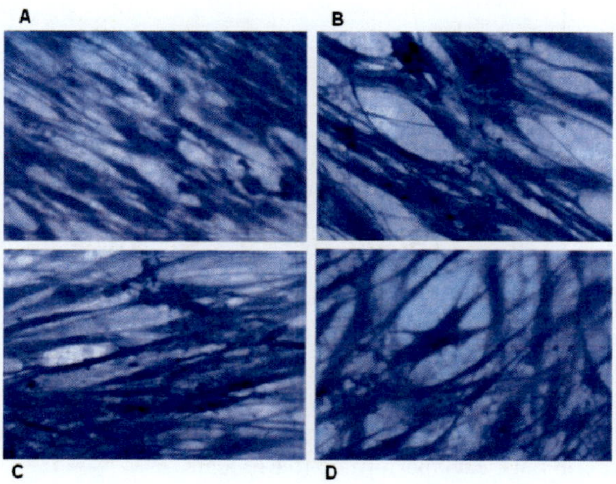

Figure 11. The microscopic appearance of MSCs cultured for 7 days on PCL-g-BMA (A), PCL-g-(BMA-AR72) (B), PCL-g-GMA (C), PCL-g-(GMA-AR72) (D), after crystal violet staining (Magnification 200x).

Figure 10 shows the macroscopic appearance of discs of PCL-g-BMA and PCL-g-GMA and of their corresponding biocomposites with functionalised nanosilica AR72. A substantial and regular cell colonization was evident on both PCL-g-BMA and PCL-g-GMA, whereas on their AR72-derivatives composites cells were not only quantitatively lesser but also unevenly distributed.

In addition, crystal violet staining performed in parallel confirmed the minor presence of the cells with respect to composites containing AR72 nanosilica (**Figure 11**).

The differentiation of MSCs toward the osteoblastic phenotype is a complex process which follows a precise temporal sequence involving several phases [132]. Alkaline phosphatase (AP), commonly considered an early marker of osteogenic differentiation, is involved in the hydroxyapatite crystal deposition and hence in mineralization of extracellular collagenic matrix [133]. Osteocalcin is a specific late marker of osteogenic differentiation, representing the major non-collagenic protein of the bone matrix, and is expressed exclusively by cells of osteoblastic lineage. Osteocalcin plays a central role in bone mineralization and calcium ion homeostasis [134]. A recent fascinating study has demonstrated that this molecule can be actually considered a hormone by which skeleton exerts a feedback endocrine regulation of energy metabolism [135].

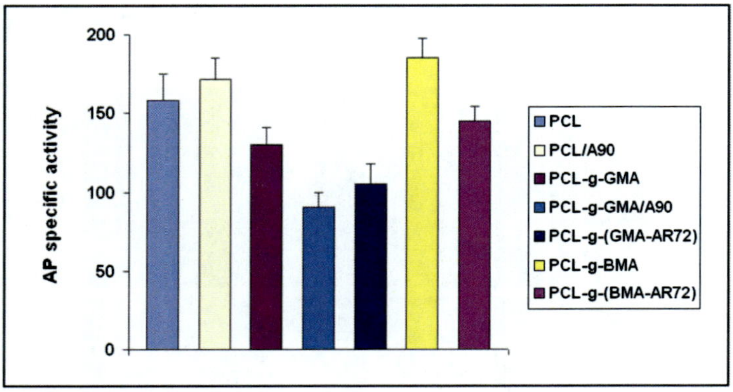

Figure 12. Alkaline phosphatase specific activity of stromal cells cultured for 7 days on PCL (control) and its new derivatives and composites. The reported values result from the average of three experiments, each performed in triplicate.

Therefore, to analyze the influence of the various materials on MSCs' differentiative pattern, the expression of the of the early and late osteoblastic markers, namely AP (**Figure 12**) and OC (**Figure 13**), at 7 and 14 days of culture, respectively, was assessed.

It was evident that PCL-g-BMA expressed the maximum value of alkaline phosphatase activity compared to PCL control, whereas its derivative composite PCL-g-(BMA-AR72) exhibited values markedly lower. This decrease was even more pronounced in the case of PCL-g-GMA and PCL-g-(GMA-AR72).

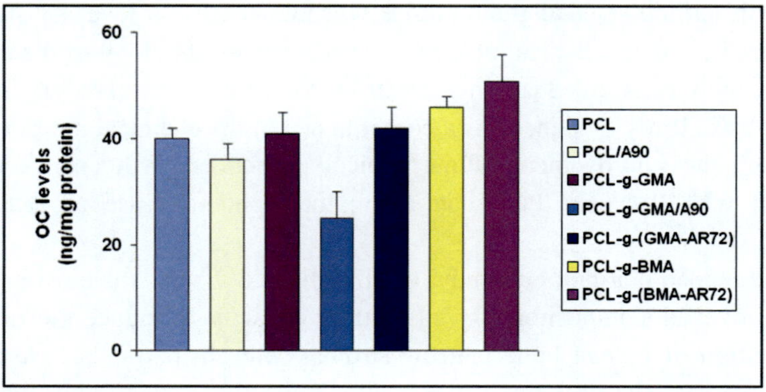

Figure 13. Osteocalcin synthesis **2 weeks after MSCs' plating on** PCL (control) and its new derivatives and composites. The last three days the cells were incubated in FCS-free Opti-MEM in presence of 0.1% bovine serum albumin and 100 nM 1,25-dihydroxycholecalciferol. The reported values result from the average of three experiments, each performed in triplicate.

On the contrary, the osteocalcin synthesis by MSCs cultured on both BMA- and GMA-derivatives composites exhibited values grater than the control PCL. In particular, it is to underline that high OC levels were expressed by cells grown on the two AR72-composites, in spite of the clearly minor presence of MSCs on these materials (Figures 9-11).

Conclusion

In this section we have reported on the preparation and characterization of new composites based on polycaprolactone reinforced with nanostructured silica, projected for biomedical applications in orthopaedic, maxillo-facial and dental surgery fields.

The chemical modification of the PCL, necessary to realize a stable interface with the nanosilica, has been carried out directly in the processing equipment and has allowed us to increase the mechanical characteristics of the polyester without worsening its biocompatibility characteristics.

The work has been carried out in two steps. In the first one, the characterization of mechanical properties performed on the new nanocomposites has demonstrated an increase of rigidity of the material without compromising the tenacity. **In particular, Young's modulus** was increased to a level far superior to the neat PCL. The mechanical properties reflect the morphology of the materials, analyzed by means of Scanning Electron Microscopy, where an interfacial adhesion was clearly evidenced as a consequence of the chemical modification. In conclusion, the improvement of mechanical properties of the nanocomposites compared with the pure PCL can allow their potential use as scaffolds or membrane barriers.

In the second phase of our study, we investigated *in vitro* the biocompatibility of these novel biocomposites as well as their capacity to induce the osteogenic differentiation of human bone marrow stromal cells. In particular, we analyzed the adhesion and the growth of these cells, being the contact with the material crucial for all the following events, including the expression of biochemical specific markers, namely alkaline phosphatase and osteocalcin. None of the new PCL-silica nanocomposites exhibited cytotoxicity, on the contrary the cells adhered and grew onto all the materials even if at a different extent.

Overall, on the basis of the biological as well as the mechanical results, the novel nanocomposites, in particular the one based on PCL-g-BMA, appear promising for bone tissue engineering applications.

Experimental

Materials

Poly-ε-caprolactone (PCL) CAPA® 650, molecular weight 50.000 m.a.u., was a gentle gift from Solvay (Italy).

Nanostructured silica Aerosil® 90, as well as functionalised nanostructured silica Aerosil® R7200, were a kind gift from Degussa.

Dibenzoylperoxide (DBPO) is a Sigma Aldrich product.

Glycidylmethacrylate (GMA) and butylmethacrylate (BMA) are Fluka reagent grade products.

All cell culture biologics were purchased from (Life Technologies, MD, USA), and all chemicals were from Sigma Chemical Co (St. Louis, MO, USA) when not otherwise specified.

Murine fibroblasts NIH 3T3 were from ATCC (LGC Milano).

Preparation of silica nanoparticles

Silica nanoparticles were prepared by using the Stöber methodology [130]. The method consists in a hydrolysis reaction followed by a polycondensation catalyzed by an alkaline water solution of tetraethoxysilane (TEOS). The process is hereafter summarized:

(a) hydrolysis:

$$Si(OEt)_4 + 4\ H_2O \xrightarrow{OH} Si(OH)_4 + 4\ EtOH$$

(b) polycondensation:

$$Si(OH)_4 \longrightarrow SiO_2 + 2\ H_2O$$

or

$$Si(OH)_4 + Si(OEt)_4 \longrightarrow SiO_2 + 4\ EtOH$$

Preparation of nanocomposites

(a) Mechanical alloys

Before the realization of the functionalized nanocomposites, mechanical alloys based on matrix PCL with a percentage of silica of 1, 3 and 5% have been prepared. The silica employed was Aerosil® 90 (A90). At first three different mixtures have been prepared in the following proportions:

- Mechanical alloy **PCL/A90 1%** = 49.5 g of PCL + 0.5 g of A90
- Mechanical alloy **PCL/A90 3%** = 48.5 g of PCL + 1.5 g of A90
- Mechanical alloy **PCL/A90 5%** = 47.5 g of PCL + 2,5 g of A90

The preparation of the material has been made in a static mixer type Brabender (Rheocord) to a temperature of 100°C and for a time of 10 minutes, 8 rpm for the first 5 minutes and then 32 rpm for the remaining time. 1,5 g of the material thus obtained have been compression moulded to a temperature of 100°C for 3 minutes and to a pressure of 5 Kg/cm^2, in order to obtain a film having the dimensions of approximately 10x10cm and a thickness between 80 and 100 μm.

(a) Functionalised nanocomposites

For allowing a direct comparison, the nanocomposites have been realized having the same percentages of nanosilica of mechanical alloys that is, 1, 3 and 5%. It has been employed the functionalised silica Aerosil® R7200 (AR72) with a vinyl end group. Moreover the PCL has been *in situ* functionalised with glycidylmethacrylate (GMA) using dibenzoylperoxide (DBPO) as radical source. The three different mixtures have been prepared in the following proportions:

- nanocomposite PCL-g-(GMA-AR72 1%) = 49.5g of PCL + 0.5g of AR72 + 0.495g DBPO + 4.95g of GMA
- nanocomposite PCL-g-(GMA-AR72 3%) = 48.5g of PCL + 1.5g of AR72 + 0.485g DBPO + 4.85g of GMA
- nanocomposite PCL-g-(GMA-AR72 5%) = 47.5g of PCL + 2.5g of AR72 + 0.475g DBPO + 4.75g of GMA

The preparation has been made in a static mixer type Brabender (Rheocord) initially to a temperature of 100°C for 10 minutes in order to obtain the formation of the GMA functionalised PCL and then for further 10 min to graft the functionalised silica on the PCL previously obtained (reactive mixing) progressively increasing the number of turns from 8 to 32 rpm. Also for these materials films have been prepared as previously described.

As reference and control materials also the following samples have been prepared:

- PCL functionalised with 1% GMA (PCL-g-GMA)
- PCL functionalised with GMA + silica A90 1% (PCL-g-GMA/A90 1%)

Differential thermal analysis (DSC)

Differential thermal analysis on polymer matrix (PCL) has been carried out using the following thermal cycle: heating of the sample from 30 to 180°C at a scanning speed of 10°C/min then cooling to - 100°C to the scanning speed of 10°C/min and successive heating till 180°C to the same scanning speed. The nanocomposites have been subjected to the same thermal cycle. The calorimeter used was Mettler DSC 30.

Mechanical tests

Tensile mechanical tests have been performed on samples in the form of dog's bone just like the norm prescribes, and tested in traction in order to obtain relevant mechanical parameters. The tests have been carried out on a Instron dynamometer model 4505 at room temperature with a traction speed of 10mm/sec. The analyses have been performed on previously described mechanical alloys and nanocomposites.

Scanning and transmission electron microscopy (SEM and TEM)

The morphology of the materials has been evaluated examining the fracture surfaces by SEM. The samples were fractured in liquid nitrogen, mounted on a support and metallized with a Au/Pd alloy. The microscope used was a Philips XL 20.

TEM allowed the observation of the as obtained silica nanoparticles. The nanoparticles were collected from a water suspension directly onto TEM supports and observed. The microscope used was a Philips EM 430.

Elution test

The samples, in form of thin layers, were first sterilized in a steam autoclave and then placed in a suitable ratio weight/volume of Dulbecco's modified Eagle's medium (0.1 g/ml medium) containing penicillin (100 units/ml), streptomycin (100 μg/ml) and fungizone (1.2 μg/ml), and left under stirring at room temperature either for 24 or for 48 hours. The corresponding extracts were added, as such (100%) or diluted 1:1 (50%), to confluent 3T3 cells, that had been plated two days before in multiwell-12. The cultures were incubated at 37°C in a 5% CO_2 humidified atmosphere and cell vitality at 24 and 48 hours was assessed by MTT test.

Preparation and characterization of bone marrow MSCs

The preparation of MSCs was performed employing heparinized human bone marrow from healthy volunteers, after informed consent, following the method of Friedenstein et al. with some modifications [12,136]. Taking into account that a number of factors, as the donor-to-donor variability, age, gender, etc, might affect the biological response, four different MSCs populations were obtained in order to enhance data reliability.

Briefly, the bone marrow sample was diluted 1:5 with Opti-MEM (Life Technologies, MD, USA), containing 10% fetal calf serum (FCS), penicillin 100 units/ml, streptomycin 100 μg/ml and sodium ascorbate 50 μg/ml (growth medium), placed in 100 mm polystyrene dishes and incubated at 37°C in a 5% CO_2 humidified atmosphere.

After 48 h, the medium was collected and centrifuged at 800 rpm. Half of the supernatant (conditioned medium, CM) was aspirated, added with an equal volume of fresh medium and placed in the first plate. The medium remaining in the tube was used to resuspend the cellular pellet and, after the addition of an equivalent volume of fresh culture medium, the entire cellular suspension was plated in a second dish. The day after, also the medium of the second plate was centrifuged: the pellet, containing all non-adherent cellular elements, was discarded while the corresponding CM was added again to the dish.

In 2-4 days, several foci of adherent spindle-like cells started to appear and to develop until to colonize the whole dish surface in the following two weeks. During this period fresh medium was added every 3 days, each time leaving one half of the old medium. After the reaching of the confluence, cells were trypsinized and amplified always diluting the conditioned medium.

MSCs were the analyzed by flow cytometry using a wide panel of monoclonal antibodies. The cells expressed specific surface markers, such as CD13, CD29, CD44, CD105, CD166, and were negative for hematopoietic cell markers CD14, CD34 and CD45.

Cultures between the 3rd and 5th passage were used in our experiments. The cells harvested from each donor were kept separately and not pooled with other preparations.

In presence of osteogenic inducers, the stromal cells go through a complex series of events, at the end of which the multipotent cells become terminally differentiated. Alkaline phosphatase and osteocalcin are phenotypic markers for early-stage differentiated and terminally differentiated osteoblasts, respectively.

Direct contact test

The samples, in form of discs (21 mm diameter and 0.1 mm thickness), were attached with biological silicone (NuSil Silicone Technology, CA, USA) to the bottom of 12-well plates and sterilized 24 h with PBS containing penicillin (1000 units/ml), streptomycin (1000 μg/ml) and fungizone (12.5 μg/ml) at 37°C in a 5% CO_2 humidified atmosphere. The materials were then placed in Opti-MEM containing antibiotics and fungizone (at concentrations ten-fold lower than those indicated above), 10% FCS and 50 μg/ml sodium ascorbate (complete medium) and kept in the incubator overnight before plating the cells.

MSCs were seeded on discs at a density of 15,000 cells/cm^2 in complete culture medium and incubated for up to 2 weeks at 37°C in a 5% CO_2 humidified atmosphere. Cell adhesion and vitality, as well as alkaline phosphatase activity and osteocalcin levels were evaluated.

MTT test

The MTT assay is a very simple and useful tool for evaluating cell vitality and proliferation. The key component is 3-(4,5-dimethylthiazol-2-yl)-2,5-diphenyltetrazolium bromide (MTT). Mitochondrial dehydrogenases of living cells reduced the tetrazolium ring, yielding a blue formazan product, which was measured spectrophotometrically. The amount of formazan produced is proportional to the number of viable cells present. MTT (5 mg/ml in DMEM without phenol red) was added to the wells in an amount equivalent to 10% of the culture medium. After an incubation of 4 h at 37°C, the liquid was aspirated and the insoluble formazan produced was dissolved in isopropanol. The optical densities were measured at 570 nm, subtracting background absorbance determined at 690 nm.

Alkaline phosphatase assay

After the removal of the medium, the wells were rinsed with 20 mM TRIS HCl-0.5 M NaCl, pH 7.4 (TBS) and the cells were solubilized with lysis buffer, that is TBS containing DTT (0.5 mM), PMSF (0.5 mM) and Triton X-100 (0.25%). After 30 min in ice, the cell lysates were centrifuged at 12,000g for 5 min, and the supernatants assayed for alkaline phosphatase activity. The protein concentration was determined using the protein assay kit (Bio-Rad Laboratories, CA, USA), according to the method of Bradford.

AP activity was determined by measuring the release of p-nitrophenol from disodium p-nitrophenyl phosphate (PNPP). The reaction mixture contained 0.1 M diethanolamine buffer, pH 10.5, 0.5mM $MgCl_2$ and 12 mM PNPP in a final volume of 0.5 ml. The reaction was initiated by the addition of the cell extract and the mixture was incubated at 37°C for 30 min. The reaction was stopped by adding 0.5 ml of 0.5 M NaOH and the absorbance was determined spectrophotometrically at 405 nm. The AP activity was normalized to the protein content and expressed as units/mg protein, where one unit was defined as the amount of enzyme that hydrolyzes 1 nmol of PNPP/min under the specified conditions.

Osteocalcin synthesis

To evaluate osteocalcin synthesis, confluent cultures grown on the different surfaces for 2 weeks were incubated in FCS-free Opti-MEM in presence of 0.1% bovine serum albumin and 100 nM 1,25- dihydroxycholecalciferol for 48 h. The levels of polypeptide secreted in the medium were measured by means of an immunoenzymatic assay (Biosource International, Camarillo, CA, USA) that utilizes monoclonal highly specific antibodies and a peroxidase as conjugated enzyme. The amount of osteocalcin was calculated in ng/ml and then normalized to the protein content.

Statistical analysis

All the experiments were performed in triplicate on at least two different cell preparations. Data are expressed as the mean ± standard deviation (SD) of values. The means of each experimental group were compared by unpaired Student's t-test, with the value of significance set at $P < 0.05$.

Novel Injectable Alginate/ N-Succinylchitosan/Calcium Sulphate Composites as Bone-Defect Fillers

A preformed scaffold might not conform to the irregular shape and size of a bone defect when implanted in the body. At the double aim to overcome this limitation and to minimize the invasivity of the surgical approach, injectable scaffolds have been designed [137].

These systems can be stored for a long time in form of soft putty-like constructs without losing their peculiarity that is just the injectability. Once injected in the defect, the material hardens making a porous and biodegradable structure that the cells at the site of bone defect can colonize finding in situ the appropriate environment in terms of cytokines and growth factors. A more promising alternative to this approach could consist in encapsulating stem cells and bioactive molecules in the injectable paste: after the injection and the hardening of the material, the cells and the growth factors are directly delivered to the site of bone repair.

At the aim to project a new injectable scaffold for bone tissue engineering application, Gomez d'Ayala *et al.* [128] developed a novel composite based on blends of alginate and a modified chitosan, N-succinylchitosan, containing calcium sulphate as mineral osteoinductive phase. It has been demonstrated that the action mechanism of calcium sulphate (CHS) occurs through the dissolution of the low soluble salt and the local precipitation of calcium phosphate (CP), due to the increase of calcium ion concentration and its contact with body fluids. The CP deposits conduce bone cells into the defect and accelerate tissue regeneration [122,124]. Therefore, it is necessary that the salt dissolution rate is compatible with bone growth inside the defect, in order to allow bone regeneration before the substitute completely disappears. Encapsulation of calcium sulphate into a slowly biodegradable polymer matrix represents a solution to the problem, as allows slowing down the salt dissolution, increasing the time of its permanence inside the bone defect.

Alginate (Alg) and chitosan (Ch) were chosen as component of the polymeric matrix because of their biocompatibility and adequate biodegradability. Unlike chitosan scaffolds, which can only be fabricated from acidic solutions, the chitosan-alginate scaffolds can be prepared from solutions at physiological or basic pH, which provide favourable environment for incorporating growth factors or proteins (as cytokines) with less risk of denaturation. Bone forming osteoblastic cells attached readily to chitosan-alginate scaffolds, proliferated well, and deposited calcified matrix [58] Moreover, in order to obtain a homogeneous paste in presence of water, chitosan was previously modified with succinic anhydride to get a water-soluble polymer, according to a previously reported procedure [85].

The succinylation reaction (illustrated in the **scheme 1**) consists of a condensation reaction between the polysaccharide amine group and the electrophilic carbonyl group of anhydride. The reaction involves the formation of an amidic bond with opening of the anhydride ring, which causes the introduction of acid carboxylic functions onto chitosan chains. It was necessary to dissolve the pristine chitosan in an acidic solution, because of its insolubility in water. However, in this condition amine groups of polymer are protonated and consequently their nucleophilicity is adversely affected; therefore, in order to preserve their nucleophilicity, a higher value of pH was necessary. As the final modified chitosan (sCh) was found to be soluble at pH 7, the pH of reaction solution was continuously adjusted to this value by a slow and simultaneous dropwise addition of NaOH solution, in order to avoid the chitosan precipitation in the course of the reaction.

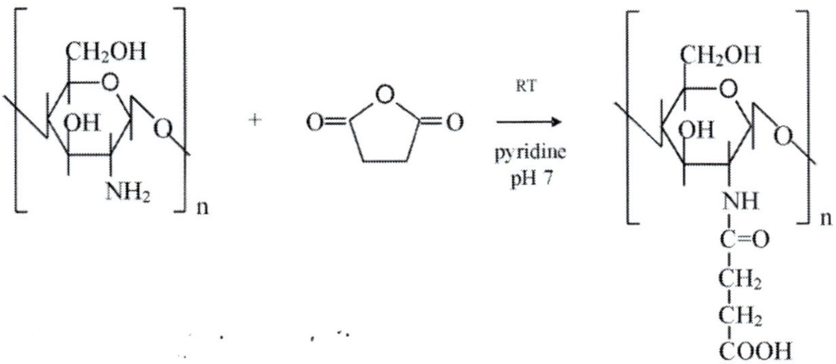

Scheme 1. Succinylation reaction of chitosan.

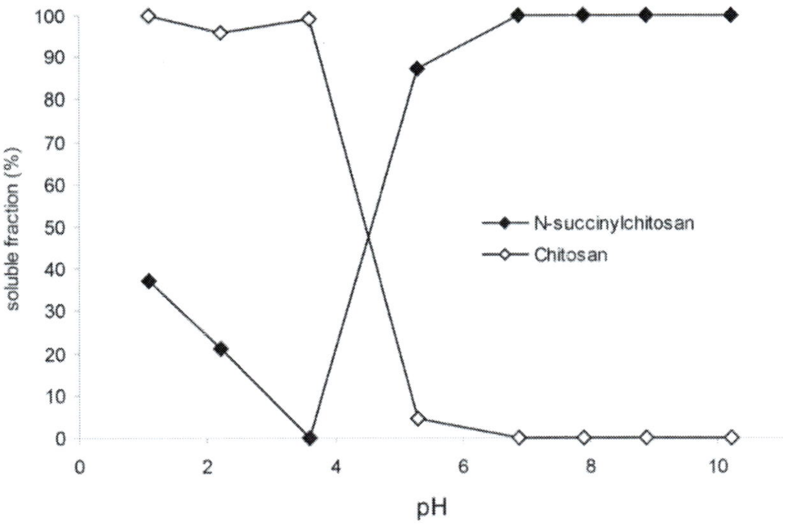

*Reprinted from *J. Biomed. Mat. Res, Part A, 81*, Development of a new calcium sulphate-based composite using alginate and chemically modified chitosan for bone regeneration, p. 817, Copyright 2007, with permission from Wiley.

Figure 14*. Solubility (% by weight) vs pH of chitosan and N-succinylchitosan.

Solubility tests were carried out in order to compare the behaviour of the two polymers at different pH values in aqueous solutions. In **Figure 14** the solubility curves of both pristine and modified polymers are shown as a function of the pH. In the investigated pH range the two polymers showed an opposite behaviour in terms of solubility, as expected. In fact, sCh is completely soluble in the range of pH between 6 and 12.

Table V. Yield strength and Young modulus (± STD) of composites with calcium sulphate and variable amounts of the polymeric components (Alg: Alginate; Ch: Chitosan; sCh: N-succinylchitosan).

Sample	Yield strength (MPa)	Young's Modulus (MPa)
CHS/Alg: 50/50	0,03 ± 0,01	20,2 ± 3,5
CHS/sCh: 50/50	0,80 ± 0,21	14,7 ± 4,3
CHS/Alg/sCh: 50/10/40	4,58 ± 1,63	427,4 ± 79,8
CHS/Alg/sCh: 50/20/30	13,62 ± 0,41	873,9 ± 145,8
CHS/Alg/sCh: 50/25/25	8,20 ± 0,15	370,8 ± 73,7
CHS/Alg/sCh: 50/30/20	9,02 ± 2,62	437,9 ± 83,5
CHS/Alg/sCh: 50/40/10	3,22 ± 0,59	444,2 ± 74,7
CHS/Alg/Ch: 50/30/20	0,04 ± 0,01	9,9 ± 0,6

To prepare the composites, finely grounded hemihydrate calcium sulphate, Alg and sCh powders were mixed in different compositions. The composite was prepared by adding dropwise distilled water to the mixture until the resulting paste became fully wet (1.8 ml of water each 1.5 g of material). The addition of water to the powder mixture promotes the transformation of calcium sulphate from the hemihydrate form to the dihydrate one. During this process calcium ions are available for the crosslinking of alginate. The setting of the materials is due to the transformation of the hemihydrate calcium sulphate into the dihydrate form and partially to the cross-linking of alginate with calcium ions. As a matter of fact, a mouldable and easy-to-spread paste, which get a hard cement in few hours, was obtained. This allows the material to be easily handled and moulded into the desired shape and to turn into a hard cement in a short setting time (2 hours).

Mechanical analyses in compression mode were performed on composites constituted by a fixed amount of CHS and variable ratios of the other two components and for sake of comparison on a composite in which N-succinylchitosan was substituted by plain chitosan. The results, shown in **Table V**, demonstrated that the blend of alginate and N-succinylchitosan causes a substantial reinforcement of composite.

In fact, by comparing mechanical properties of CHS/Alg/sCh composite with those of plain chitosan-based composite, it has been observed that in presence of the modified chitosan the mechanical performance is largely improved. It seems that the carboxylic groups grafted onto the chitosan modified chain, in addition to those of alginate, enhance the chelating power of the polysaccharide mixture because of an additional synergistic effect with respect to the classic "egg-box" effect of alginate (**Figure 15**) [138].

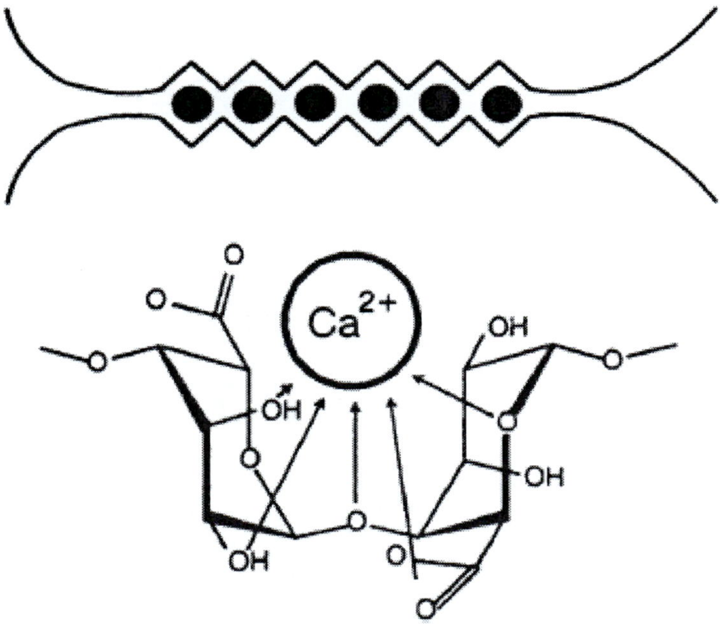

Figure 15. Egg-box model for the cross-linking of alginate with calcium ions.

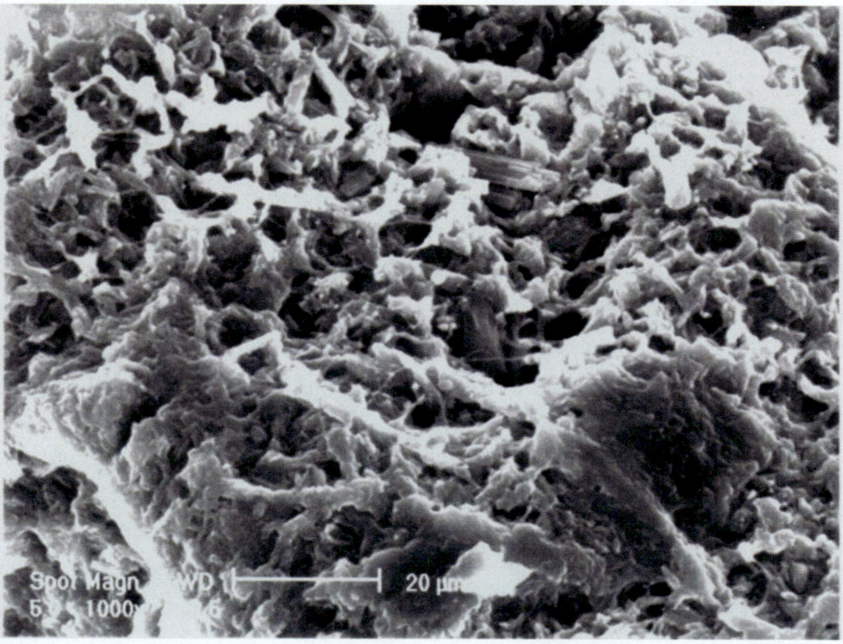

*Reprinted from *J. Biomed. Mat. Res, Part A, 81*, Development of a new calcium sulphate-based composite using alginate and chemically modified chitosan for bone regeneration, p. 817, Copyright 2007, with permission from Wiley.

Figure 16*. Scanning electron micrographs of CHS/Alg/sCh composite.

This was demonstrated also by a rheological study of a blend of plain alginate and N-succinyl-chitosan polymers in presence of a slow release calcium ions source [139]. This study showed that, in favourable conditions of concentration and composition, most of the chitosan molecules can effectively interact with the alginate network and might co-crosslink with alginate through calcium ions. A significant acceleration of the gelation kinetic observed for a given alginate/N-succinylchitosan composition, with respect to the case of the plain alginate gel, confirmed the hypothesis of a synergistic effect of the N-succinylchitosan in chelating calcium ions during the alginate gelation process.

Besides, morphological analyses showed that calcium sulphate crystals are embedded into the cross-linked polymer matrix with scarce evidence of discontinuity between the polymeric components and the inorganic filler (**Figure 16**). This could depend on the presence on both polymer chains of negatively charged carboxylic groups, which allow a stronger interaction with calcium ions.

Biological Evaluation

Biological assays were performed on a CHS/Alg/sCh 50/30/20 composite as such or with the addition of 10% demineralized bone matrix (DBM) (CHS/DBM/Alg/sCh 40/10/30/20).

To test *in vitro* the effect of this material on the biological response of MSCs, we made it to harden and then analyzed its influence on the growth and differentiation of bone marrow stromal cells. The samples, in form of little chips, derived from the mincing of the materials, were first autoclaved and then placed in a suitable ratio weight/volume of Opti-MEM (4 mg/ml medium) containing penicillin (100 units/ml), streptomycin (100 μg/ml) and fungizone (1.2 μg/ml), and left under agitation at room temperature for 24 hours.

This remarkable difference between the ratio weight/volume selected for the blends composites, compared to the above mentioned PCL-nanocomposites, is justified by the fast breaking up of the blends in aqueous environment forming a lot of tiny fragments that, in turn, are responsible for a large surface exchange between the materials and the medium.

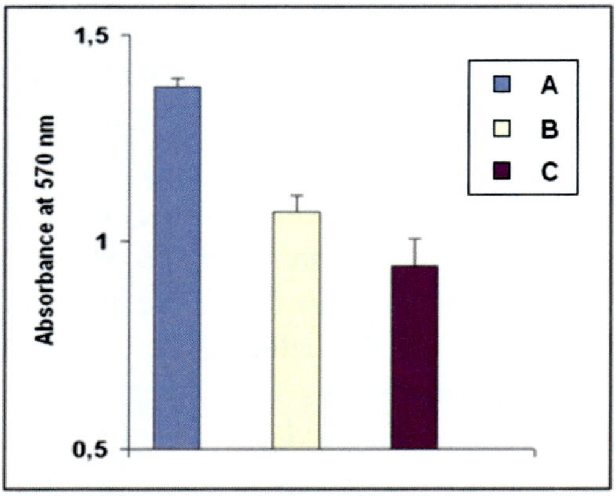

Figure 17. MTT test performed on MSCs incubated for 7 days in presence of complete medium added with: A = 1.8 mg/ml calcium sulphate; B = 4 mg/ml CHS/Alg/sCh (50/30/20); C = 4 mg/ml CHS/DBM/Alg/sCh (40/10/30/20). The reported values result from the average of three experiments, each performed in triplicate.

The corresponding extracts were centrifuged at 15,000xg for 20 min and the supernatants added (100%) to semiconfluent cultures of bone marrow MSCs. As control extract we used medium complete added with calcium sulphate (1,8 mg/ml). After 7 day-culture at 37°C in a 5% CO_2 humidified atmosphere, MSCs vitality was assessed by MTT assay (**Figure 17**) and cell morphology was evaluated by crystal violet staining (**Figure 18**).

As shown in Fig. 17, the proliferation of cells in presence of each extract was clearly lesser than control, especially with respect to DBM-containing material. In the latter case, to the minor cell growth corresponded a more differentiated morphology, namely a polygonal shape of the MSCs.

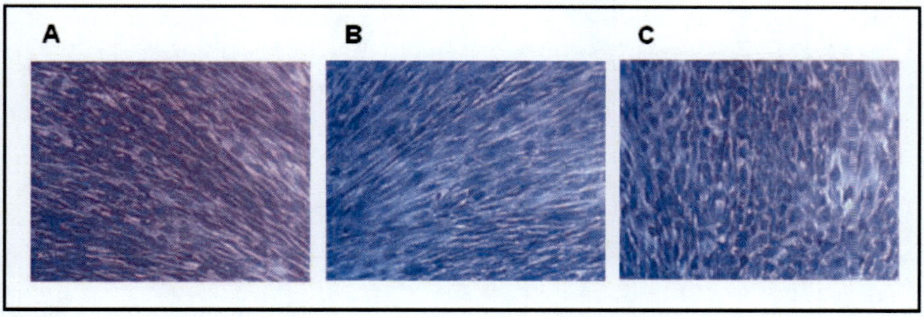

Figure 18. The microscopic appearance of MSCs cultured for 7 days in complete medium added with: A = 1.8 mg/ml calcium sulphate; B = 4 mg/ml CHS/Alg/sCh (50/30/20); C = 4 mg/ml CHS/DBM/Alg/sCh (40/10/30/20), after crystal violet staining (Magnification 100x).

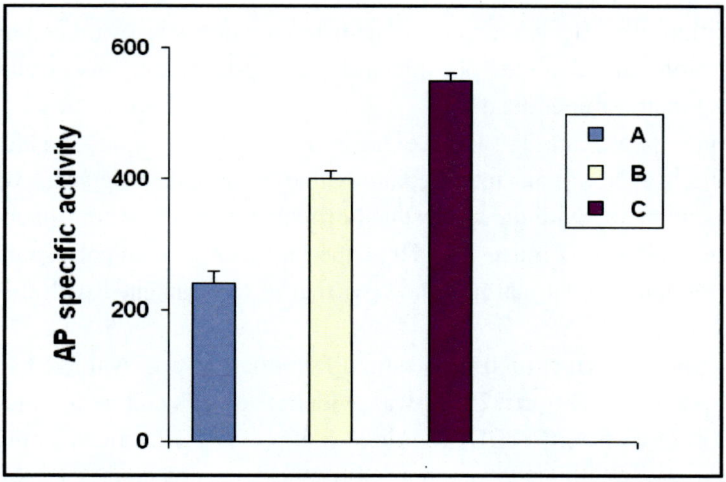

Figure 19. Alkaline phosphatase specific activity of stromal cells cultured for 7 days in complete medium added with: A = 1.8 mg/ml calcium sulphate; B = 4 mg/ml CHS/Alg/sCh (50/30/20); C = 4 mg/ml CHS/DBM/Alg/sCh (40/10/30/20). The reported values result from the average of three experiments, each performed in triplicate.

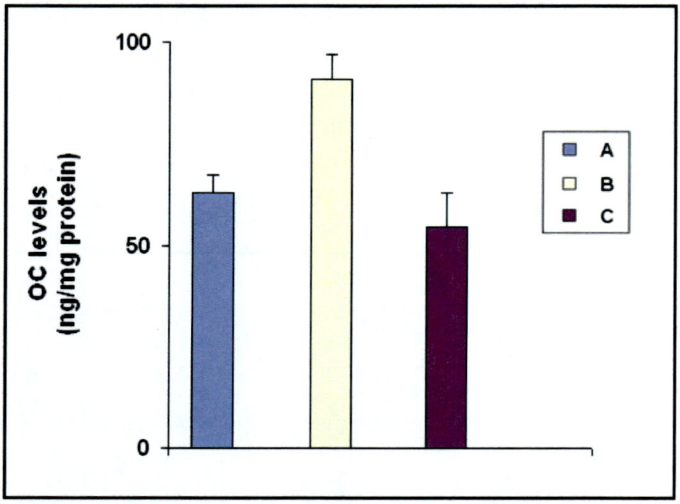

Figure 20. Osteocalcin synthesis **after 2 weeks MSCs' culture** in medium added with: A = 1.8 mg/ml calcium sulphate; B = 4 mg/ml CHS/Alg/sCh (50/30/20); C = 4 mg/ml CHS/DBM/Alg/sCh (40/10/30/20). The last three days the cells were incubated in FCS-free Opti-MEM in presence of 0.1% bovine serum albumin and 100 nM 1,25-dihydroxycolecalciferol. The reported values result from the average of three experiments, each performed in triplicate.

In addition, the effects on osteogenic differentiation was evaluated analyzing the expression of alkaline phosphatase and osteocalcin, as well as the extracellular matrix mineralization.

Alkaline phosphatase was expressed at the highest value with CHS/DBM/Alg/sCh, at an intermediate value with CHS/Alg/sCh, while the lowest value matched with the highest cell growth extent of the control containing only calcium sulphate (**Figure 19**). Thus, the least cell growth corresponded a to more differentiated polygonal morphology, that in turn matched with the highest value of expression of AP.

In addition, the effect on osteoblastic differentiation was evaluated analyzing the osteocalcin levels (**Figure 20**). It was evident that cells cultured in presence of the extract derived from CHS/DBM/Alg/sCh expressed OC values similar to the control, whereas the highest amount of osteocalcin was expressed by MSCs grown in the presence of CHS/DBM/Alg/sCh extract.

After 3 week-culture in the presence of the extracts (changing the medium every 3 days), a staining with crystal violet was performed (**Figure 21**). It was apparent that, parallel to the higher proliferation, a dense network of collagen fibers covered up and hidden the control cells, while the cellular elements were more clearly distinguishable in the presence of each extract.

Finally, the extracellular matrix mineralization was tested by the quantification of the calcium levels. MSCs cultures were incubated for 2 weeks in the presence of the extracts and for the following 14 days in the presence an osteogenic medium composed of 100 nM dexamethasone and 10 mM β-glycerophosphate. Then, Alizarin Red assay showed a remarkably more intense and spread staining in control culture, whereas separate calcification nodules were apparent in extract-treated cells (**Figure 22**).

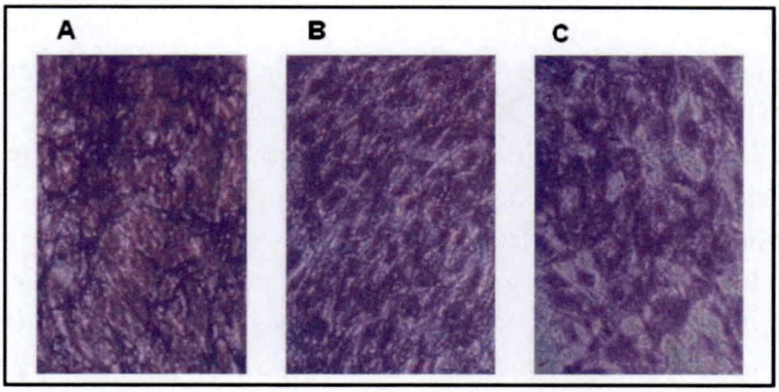

Figure 21. The microscopic appearance of MSCs cultured for 3 weeks in complete medium added with: A = 1.8 mg/ml calcium sulphate; B = 4 mg/ml CHS/Alg/sCh (50/30/20); C = 4 mg/ml CHS/DBM/Alg/sCh (40/10/30/20), after crystal violet staining (Magnification 100x).

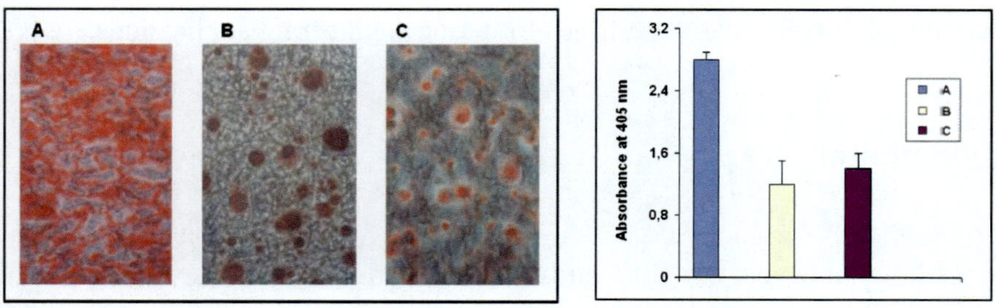

Figure 22. Alizarin Red staining. MSCs were cultured for 14 days in complete medium added with: A = 1.8 mg/ml calcium sulphate; B = 4 mg/ml CHS/Alg/sCh (50/30/20); C = 4 mg/ml CHS/DBM/Alg/sCh (40/10/30/20). For the following 2 weeks the cells were incubated in the presence of an osteogenic medium (100 nM dexamethasone - 10 mM β-glycerophosphate). First panel: microscopic appearance of cells (Magnification 100x). Second panel: quantification of matrix mineralization.

Conclusion

Calcium sulphate was encapsulated in a biodegradable polymeric matrix in order to prolong its structural integrity and slow-down its resorption. The analyses performed on the new material demonstrate that it should be suitable for biomedical applications and easy to use for the clinicians. In fact, the paste obtained by adding water to the powder mixture of the components shows a putty-like consistency and can be easily moulded to the desired shape.

Our results have demonstrated that the blend of alginate and N-succinylchitosan exhibited a substantial reinforcement of the corresponding obtained composite CHS/Alg/sCh, as a remarkable improvement of the mechanical performance compared with that of the plain alginate/chitosan- based composite (CHS/Alg/Ch) was found.

The addition of 10% demineralised bone matrix was also considered, in order to improve the osteoinductive potential of the composite.

As regards the biological features, assessed by elution test, it was evident that both formulations (Alg/sCh/CHS and Alg/sCh/DBM/CHS/) decreased cell growth, but in parallel greatly increased the osteogenic expression. In particular, the cells cultured in the presence of Alg/sCh/CHS, compared with control cells, expressed higher levels of the early marker AP, and levels even greater than the late marker OC. Furthermore, the extracellular matrix, even if apparently less rich of collagen fibers, was able to mineralize and to produce calcific nodules.

These encouraging results support the potential applications of the novel N-succinylchitosan/alginate-based injectable scaffolds as an improved alternative to other natural polymer-based scaffolds for tissue engineering applications.

Experimental

Materials

Medium molecular weight Chitosan (Ch), medium molecular weight Sodium Alginate (Alg) and Succinic Anhydride (SA) were all purchased from Fluka-Aldrich and used as received.

Hemihydrate calcium sulphate (CHS) was kindly provided by ClassImplant s.r.l., Rome, Italy.

Demineralized bone matrix (DBM) was purchased from Integra Orthobiologics, Irvine (CA), USA.

Pyridine, purchased from Fluka, was distilled under reduced pressure prior to use.

All the solvents were of analytical grade.

All cell culture biologics were purchased from (Life Technologies, MD, USA), and all chemicals were from Sigma Chemical Co (St. Louis, MO, USA) when not otherwise specified.

Preparation of N-succinylchitosan

To a 2% chitosan solution (w/v) in aqueous HCl (0,37%) a solution of succinic anhydride (SA) in pyridine (12,5% w/v) was dropwise added under vigorous stirring at room temperature (RT). SA was used in a 2:1 molar ratio with respect to chitosan. The pH of the reaction mixture was adjusted to 7,0 by simultaneously adding a NaOH solution (1,0M) dropwise. NaOH addition was continued until the pH was stabilized [52]. After 40 minutes of reaction, the modified polymer obtained was precipitated in a large excess of methanol, washed with acetone and desiccated. The as obtained product was finely grounded. In order to remove ungrafted SA, the resulting powder was dissolved and dialysed in a pH=10 NaOH solution for 48h using a dialysis membrane bag with a molecular weight cut-off of 10 kDa.

Mechanical tests

Mechanical tests were performed on cylindrical samples (diameter 13 mm, height 15 cm) obtained by compression moulding the wet powder mixture at room temperature and by leaving it at room temperature for setting. Samples were then analysed in compressive mode at room temperature on an Instron dynamometer model 4505 at a cross-head speed of 1 mm/min.

Osteogenic markers (AP, OC) evaluation

The same protocol described in the case of PCL-silica nanocomposites has been adopted for these novel alginate/N-succinylchitosan/calcium sulphate composites.

Extracellular matrix mineralization

In order to induce extracellular matrix mineralization, control cells were cultured for 3 weeks in the presence of an osteogenic medium (culture medium supplemented with 10 mM β-glycerophosphate, 100 nM dexamethasone). At the end of this period, the calcium levels were measured colorimetrically using Alizarin Red Staining (ARS). Briefly, cell layers were washed with PBS, fixed in

10% formaldehyde for 15 min and then washed with dH_2O prior to adding 1 ml of 40 mM ARS (pH 4.1) per well. The plate was incubated for 20 min and then washed four times with dH_2O. For quantification of staining, 800 μl 10% acetic acid was added to each well, and the plate was incubated at room temperature for 30 min with shaking. The cell layer was then scraped from the plate, collected together with the acetic acid and vortexed for 30 sec. The slurry was centrifuged at 20,000xg for 15 min and 500 μl of the supernatant was transferred into a new 1.5-ml microfuge tube. Then 200 μl of 10% ammonium hydroxide was added to neutralize the acid. Aliquots of the obtained solution were read in triplicate at 405 nm in 96-well.

References

[1] Langer, R. & Vacanti, J. P. (1993). Tissue engineering. *Science*, *260*, 920-6.

[2] MacArthur, B. & Oreffo, R. O. C. (2005). Bridging the gap. *Nature*, *433*, 19.

[3] Friedenstein, A. J., Chailakhyan, R. K. & Gerasimov, U. V. (1987). Bone marrow osteogenic stem cells: in vitro cultivation and transplantation in diffusion chambers. *Cell Tissue Kinet*, 20, 263-72.

[4] Prockop, D. J. (1997) Marrow stromal cells as stem cells for nonhematopoietic tissues. *Science*, 27, 671-74.

[5] Pittenger, M. F., Mackay, A. M., Beck, S. C., Jaiswal, R. K., Douglas, R., Mosca, J. D., Moorman, M. A., Simonetti, D. W., Craig, S. & Marshak, D. R. (1999). Multilineage potential of adult human mesenchymal stem cells. *Science*, *284*, 143-147.

[6] Toma, C., Pittenger, M. F., Cahill, K. S., Byrne, B. J. & Kessler, P. D. (2002). Human mesenchymal stem cells differentiate to a cardiomyocyte phenotype in the adult murine heart. *Circulation*, *105*, 93-98.

[7] Kopen, G. C., Prockop, D. J. & Phinney, D. G. (1999). Marrow stromal cells migrate throughout forebrain and cerebellum, and they differentiate into astrocytes after injection into neonatal mouse brains. *Proc Natl Acad Sci*, USA, 96, 10711-16.

[8] Woodbury, D., Schwarz, E. J., Black, I. B. & Prockop, D. J. (2000). Adult rat and human bone marrow stromal cells differentiate into neurons. *Neurosci Res*, 61, 364-70.

[9] Seshi, B., Kumar, S. & Sellers, D. (2000). Human bone marrow stromal cell: coexpression of markers specific for multiple mesenchymal cell lineages. *Blood Cells Mol Dis*, 26, 234-46.

[10] Herzog, E. L., Chai, L. & Krause, D. S. (2003). Plasticity of marrow-derived stem cells. *Blood*, *102*, 3483-93.

[11] Di Nicola, M., Carlo-Stella, C., Magni, M., Milanesi, M., Longoni, P. D., Matteucci, P., Grisanti, S. & Gianni, A. M. (2002). Human bone marrow stromal cells suppress T-lymphocyte proliferation induced by cellular or nonspecific mitogenic stimuli. *Blood*, *99*, 3838-3843.

[12] Oliva, A., Passaro, I., Di Pasquale, R., Di Feo, A., Criscuolo, M., Zappia, V., Della Ragione, F., D'Amato, S., Annunziata, M. & Guida, L. (2005). Ex vivo expansion of bone marrow stromal cells by platelet-rich plasma: a promising strategy in maxillo-facial surgery. *Int J Immunopathol Pharmacol*, *18*, 47-53

[13] Meijer, G. J., de Bruijn, J. D., Koole, R. & van Blitterswijk, C. A. (2007). Cell-based bone tissue engineering. *PLoS Med*, *4*, 260-264.

[14] Caplan, A. I. (2007). Adult mesenchymal stem cells for tissue engineering versus regenerative medicine. *J Cell Physiol*, *213*, 341-347.

[15] Wu, Y., Chen, L., Scott, P. G. & Tredget, E. E. (2007). Mesenchymal stem cells enhance wound healing through differentiation and angiogenesis. *Stem Cells*, *25*, 2648-2659.

[16] Boyle, W. J., Simonet, W. S. & Lacey, D. L. (2003). Osteoclast differentiation and activation. *Nature*, *423*, 337-342.

[17] Ilan, D. I. & Ladd, A. L. (2003). Bone graft substitutes. *Operat Tech Plast Reconstr Surg*, 9,151-160.

[18] Damien, C. J. & Parsons, J. R. (1991). Bone grafts and bone graft substitutes: A review of current technology and applications. *J Appl Biomater*, *2*, 187-208.

[19] Murugan, R. & Ramakrishna, S. (2005) Development of nanocomposites for bone grafting. *Composites Science and Technology*, *65*, 2385-2406.

[20] de Groot, K. (1983). Ceramic of calcium phosphate: Preparation and properties. In: K. de Groot (Ed), *Bioceramics of Calcium Phosphate* (100-114). Boca Raton, FL: CRC.

[21] Sato, S., Koshino, T. & Saito, T. (1998). Osteogenic response of rabbit tibia to hydroxyapatite particle–plaster of Paris mixture. *Biomaterials*, *19*, 1895-1900.

[22] Jarcho, M. (1981). Calcium phosphate ceramics as hard tissue prosthetics. *Clin Orthop Relat Res*, *157*, 259-278.

[23] Ogiso, M. (1998). Reassessment of long-term use of dense HA as dental implant: Case report. *J Biomed Mater Res B Appl Biomater*, *43*, 318-320.

[24] Hench, L. L. (1991). Bioceramics: From concept to clinic. *J Am Ceram Soc*, *74*, 1487-1510.

[25] Bajpai, P. K. (1992) Ceramics: a novel device for sustained long-term delivery of drugs. In: J. E. Hulbert, & S. F. Hulbert (Eds.), *Bioceramics* (87-99). Terre Haute.

[26] Murugan, R. & Ramakrishna, S. (2004). Bioresorbable composite bone paste using polysaccharide based nano hydroxyapatite. *Biomaterials*, *25*, 3829-3835.

[27] Komath, M. & Varma, H. K. (2003). Development of a full injectable calcium phosphate cement for orthopedic and dental applications. *Bull Mater Sci*, *26*, 415-422.

[28] Ural, E., Kesenci, K., Fambri, L., Migliaresi, C. & Piskin, E. (2000). Poly(D,Llactide/e-caprolactone)/hydroxyapatite composites. *Biomaterials*, *21*, 2147-2154.

[29] Walsh, D., Furuzono, T. & Tanaka, J. (2001). Preparation of porous composite implant materials by in situ polymerization of porous apatite containing e-caprolactone or methyl methacrylate. *Biomaterials*, *22*, 1205-1211.

[30] Coombes, A. G. A., Rizzi, S. C., Williamson, M., Barralet, J. E., Downes, S. & Wallace, W. A. (2004). Precipitation casting of polycaprolactone for applications in tissue engineering and drug delivery. *Biomaterials*, *25*, 315-325.

[31] Marra, K. G., Szem, J. W., Kumta, P. N., DiMilla, P. A. & Weiss, L .E. (1999). In vitro analysis of biodegradable polymer blend/hydroxyapatite composites for bone tissue engineering. *J Biomed Mater Res*, *47*, 324-335.

[32] Reis, M. C., Reis, R. L., Feijen, M. B., Feijen, D. W. & Feijen, J. (2003). Development and properties of polycaprolactone/hydroxyapatite composite biomaterials. *J Mater Sci-Mater M*, *14*, 103-107.

[33] Rizzi, S. C., Heath, D. J., Coombes, A. G., Bock, N., Textor, M. & Downes, S. (2001). Biodegradable polymer/hydroxyapatite composites: Surface analysis and initial attachment of human osteoblasts. *J Biomed Mater Res*, *55*, 475-486.

[34] Calandrelli, L., Immirzi, B., Malinconico, M., Volpe, M. G., Oliva, A. & Della Ragione, F. (2000). Preparation and characterisation of composites based on biodegradable polymers for “in vivo” application. *Polymer*, *41*, 8027-8033.

[35] Itoh, S., Kikuchi, M., Takakuda, K., Nagaoka, K., Koyama, Y., Tanaka, J. & Shinomiya, K. (2002). Implantation study of a novel hydroxyapatite/collagen (HAp/col) composite into weight-bearing sites of dogs. *J Biomed Mater Res*, *63*, 507-15.
[36] Zhao, F., Yin, Y., Lu, W. W., Leong, J. C., Zhang, W., Zhang, J., Zhang, M. & Yao, K. (2002). Preparation and histological evaluation of biomimetic three-dimensional hydroxyapatite/chitosan-gelatin network composite scaffolds. *Biomaterials*, *23*, 3227-34.
[37] Murugan, R. & Panduranga Rao, K. (2002). Biodegradable coralline hydroxyapatite composite gel using natural alginate. *Bioceramics*, *15*, 407-10.
[38] Murugan, R. & Panduranga Rao, K. (1998). Synthesis and characterization of bioactive chitosan bonded bone cement for hard tissue replacement. In: Srinivasan KSV (Ed). *Proceedings of the IUPAC Macromolecules*. (638-641). New Delhi, Allied Publication.
[39] Lakes, R. (1993). Materials with structural hierarchy. *Nature*, *361*, 511-515.
[40] Hartgerink, J. D., Beniash, E. & Stupp, S. I. (2001). Self-assembly and mineralizationùof peptide-amphiphile nanofibers. *Science*, *294*, 1684-1688.
[41] Suchanek, W. & Yoshimura, M. (1998). Processing and properties of hydroxyapatite-based biomaterials for use as hard tissue replacement implants. *J Mater Res*, *13*, 94-117.
[42] TenHuisen, K. S., Martin, R. I., Klimkiewicz, M. & Brown, P. W. (1995). Formation and properties of a synthetic bone composite: hydroxyapatite-collagen. *J Biomed Mater Res*, *29*, 803-810.
[43] Webster, T. J., Siegel, R. W. & Bizios, R. (2000). Enhanced functions of osteoblasts on nanophase ceramics. *Biomaterials*, *21*, 1803-1810.
[44] Webster, T. J., Ergan, C., Doremus, R. H., Siegel, R. W. & Bizios, R. (2000). Specific proteins mediate enhanced osteoblast adhesion on nanophase ceramics. *J Biomed Mater Res*, *51*, 475-483.
[45] Webster, T. J., Siegel, R. W. & Bizios, R. (1999). Osteoblast adhesion on nanophase ceramics. *Biomaterials*, *20*, 1221-1227.
[46] Rhee, S. H. & Choi, J. Y. (2002). Preparation of a Bioactive Poly(methyl methacrylate)/Silica Nanocomposite *J Am Ceram Soc*, *85*, 1318-1320.
[47] Rhee, S. H. (2003). Effect of calcium salt content in the poly(ε-caprolactone)/silica nanocomposite on the nucleation and growth behavior of apatite layer. *J Biomed Mater Res Part A*, *67A*, 1131-1138;

[48] Yoo, J. J. & Rhee, S. H. (2004). Evaluations of bioactivity and mechanical properties of poly (ε-caprolactone)/silica nanocomposite following heat treatment *J Biomed Mater Res Part A*, *68A*, 401-410.

[49] Pecora, G., Andreana, S., Margarine, J. E., Covani, U. & Sottosanti, J. S. (1997). Bone regeneration with a calcium sulphate barrier. *Oral Surg Oral Med Oral Pathos*, *84*, 424-429.

[50] Pecora, G., De Leonardis, D., Ibrahim, N., Bovi, M. & Cornelini, R. (2001). The use of calcium sulphate in the surgical treatment of a "through and trough" periradicular lesion. *Int Endod J*, *34*, 189-196.

[51] Liu, Q., de Wijn, J. R., de Groot, K. & Van Blitterswijk, C. A. (1998). Surface modification of nano-apatite by grafting organic polymer. *Biomaterials*, *19*, 1067-1072.

[52] Jagur-Grodzinski, J. (2006). Polymers for tissue engineering, medical devices, and regenerative medicine. Concise general review of recent studies. *Polym Adv Technol*, *17*, 395-418

[53] Li, W. J. & Tuan, R. S. (2005). Polymeric scaffolds for cartilage tissue engineering. *Macromol Symp*, *227*, 65-75.

[54] Jarcho, M. (1981) Calcium phosphate ceramics as hard tissue prosthetics. *Clin Orthop Relat Res*, *157*, 259-278.

[55] Mori, M., Yamaguchi, M., Sumitomo, S. & Takai, Y. (2004). Hyaluronan-based biomaterials for tissue-engineering. *Acta Histochem Cytochem*, *37*, 1-5.

[56] Giavaresi, G., Tschon, M., Daly, J. H., Liggat, J. J., Fini, M., Torricelli, P. & Giardino, R. (2004). Natural and synthetic polyesters for musculoskeletal tissue repair: Experimental in vitro and in vivo evaluation. *Int J Artificial Organs*, *27*, 796-805.

[57] Bačáková, L., Filová, E., Rypáček, F., Švorčík, V. & Starý, V. (2004). Cell adhesion on artificial materials for tissue engineering. *Physiological Research*, *53*, S35-S45.

[58] Li, Z., Ramay, H. R., Hauch, K. D., Xiao, D. & Zhang, M. (2005). Chitosan-alginate hybrid scaffold for bone tissue engineering. *Biomaterials*, *26*, 3919-3928.

[59] Hench, L. L. & Wilson, J. (1984). Surface-active biomaterials. *Science*, *226*, 630-636.

[60] Cowin, S. C., Vanburskirk, W. C. & Ashaman, R. B. (1987). In: R. Skalak, & S. Chien (Eds.), *Handbook of Bioengineering* (97). New York: McGraw Hill.

[61] Sato, T., Chen, G., Ushida, T., Ishii, T., Ochiai, N., Tateishi, T. & Tanaka, J. (2004). Evaluation of PLLA-collagen hybrid sponge as a scaffold for cartilage tissue engineering. *Mater Sci Eng C*, *24*, 365-372.

[62] Schantz, J. T., Brandwood, A., Hutmacher, D. W., Khor, H. L. & Bittner, K. (2005). Osteogenic differentiation of mesenchymal progenitor cells in computer designed fibrin-polymer-ceramic scaffolds manufactured by fused deposition modeling. *J Mater Sci-Mater Med*, *16*, 807-819.

[63] Shogren, R. L. & Bagley, E. B. (1999). Natural polymer as advanced materials: some research needs and directions. In S. H. Imam, R. V. Greene, & B. R. Zaidi (Eds.), *Biopolymers - utilizing nature's, advanced materials, ACS symposium series*, *723* (2-11). Washington, Oxford University Press.

[64] Kaplan, D. L. (1998). *Biopolymers from renewable resources: Introduction to polymers from renewable resources*. In: D. L. Kaplan (Ed). Berlin, Germany: Springer Verlag.

[65] Yannas, I. V. (1996). *Biomaterials science: An introduction to materials in medicine. Natural materials.* In: B. D. Ratner, A. S. Hoffman, F. J, Schoen, & J. E. Lemons (Eds.), California, USA: Academic Press.

[66] Miyamoto, T., Takahashi, S., Ito, H., Inagaki, H. & Noishiki, Y. (1989). Tissue biocompatibility of cellulose and its derivatives. *J Biomed Mater Res*, *23*, 125-133.

[67] Hayashi, T. (1994). Biodegradable polymers for biomedical uses. *Prog Polym Sci*, *19*, 663-702.

[68] Ravi Kumar, M. N. V., Muzzarelli, R. A. A., Muzzarelli, C., Sashiwa, H. & Domb, A. J. (2005). Chitosan chemistry and pharmaceutical perspectives. *Chem Rev*, *104*, 6017-6083.

[69] Jayakumar, R., Prabaharan, M., Reis, R. L. & Mano, J. F. (2005). Graft copolymerized chitosanpresent status and applications. *Carbohydr Polym*, *62*, 142-158.

[70] Bernkop-Schnurch, A., Hornof, M. & Guggi, D. (2004). Thiolated chitosans. *Eur J Pharm Biopharm*, *57*, 9-17.

[71] Ishihara, M., Ono, K., Sato, M., Nakanishi, K., Saito, Y., Yra, H., Matsui, T., Hattori, H., Fujita, M., Kikuchi, M. & Kurita, A. (2001). Acceleration of wound contraction and healing with a photocrosslinkable chitosan hydrogel. *Wound Rep Reg*, *9*, 513-21.

[72] Razania, A. & Healy, K. E. (1999). Biomimetic peptide surfaces that regulate adhesion, spreading, cytoskeletal organization, and mineralization of the matrix deposited by osteoblast-like cells. *Biotechnol Prog*, *15*, 19-32.

[73] Park, Y. J., Lee, Y. M., Park, S. N., Sheen, S. Y., Chung, C. P. & Lee, S. J. (2000). Platelet derived growth factor releasing chitosan sponge for periodontal bone regeneration. *Biomaterials*, *21*, 153-159.

[74] Klokkevold, P. R., Vandemark, L., Kenney, E. B. & Bernard, G. W. (1996). Osteogenesis enhanced by chitosan poly(N-acetyl glucosaminoglycan) in vitro. *J Periodontol*, *67*, 1170-175.

[75] Muzzarelli, R. A., Mattioli-Belmonte, M., Tietz, C., Biagini, R., Ferioli, G., Brunelli, M. A., Fini, M., Giardino, R., Ilari, P. & Biagini, G. (1994). Stimulatory effect on bone formation exerted by a modified chitosan. *Biomaterials*, *15*, 1075-1081.

[76] Chandy, T. & Sharma, C. P. (1993). Chitosan matrix for oral sustained delivery of ampicillin. *Biomaterials*, *14*, 939-44.

[77] Hirano, S., Tsuchida, H. & Nagao, N. (1989). N-acetylation in chitosan and the rate of its enzymatic hydrolysis. *Biomaterials*, *10*, 574-576.

[78] Kuen, Y. L., Wuan, S. H. & Won, H. P. (1995). Blood compatibility and biodegradability of partially N-acylated chitosan derivatives. *Biomaterials*, *16*, 1211-216.

[79] Park, Y. J., Lee, Y. M., Park, S. N., Sheen, S. Y., Chung, C. P. & Lee, S. J. (2000). Platelet derived growth factor releasing chitosan sponge for periodontal bone regeneration. *Biomaterials*, *21*, 153-159.

[80] Klokkevold, P. R., Vandemark, L., Kenney, E. B. & Bernard, G. W. (1996). Osteogenesis Enhanced by Chitosan in vitro. *J Periodontol*, *67*, 1170-1175.

[81] Gutowska, A., Jeong, B. & Jasionowski, M. (2001). Injectable gels for tissue engineering. *Anatomical Record*, *263*, 342-349.

[82] Malette, W. G., Quigley, H. & Adickes, E. D. (1986). *Chitosan Effect in Nature and Technology: Chitin in nature and technology*. In: R. Muzzarelli, C. Jeuiaux, & Gooddy, G. W. (Eds.), New York, Plenum Press.

[83] Somashekar, D. & Joseph, R. (1996). Chitosanase-properties and applications: a review. *Bioresour Technol*, *55*, 35-45.

[84] Murugan, R. & Ramakrishna, S. (2004). Bioresorbable composite bone paste using polysaccharide based nano hydroxyapatite. *Biomaterials*, *25*, 3829-3835.

[85] Yang, D., Jin, Y., Ma, G., Chen, X., Lu, F. & Nie, J. (2008). Fabrication and characterization of chitosan/PVA with hydroxyapatite biocomposite nanoscaffolds. *J Appl Polym Sci*, *110*, 3328-3335.

[86] Aiedeh, K. & Taha, M. O. (1999). Synthesis of chitosan succinate and chitosan phthalate and their evaluation as suggested matrices in orally

administered, colon-specific drug delivery systems. *Arch Pharm*, *332*, 103-107.

[87] Liu, Y. F., Huang, K. L., Peng, D. M., Ding, P. & Li, G. Y. (2007). Preparation and characterization of glutaraldehyde cross-linked O-carboxymethylchitosan microspheres for controlled delivery of pazufloxacin mesilate. *Int J Biol Macromol*, *41*, 87-93.

[88] Wang, X. (1998). Calcium alginate gels: formation and stability in the presence of an inert electrolyte. *Polymer*, *39*, 2759-2764.

[89] LeRoux, M. A. (1999). Compressive and shear properties of alginate gel: effects of sodium ions and alginate concentration. *J Biomed Mater Res*, *47*, 46-53.

[90] Mumper, R. J. (1994). Calcium alginate beads for the oral delivery of TGF-beta1. *J Control Release*, *30*, 241-251.

[91] Zhang, S. M., Cui, F. Z., Liao, S. S., Zhu, Y. & Han, L. (2003). Synthesis and biocompatibility of porous nano-hydroxyapatite/ collagen/ alginate composite *J Mater Sci: Mater Med*, *14*, 641-645.

[92] Parhi, P., Ramanan, A. & Ray, A. R. (2006). Preparation and characterization of alginate and hydroxyapatite-based biocomposite. *J Appl Polym Sci*, *102*, 5162-5165.

[93] Tampieri, A., Sandri, M., Landi, E., Celotti, G., Roveri, N., Mattioli-Belmonte, M., Virgili, L., Gabbanelli, F. & Biagini, G. (2005). HA/alginate hybrid composites prepared through bio-inspired nucleation. *Acta Biomaterialia*, *1*, 343-351.

[94] Bottini, M., Cerignoli, F., Dawson, M. I., Magrini, A., Rosato, N. & Mustelin, T. (2006). Full-Length Single-Walled Carbon Nanotubes Decorated with Streptavidin-Conjugated Quantum Dots as Multivalent Intracellular Fluorescent Nanoprobes. *Biomacromolecules*, *7*, 2259-2263.

[95] Bottini, M., Magrini, A., Di Venere, A., Bellucci, S., Dawson, M. I., Rosato, N., Bergamaschi, A. & Mustelin, T. (2006). Synthesis and Characterization of Supramolecular Nanostructures of Carbon Nanotubes and Ruthenium-Complex Luminophores. *J Nanosci Nanotechnol*, *6*, 1381-1386.

[96] Wang, J., Zhang, K. & Zhu, Y. (2005). Synthesis of SiO_2-Coated $ZnMnFe_2O_4$ nanospheres with Improved Magnetic Properties. *J Nanosci Nanotechnol*, *5*, 772-775.

[97] Lin, Y. W., Liu, C. W. & Chang, H. T. (2006). Synthesis and Properties of Water-Soluble Core-Shell-Shell Silica-CdSe/CdS-Silica Nanoparticles. *J Nanosci Nanotechnol*, *6*, 1092-1100.

[98] Lin, Y. S., Tsai, C. P., Huang, H. Y., Kuo, C. T., Hung, Y., Huang, D. M., Chen, Y. C. & Mou, C. Y. (2005). Well-Ordered Mesoporous Silica Nanoparticles as Cell Markers. *Chem Mater*, *17*, 4570-4573.

[99] Bharali, D. J., Klejbor, I., Stachowiak, E. K., Dutta, P., Roy, I., Kaur, N., Bergey, E. J., Prasad, P. N. & Stachowiak, K. M. (2005). Organically modified silica nanoparticles: A nonviral vector for *in vivo* gene delivery and expression in the brain. *PNAS*, *102*, 11539-11544.

[100] Yang, J. M., Lu, C. S., Hsu, Y. G. & Shih, C. H. (1997). Mechanical Properties of Acrylic Bone Cement Containing PMMA-SiO2 Hybrid Sol-Gel Material, *J Biomed Mater Res*, *Appl Biomater*, *38B*, 143-54.

[101] Wei, Y. & Jin, D. (1997). A New Class of Organic-Inorganic Hybrid Dental Materials. *Polym Prepr*, *38*, 122-123.

[102] Paul, P. P., Timmons, S. F. W. & Machowski, J. (1997). Organic-Inorganic Hybrid Dental Restorative Composites. *Polym Prepr*, *38*, 124-125.

[103] Yoo, J. J. & Rhee, S. H. (2004). Evaluations of bioactivity and mechanical properties of poly (ε-caprolactone)/silica nanocomposite following heat treatment. *J Biomed Mater Res Part A*, *68A*, 401-410.

[104] Wei, J., Heo, S. J., Liu, C., Kim, D. H., Kim, S. E., Hyun, Y. T., Shin, J. W. & Shin, J. W. Preparation and characterization of bioactive calcium silicate and poly(ε-caprolactone) nanocomposite for bone tissue regeneration. *J Biomed Mater Res Part A*, *DOI:* 10.1002/jbm.a.32139

[105] Xu, T., Zhang, N., Nichols, H. L., Shi, D. & Wen, X. (2007). Modification of nanostructured materials for biomedical applications. *Mater Sci Eng C*, *27*, 579-594.

[106] Redhead, H. M., Davis, S. S. & Illum, L. (2001). Drug delivery in poly(lactide-co-glycolide) nanoparticles surface modified with poloxamer 407 and poloxamine 908: in vitro characterisation and in vivo evaluation. *J Control Release*, *70*, 353-363.

[107] Neal, J. C., Stolnik, S., Garnett, M. C., Davis, S. S. & Illum, L. (1998). Modification of the copolymers poloxamer 407 and poloxamine 908 can affect the physical and biological properties of surface modified nanospheres. *Pharmaceutical Research*, *15* 318-324.

[108] Park, Y. J., Nah, S. H., Lee, J. Y., Jeong, J. M., Chung, J. K., Lee, M. C., Yang, V. C. & Lee, S. J. (2003). Surface-modified poly(lactide-co-glycolide) nanospheres for targeted bone imaging with enhanced labeling and delivery of radioisotope. *J Biom Mater Res Part A*, *67A*, 751-760.

[109] Hrapovic, S., Liu, Y. L., Male, K. B. & Luong, J. H. T. (2004). Electrochemical biosensing platforms using platinum nanoparticles and carbon nanotubes. *Analytical Chemistry*, *76*, 1083-1088.

[110] Lin, Y. H., Lu, F., Tu, Y. & Ren, Z. F. (2004). Glucose biosensors based on carbon nanotube nanoelectrode ensembles. *Nano Letters*, *4*, 191-195.

[111] Artyukhin, A. B., Bakajin, O., Stroeve, P. & Noy, A. (2004). Layer-by-layer electrostatic self-assembly of polyelectrolyte nanoshells on individual carbon nanotube templates. *Langmuir*, *20*, 1442-1448.

[112] Islam, M. F., Rojas, E., Bergey, D. M., Johnson, A. T. & Yodh, A. G. (2003). High weight fraction surfactant solubilization of single-wall carbon nanotubes in water. *Nano Letters*, *3*, 269-273.

[113] Nakashima, N., Tomonari, Y. & Murakami, H. (2002). Water-soluble single-walled carbon nanotubes via noncovalent sidewall-functionalization with a pyrene-carrying ammonium ion. *Chemistry Letters*, *31*, 638-639.

[114] Qhobosheane, M., Santra, S., Zhang, P: & Tan, W. H. (2001). Biochemically functionalized silica nanoparticles *Analyst*, *126*, 1274-1278.

[115] Moghimi, S. M., Hunter, A. C. & Murray, J. C. (2001). Long-Circulating and Target-Specific Nanoparticles: Theory to Practice. *Pharmacol Rev*, *53*, 283-318.

[116] Xu, H., Yan, F., Monson, E. E. & Kopelman, R. (2003). Room-temperature preparation and characterization of poly (ethylene glycol)-coated silica nanoparticles for biomedical applications *J Biom Mater Res Part A*, *66A*, 870-879.

[117] Butterworth, M. D., Illum, L. & Davis, S. S. (2001). Preparation of ultrafine silica- and PEG-coated magnetite particles *Colloids and Surfaces A: Physicochem Eng Aspects*, *179*, 93-102.

[118] Li, S., Shah, A., Hsieh, A. J., Haghighat, R., Solomon Praveen, S., Mukherjee, I., Wei, E., Zhang, Z. & Wei, Y. (2007). Characterization of poly(2-hydroxyethyl methacrylate-silica) hybrid materials with different silica contents. *Polymer*, *48*, 3982-3989.

[119] Kang Y. M., Kim, K. H., Seol, Y. J. & Rhee, S. H. (2009). Evaluations of osteogenic and osteoconductive properties of a non-woven silica gel fabric made by the electrospinning method. *Acta Biomaterialia*, *5*, 462-469.

[120] Xua, H. H. K., Smith, D. T. & Simon, C. G. (2004). Strong and bioactive composites containing nano-silica-fused whiskers for bone repair. *Biomaterials*, *25*, 4615-4626.

[121] Beletskii, B. I., Shumskii, V. I., Nikitin, A. A. & Vlasova, E. B. (2000). Biocomposite calcium-phosphate materials used in osteoplastic surgery. *Glass and Ceramics*, *57*, 322-325.
[122] Coetzee, A. S. (1980). Regeneration of bone in the presence of calcium sulphate. *Arch Otolaryngol*, *106*, 405-409.
[123] Carinci, F., Piattelli, A., Stabellini, G., Palmieri, A., Scapoli, L., Laino, G., Caputi, S. & Pezzetti, F. (2004). Calcium sulfate: Analysis of MG63 osteoblast-like cell response by means of a microarray technology. *J Biomed Mater Res*, *71B*, 260-267.
[124] Walsh, W. R., Morberg, P., Yu, Y., Yang, J. L., Haggard, W., Sheat, P. C., Svehla, M. & Bruce, J. M. (2002). Response of a calcium sulphate bone graft substitute in a confined cancellous defect. *Clin Orthop Relat Res*, *406*, 228-236.
[125] Ricci, J. L., Alexander, H., Nadkarni, P., Hawkins, M., Turner, J., Rosenblum, S., Brezenoff, L., De Leonardis, D. & Pecora, G. (2000). *Bone engineering: Biological mechanisms of calcium sulfate replacement by bone.* In: J. E. Davies (Editor), Toronto, Canada: Em Squared Inc.
[126] Doadrio, J. C., Arcos, D., Cabanas, M. V. & Vallet-Regi, M. (2004). Calcium sulphate-based cements containing cephalexin. *Biomaterials*, *25*, 2629-2635.
[127] La Gatta, A., De Rosa, A., Laurienzo, P; Malinconico, M., De Rosa, M. & Schiraldi, C. (2005). A Novel Injectable Poly(epsilon-caprolactone)/Calcium Sulphate System for Bone Regeneration: Synthesis and Characterization. *Macromol Biosci*, *5*, 1108-1117.
[128] Gomez d'Ayala, G., De Rosa, A., Laurienzo, P. & Malinconico, M. (2007). Development of a new calcium sulphate-based composite using alginate and chemically modified chitosan for bone regeneration. *J Biomed Mater Res*, *81A*, 811-820.
[129] Sosnik, A. & Cohn, D. (2003). Poly(ethylene glycol)-poly(epsilon-caprolactone) block oligomers as injectable materials. *Polymer*, *44*, 7033-7042.
[130] Kweon, H.Y.; Yoo, M.K.; Park, I.K.; Kim, T.H.; Lee, H.C.; Lee, H.S.; Oh, J.-S.; Akaike, T. & Cho, C.-S. (2003). A novel degradable polycaprolactone networks for tissue engineering *Biomaterials*, 24, 801-808.
[131] Stöber, W., Fink, A. & Bohn, E. (1968). Controlled growth of monodisperse silica spheres in the micron size range. *J Colloid Interf Sci*, *26*, 62-69.

[132] Lian, J. B. & Stein, G. S. (1992). Concepts of osteoblast growth and differentiation: basis for modulation of bone cell development and tissue formation. *Crit Rev Oral Biol Med*, *3*, 269-305.

[133] Majeska, R. J. & Rodan, G. A. (1982). The effects of 1,25(OH)2D3 on alkaline phosphatase in osteoblastic osteosarcoma cells. *J Biol Chem*, *257*, 3362-67.

[134] Harada, S. & Rodan, G. A. (2003). Control of osteoblast function and regulation of bone mass. *Nature*, *423*, 349-355.

[135] Lee, N. K., Sowa, H., Hinoi, E., Ferron, M., Ahn, J. D., Confavreux, C., Dacquin, R., Mee, P. J., McKee, M. D., Jung, D. Y., Zhang, Z., Kim, J. K., Mauvais-Jarvis, F., Ducy, P. & Karsenty, G. (2007). Endocrine regulation of energy metabolism by the skeleton. *Cell*, *130*, 456-69.

[136] Friedenstein, A. J., Deriglasova, U. F., Kulagina, N. N., Panasuk, A. F., Rudakowa, S. F., Luria, E. A. & Ruadkow, I. A. (1974). Precursors for fibroblasts in different populations of hematopoietic cells as detected by the in vitro colony assay method. *Exp Hematol*, *2*, 83-92.

[137] Hou, Q. P., De Bank, P. A. & Shakesheff, K. M. (2004). Injectable scaffolds for tissue regeneration *J Mater Chem*, *14*, 1915-1923

[138] Grant, G. T., Morris Edwin, R., Rees David, A., Smith Peter, J. C. & Thom, D. (1973). *FEBS Lett*, *32*, 195.

[139] Nobile, M. R., Pirozzi, V., Somma, E., Gomez d'Ayala, G. & Laurienzo, P. (2008). Development and rheological investigation of novel alginate/N-succinylchitosan hydrogels. *J. Polym. Sci. B*, *46*, 1167-1182.

In: Bone Regeneration: Growth Factors... ISBN: 978-1-60876-621-5
Editors: V. Legard et al., pp. 59-91

Chapter II

BMP Signaling and Bone Formation

Hue H. Luu*[*], *Eric Wagner, Rex C. Haydon, Russell R. Reid and Tong-Chuan He
The University of Chicago Medical Center, The Molecular Oncology Laboratory, Department of Surgery, Section of Orthopaedics, 5841 South Maryland MC3079, Chicago, IL 60637.

Abstract

Bone formation is a complex and coordinated series of events resulting in a macro-structure that has the ability to continually repair and remodel throughout the life of a human. The human skeleton not only provides structural support but is also a complex network involving multiple organ systems such as the hematopoietic and endocrine systems. Integral in the steps of bone formation is the highly regulated process of osteoblastic differentiation. This process is critical for the normal development of bone, physiologic remodeling, and repair in an injured state. Over the past decade, we have begun to dissect the important regulatory elements of this cascade of events for osteogenesis. These key elements include morphogens, signal

[*] Corresponding author: Department of Surgery, Section of Orthopaedic Surgery, 5841 South Maryland, MC3079, Chicago, IL 60637, E-mail: hluu@surgery.bsd.uchicago.edu, Phone: 773-702-6216, 773-702-4765 (FAX).

transduction proteins, and transcription factors. Aberrations in key regulatory genes result in a number of human pathologic musculoskeletal conditions. This chapter will discuss normal bone development, mesenchymal stem cells in osteogenesis, and essential elements that regulate osteogenesis at the extracellular, intracellular, and nuclear levels. Focus will be directed to Bone Morphogenetic Protein (BMP) signaling in osteogenesis. Like TGFβ, BMP signaling is through the Smad pathway and regulates a number of downstream effecter genes that leads to bone formation. In the past five to ten years, we have begun to use recombinant human BMPs in the clinical setting to augment bone healing in craniofacial surgeries, spinal fusions, and fracture care. As we further understand the normal processes of osteogenesis and bone formation, we can potentially identify new therapies for bone regeneration in an injured or diseased state, as well as for bone loss in aging.

Introduction

Bone is one of a few organs in the human body that has regenerative capacity. Bone formation is a complex process that includes cells from the mesodermal lineage. The end result is a skeletal system that has tremendous reparative ability, provides mechanical support for ambulation and other functions (e.g. mastication), and is a reservoir for calcium and phosphate homeostasis. Bone remodeling is essential in maintaining a healthy skeletal system and in the healing process of a fracture. Many factors, especially Bone Morphogenetic Proteins (BMPs), play an important role in bone regeneration. Several BMPs are among most potent stimulators of osteoblastic differentiation, which is one of early steps in bone regeneration.

In the last 5 years, recombinant BMPs have been used in the surgical arena to help promote bone healing. Conversely, disruption of BMP signaling can lead to a number of rare human developmental or skeletal diseases. Similar to TGFβ, BMP signaling occurs through the Smad pathway which subsequently regulates a number of downstream effector genes. Activation of this pathway leads to the differentiation of mesenchymal stem cells (MSCs) into mature osteoblasts, and thus bone formation. As we understand this complex and regulated process, we can begin to reap some of the benefits of this knowledge and apply it clinically to improve the health and quality of life for humans.

Bone Formation and Regeneration

As part of the connective tissue system, bone is derived from the embryonic mesodermal layer. It retains the potential to regenerate and remodel throughout a human's life. During development, bone can form in two different mechanisms, intramembranous ossification or endochondral ossification. Intramembranous ossification occurs by direct condensation of mesenchymal cells. Intramembranous ossification is observed in flat bones of the skull, the mandible, and part of the clavicle. In contrast, endochondral ossification relies on the formation of a cartilaginous anlage and subsequent ossification. This occurs at the growth plates of long bones. The process of bone formation at a growth plate is well coordinated and involves proliferation and hypertrophy of cartilaginous cells followed by calcification and ossification. Closure of the growth plates signifies the cessation of skeletal growth, which occurs around age 13-15 for girls and age 15-18 for boys.

Although longitudinal growth has stopped, the bone continues to have the capacity to remodel and regenerate in the setting of injury. The bone regeneration that occurs in a fracture callous is identical to the process at a growth plate. Classically, bone regeneration relies on osteoinduction, osteoconduction, and mechanical stimulation for remodeling. Osteoinduction implies the presence of cytokines or growth factors to induce mesenchymal stem cells to differentiate into osteoblasts and lay down the osteoid matrix that eventually becomes bone. The initial phase of a fracture is the formation of a hematoma and an intense inflammatory response. It is in this milieu that cytokines and growth factors are released for the chemotaxis and proliferation of mesenchymal stem cells. Some of these factors include members of the Bone Morphogenetic Protein (BMP) family. This is the osteoinductive phase of fracture healing. By the third day, the mesenchymal stem cells have become adherent and have reached their maximal proliferation (Reddi 2001). By the fifth day, the mesenchymal stem cells have differentiated into chondroblasts and reach maximal chondrogenesis by day eight. Calcification begins to form, and by day 11, the osteoblasts have differentiated and ossification begins.

The presence of an osteoconductive material (e.g. pieces of bone or bone graft) helps facilitate the process. Osteoconductive material essentially is a lattice work along which new bone forms. Once the new bone has been created, it requires remodeling. Remodeling of bone abides by the principles of Wolff's law, in which bone that sees a mechanical stress will respond to that stress by laying down new bone (Wolff 1892). The mechanical strain that is exerted on osteoblasts and osteoclasts will promote bone formation, while the lack of strain will stimulate bone resorption (Frost 2001).

Bone Morphogenetic Proteins and Their Receptors

Bone Morphogenetic Proteins are potent osteoinductive factors that promote osteoblastic differentiation and bone formation. BMPs are secreted proteins that are abundantly found in bone and are part of the Transforming Growth Factor β (TGFβ) superfamily. Marshall Urist first reported on the discovery of these proteins in the 1960s when he observed that demineralized bone matrix was able to induce bone formation in animal models (Urist 1965). From the demineralized bone matrix, he purified and identified bone morphogenetic proteins. At least 15 BMPs have been identified. While many of the BMPs, such as BMP2, BMP6, BMP7, and BMP9 are potent stimulators of bone formation, there are several BMPs whose known functions have little to do with osteogenesis. Members of the BMP family and their functions are shown in **Table 1**.

Table 1: Members of the BMP Family

BMP	Other Names	Known Skeletal Function(s)	Other Functions	References
BMP2	BMP2A	Bone and cartilage formation	Heart development	Kang 04, Cheng 03, Zhang 1996
BMP3	osteogerin	Inhibitor of bone formation		Luu 2007, Daluiski 2001, Bahamonde 2001
BMP4	BMP2B	Bone formation	Spermatogenesis; allantois development	Hu 2004, Winnier 1995, Gysin 2002, Wright 2002, Yamaguchi 1996
BMP5		Bone formation, chondrocyte proliferation	Ovarian folliculogenesis	Ho 2008, Mailhot 2008, Bailon 1999, Pierre 2005
BMP6	Vgr-1	Bone and cartilage formation	Skin epithelial growth and wound healing	Kang 2004, Cheng 2003, Gosselet 2007, Kaiser 1998
BMP7	OP-1	Bone formation	Kidneyand eye development; spermatogenesis	Kang2004, Cheng 2003, Hu 2004, Zhao 2001, Mitu 2008
BMP8	OP-2	Bone formation	Spermatogenesis and epididymis function	Hu 2004, Zhao 1998
BMP9	GDF-2	Bone formation	Neuronal development, hepatocyte and hematopoetic growth	Kang 2004, Cheng 2003
BMP10		Unknown	Expression in cardiac myocytes	Chen 2006, Chen 2004
BMP11	GDF-11	Axial skeletal patterning	Limb development	McPherron 1999
BMP12	GDF-7, CDMP-3	Tendon and ligament development	Seminal vessicle and spinal neuron development	Lee 1998
BMP13	GDF-6, CDMP-2	Tendon, ligament, and cartilage development		Settle 2003
BMP14	GDF-5, CDMP-1	Chondrogenesis and joint formation, tendon healing	Limb development	Storm 1994, Bolt 2007
BMP15	GDF-9B	Unknown	Oogenesis and fertility	Dong 1996

Note: BMP-1 is a protease and therefore not included.

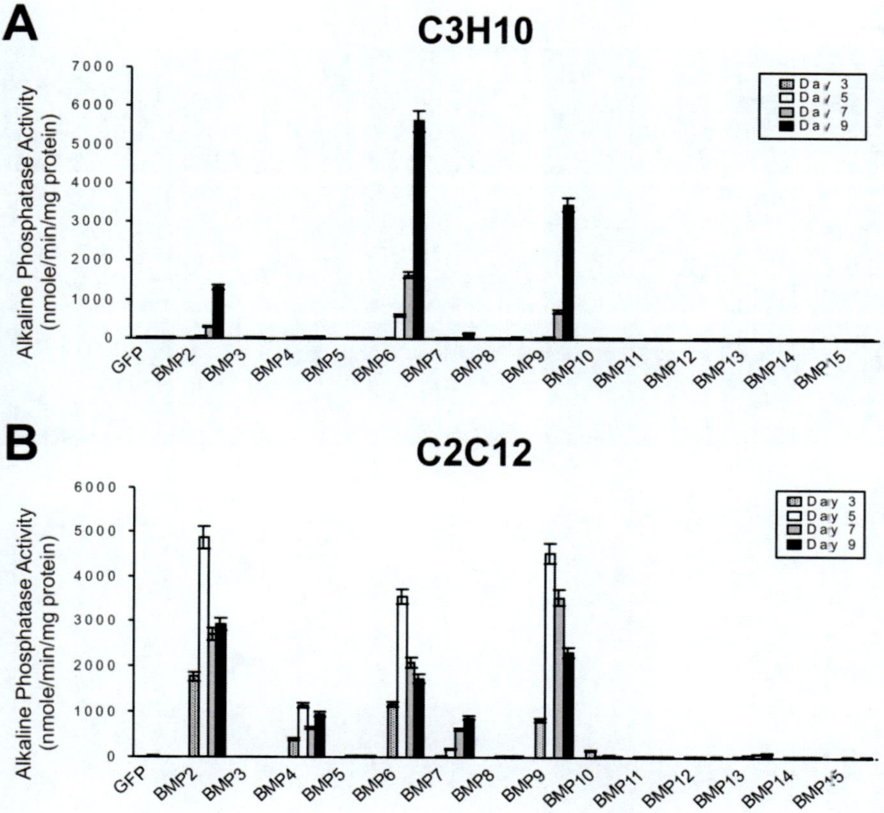

Figure 1. BMP induction of osteogenic differentiation marker alkaline phosphatase activity in pleuripotent C3H10T1/2 (A) and C2C12 (B) cells. Cells were infected with adenovirus containing BMP2-BMP14 and alkaline phosphatase activity was assayed at multiple time points. Reproduced with permission. Cheng et al. J. Bone and Joint Surgery 2003.

The biologic functions of some of the BMPs are non-osteogenic. BMP 12 and 13 are believed to be important for ligament and tendon development while BMP14 may be important for chondrogenesis (Bolt, Clerk et al. 2007). BMP15 has been shown to have an important role in ovulation and female fertility. In vertebrates, BMP2 acts to direct the development of neural crest cells into a neuronal lineage (Christiansen, Coles et al. 2000). BMP4, BMP7, and BMP8 appear to play a role in spermatogenesis (Hu, Chen et al. 2004). BMP 7 has been shown to be important for renal development as well as chronic renal disease (Mitu and Hirschberg 2008). Despite its name, BMP1 functions as a protease.

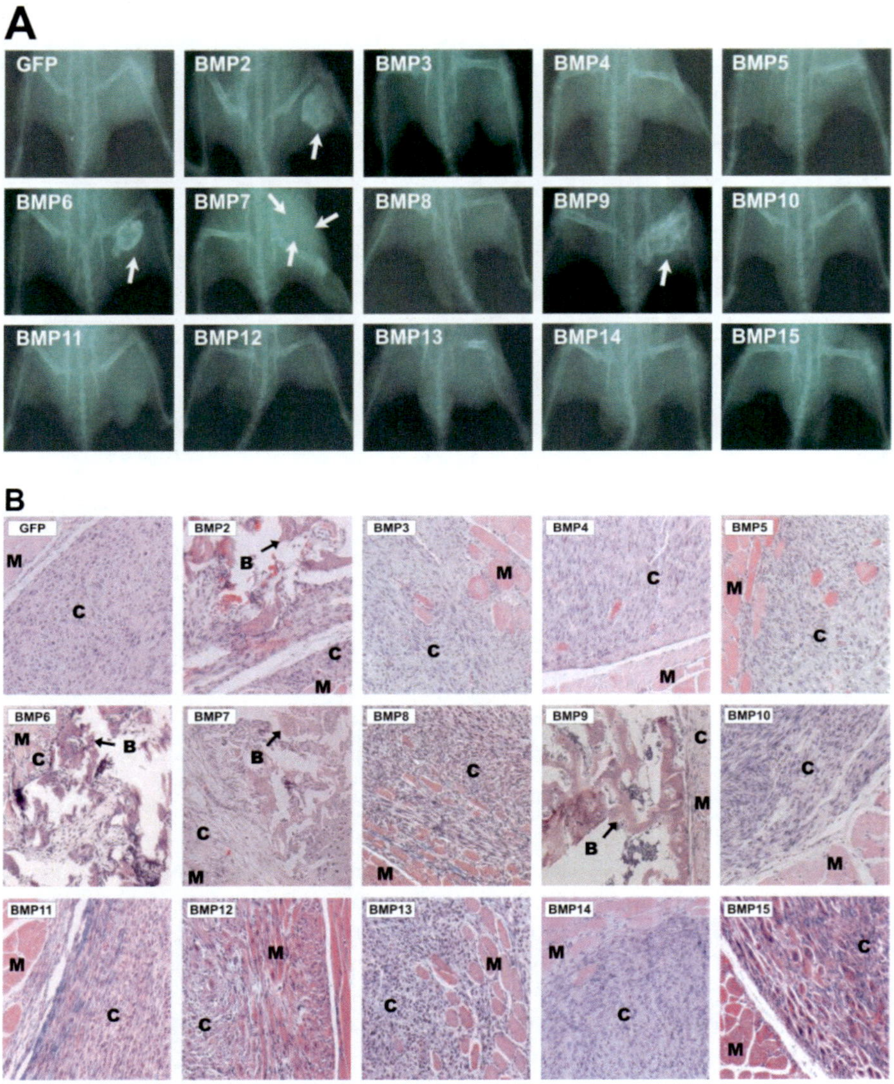

Figure 2. Comparison of heterotopic bone formation by 14 different BMPs. Pleuripotent C2C12 cells were infected with adenovirus containing BMP2-BMP14. The cells were harvested and injected into the quadriceps muscle of athymic nude mice. (A) Radiographs taken at 5 weeks demonstrated the ability for each of the BMPs to induce heterotopic bone formation (arrow). (B) The sites of injection or heterotopic bone were harvested for histologic evaluation. Magnification 200X. b, hetertopic bone with osteoid matrix (arrows); c, injected C2C12 cells; and m, muscle. Reproduced with permission. Kang et al. Gene Therapy 2004.

All members of the BMP family bind to the BMP receptor complex which is composed of the Type I and Type II receptors. There are three Type I receptors (ALK-2, ALK-3, and ALK-6). The ALK-2 receptor is also known as ActR-1A or the Type IA activin receptor. ALK-3 is also called BMPR-IA while ALK-6 is also known as BMPR-IB. Similarly, there are three Type II receptors (BMPR-II, ActRII, and ActRIIB). Upon binding with the BMP ligand, these serine/threonine kinases form a heterotetrameric complex. This complex is composed of pairs of Type I and Type II receptors. The crystal structures of the majority of the BMPs with or without their receptors have been characterized. The essential domains for binding of BMPs to the receptors are the "wrist" and "knuckle" epitopes. The "wrist" epitope of BMP binds to the Type I receptor while the "knuckle" epitope binds to the Type II receptor.

In vitro studies have shown that BMP2, BMP6, and BMP9 are able to effectively induce alkaline phosphatase activity, a marker for osteoblastic differentiation, in mouse embryonic fibroblast C3H10T1/2 cells (Cheng, Jiang et al. 2003; Luu, Song et al. 2007) (**Figure 1A**). Similarly BMP2, BMP4, BMP6, BMP7, and BMP9 are able to induce alkaline phosphatase activity in mouse myoblast C2C12 cells (Cheng, Jiang et al. 2003; Luu, Song et al. 2007) (**Figure 1B**). In vivo experiments have demonstrated that BMP2, BMP6, BMP7, and BMP9 are all able to induce heterotopic bone formation (**Figure 2**) (Kang, Sun et al. 2004). Currently, recombinant BMP2 and BMP7 are used clinically in humans to promote bone healing.

Among the BMPs, BMP3 is most unique in that it has antagonistic activities. BMP3 has been shown to inhibit the osteogenic effects of other BMPs when they are co-expressed in osteoblast precursor cell lines (Bahamonde and Lyons 2001; Cheng, Jiang et al. 2003). Furthermore, recombinant human BMP3 alone fails to demonstrate any osteogenic activity (Takao, Hino et al. 1996). In the maturing rat skull, BMP3 mRNA is downregulated in a fusing (or ossifying) suture, whereas it remains stably expressed or upregulated in patent cranial sutures (Nacamuli, Fong et al. 2005). It appears that the antagonistic activity of BMP3 is not at the receptor level. Instead, BMP3 likely activates a pathway whose downstream genes antagonize the Smad pathway. BMP3 knockout mice (BMP3 -/-) are viable and do not demonstrate any obvious skeletal phenotypes as embryos or neonates. Interestingly, adult mice exhibit twice as much trabecular bone as their wild-type littermates. Therefore, BMP3 is an inhibitor of osteogenesis both in vitro and in vivo.

BMP Signaling

BMPs are member of the Transforming Growth Factor β (TGFβ) superfamily, with more than 30 members identified in mammals (Rosen and Thies 1992; Hogan 1996; Hoffmann and Gross 2001; Luo, Sun et al. 2005; Luu, Song et al. 2007). As described, the BMP ligand binds to a tetrameric receptor complex of two pairs of transmembrane serine-threonine kinases (**Figure 3**). Upon binding, the activated receptors phosphorylate the Receptor-regulated Smads or R-Smads (Smads 1, 2, 3, 5, and 8). Phosphorylation of R-Smads is mediated by the Type I receptor. The R-Smads then dimerize with Smad 4, also known as co-Smad, and the dimer translocates into the nucleus. The Smad complex acts as a transcription factor to regulate a number of down stream genes, such as Runx2.

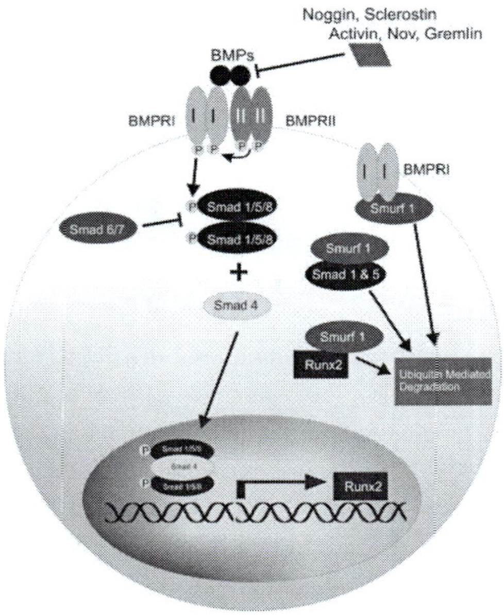

Figure 3. Skematic diagram of BMP signaling. The BMP ligand binds to a tetrameric receptor complex of two pairs of transmembrane serine-threonine kinases. Upon binding, the activated Type II receptor phosphorylates the Type I receptor. The Type I receptor then phosphorylates the Receptor-regulated Smads or R-Smads (Smads 1, 2, 3, 5, and 8). The R-Smads then dimerize with Smad 4, also known as co-Smad, and the dimer translocates into the nucleus. The Smad complex acts as a transcription factor to regulate a number of down stream genes, such as Runx2. The pathway is negatively regulated by a number of secreted antagonists (e.g. noggin, sclerostin, activin, NOV, and Gremlin) as well as intracellular inhibitory Smads (Smad 6 and 7) or Smurf 1.

Table 2: Mammalian Inhibitors of BMP Signaling

Gene	Function/Mechanism	References
Secreted Inhibitors		
Noggin	Blocks BMP binding to Type I and Type II BMP receptors	Groppe 2002
Sclerostin	Binds BMP5, 6, & 7 to prevent binding to the BMP receptors	Chen 2004
Gremlin	Binds BMP2, 4, & 7 to prevent binding to the BMP receptors	Canalis 2003, Westendorf 2004, Gazzero 2006, Kawano 2003
NOV	Binds to BMP2 to inhibit binding to the BMP receptors	Canalis 2007
Dan	Binds to BMP 12, BMP 13, & BMP14 to inhibit binding to the BMP receptors	Gerlach 2002, Dione 2001
Cerebrus	Binds to Wnt and BMPs	Piccolo 1999
Cer1	Binds to BMPs to inhibit binding to the BMP receptors	Belo 2000
Coco	Maternal BMP, Wnt, and TGFβ antagonist	Bell 2003
Caronte	Antagonizes BMPs	Rodriguez 1999, Yokouchi 1999
Dante	Blockade at the BMP receptor level.	Katoh 2004, Katoh 2006
WISE	Binds to BMP4 or Wnt receptor LRP6	Lintern 2009
Chordin	Binds to BMP2 & BMP4 to inhibit binding to the BMP receptors	Larrain 2000
Tsg (Twisted gastrulation)	Binds to BMP2 & BMP4 to inhibit binding to the BMP receptors; Binds to RANKL; Can also be a BMP agonist	Mahoney 2008, Chang 2001, Ross 2001, Nosaka 2003, Oelgeschlager 2000
BMPER	Binds to BMP 2, BMP4, & BMP6 to block binding to the BMP receptor	Moser 2003
Activin	Competes for binding to type II BMP receptor ActRII	Candia 1997, Kawabata 1998
Transmembrane Inhibitors		
CRIM1 (CR-rich motor neuron)	Transmembrane protein with an extracellular domain that gets cleaved and binds to BMP4 & BMP7	Wilkinson 2003, Kolle 2000
Amnionless	Transmembrane protein with an extracellular domain that competitively binds to BMP2 & BMP7	Kalantry 2001, Zhang 1996, Coucouvanis 1999
Intracellular Inhibitors		
Smad6 and Smad7	Binds to cytoplasmic domain of Type I BMP receptor to prevent phosphorylation of stimulatory Smads	Imamura 1997, Hanyu 2001
Smurf1	Mediates degradation of Smads 1, 5, & 6, Runx2, & Type I BMP receptor	Zhu 1999, Zhao 2003, Chen 2004, Murakami 2003

BMP signaling is modulated at multiple levels. Bone formation involves the balanced coordination of stimulus signals (BMP) and BMP antagonists at multiple levels. In the extracellular compartment, several BMP antagonists have been identified (**Table 2**). These antagonists, such as Noggin, work by binding to the BMP ligand to prevent the binding of the BMPs to their receptors (Canalis 2007). In the cytoplasmic compartment, I-Smads (Smad 6 and Smad 7) are inhibitors of BMP signaling. They compete with the R-Smads for binding to the Type I BMP receptor, thereby inhibiting transduction of the BMP signal

(Imamura, Takase et al. 1997; Hanyu, Ishidou et al. 2001). Smurfl (Smad ubiquitin regulatory factor 1) is also another negative regulator of BMP signaling. Smurf 1 is a E3 ubiquitin ligase that interacts with both the BMP receptor as well as Smad 1 and Smad 5, leading to their degradation (Murakami, Watabe et al. 2003).

Mutations in BMP Signaling

Many investigators have created null mutations in the BMPs, BMP receptors, or BMP signaling effector genes, such as Smads. BMP2 null mice are embryonic lethal, with heart defects, lack of amnion, and delayed formation of the primitive streak (Zhang and Bradley 1996). BMP3 is a negative regulator of osteogenesis and a null mutation of BMP3 leads to increased bone density (Daluiski, Engstrand et al. 2001). BMP4 null mice are also embryonic lethal with absence of the allantois, as well as head defects (Winnier, Blessing et al. 1995). BMP5 null mice have a short ear phenotype and have multiple defects in the skeleton, in addition to lung and kidney defects (Kingsley, Bland et al. 1992; Kingsley 1994). These mice have an overall lower bone mass. BMP 6 null mice are viable with no obvious skeletal abnormalities except for delayed ossification of the sternum (Solloway, Dudley et al. 1998).

BMP7 null mice have multiple minor skeletal defects and die shortly after birth due to kidney agenesis (Dudley, Lyons et al. 1995; Luo, Hofmann et al. 1995). In addition, these null mice have defects in their eyes. BMP8 mutants are viable but infertile due to deficiencies in spermatogenesis as well as abnormal development of the epididymis (Zhao, Liaw et al. 1998). There are no obvious skeletal defects. To date, there are no knock out mice for BMP9 and BMP10. Homozygous mutant BMP11 mice have defects in the Anterior-Posterior patterning in skeletal development (McPherron, Lawler et al. 1999). There is a more posterior displacement of the hind limbs. BMP12 null mice have defects in neuronal development, specifically the D1A neurons (Lee, Mendelsohn et al. 1998). Homozygous BMP13 mutant mice have defects in joint, cartilage, and ligament formation (Settle, Rountree et al. 2003). Mutations in BMP14 results in brachypodism, which results in shortened limbs and reduced number of bones (Storm, Huynh et al. 1994). Finally, BMP15 deficient mice are sterile due to defects in oogenesis.

BMPR-IA null mice have an embryonic lethal phenotype with death occurring at embryonic day 9.5 (Mishina, Suzuki et al. 1995). Upon examination, embryos are noted to have no mesoderm, suggesting that the BMPR-IA is essential for the fate of the mesodermal layer during gastrulation. In contrary, mice deficient in BMPR-IB are viable but have defects in the appendicular skeleton (Yi, Daluiski et al. 2000). These mice have deformities in the proximal phalanges, metacarpals, and metatarsals. The resultant deformity is likely due to reduction in the proliferation of chondrocyte precursors as well as chondrocytes. Interestingly, mice with mutation in both the BMPR-IB receptor as well as BMP7 have severe appendicular skeletal defects.

Antagonist to BMP Signaling

Noggin is a secreted BMP antagonist that is important for skeletal development. Mice with a noggin null mutation have increased BMP signaling and have a number of skeletal abnormalities (Brunet, McMahon et al. 1998). Mice that are heterozygous for noggin are normal, while the homozygous mice have defects in the vertebrae, ribs, and limbs. Interestingly, the severity of the axial defects increases the further caudal the segment is located. The skull and cervical vertebrae are nearly normal, while the thoracic vertebrae are fused and have incomplete closure of the posterior vertebral elements. The lumbar and sacral vertebrae are absent, as well as some of the ribs. The limbs are shorter than the wild type littermates, with ossification and fusion of some of the joints.

Sclerostin is another secreted antagonist of BMP signaling by binding to BMP5, BMP6, and BMP7 with high affinity (Chen, Zhao et al. 2004). Over-expression of sclerostin in mouse pluripotent C3H10T1/2 cells results in blockage of osteoblastic differentiation and subsequent osteoporosis (Winkler, Sutherland et al. 2003). Sclerostin knockout mice have increased bone mineral density (BMD) throughout the skeleton (Li, Ominsky et al. 2008). There are not obvious skeletal defects in both the homozygous and heterozygous knockout mice While the homozygous knockout mice have increased bone density, the heterozygous knockout mice have a similar bone density to the wild type. Mechanical testing demonstrates that the sclerostin null mice have increased bone strength compared to the wild type.

Gremlin is a secreted protein that serves as a BMP antagonist by binding to BMP 2, BMP 4, and BMP 7, and preventing their activation through an alteration in the Wnt signaling pathway (Gazzerro, Smerdel-Ramoya et al. 2007). Overexpression in transgenic mice behind the osteocalcin promoter leads to osteopenia, with diffusely reduced trabecular and woven bone (Gazzerro, Pereira et al. 2005). Mice with a homozygous null mutation for Gremlin are characterized by abnormalities of the limbs and lungs, the absence of kidneys, and intrauterine fatality (Khokha, Hsu et al. 2003; Michos, Panman et al. 2004). Kidney agenesis results from failure of metanephric kidney development initiated by ureter growth. Conditional knockout of Gremlin leads to increased trabecular bone volume from increased osteoblastic activity stemming from BMP-2 stimulation, with Runx2 and osteocalcin expression (Gazzerro, Smerdel-Ramoya et al. 2007). Thus, it appears Gremlin affects early skeletal, lung, and kidney development associated with BMPs.

Smad proteins 1, 5, and 8 are receptor-activated after ligand binding induces the phosphorylation of BMP type I and II receptors (Massague, Seoane et al. 2005). After activation, they complex with Smad 4 and translocate to the nucleus to bind to cofactors and activate transcription (Chen, Zhao et al. 2004). Although individual Smad loss leads to minimal cartilagenous defects and viable mice, other reports have shown that the null mutations in Smad 1 and 5 lead to intrauterine death (Retting, Song et al. 2009). The Smad 1 mutants fail to establish the umbilical-placental connection, while the Smad 5 mutants display abnormal angiogenesis leading to irregular vasculature (Yang, Castilla et al. 1999; Tremblay, Dunn et al. 2001). Furthermore, the combination of Smad 1 and 5 null mutations leads to chrondrodysplasia from alteration in the BMP associated chrondrocyte differentiation pathway (Retting, Song et al. 2009). This results in shorter long bones and absent hypertrophic zones in growth plates. The knockout of the Smad antagonist, Smad 6, results in multiple cardiovascular abnormalities, including aortic ossification and valvular hyperplasia (Galvin, Donovan et al. 2000). Thus, Smads are thought to play a crucial role in BMP-related fetal development, angiogenesis, endochondral ossification, and growth plate bone formation.

Activin serves as a BMP signaling antagonist through competition for binding to the type II BMP receptor ActRII (Candia, Watabe et al. 1997; Kawabata, Imamura et al. 1998). Overexpression of ActRII can overcome Activin inhibition of the BMP7 signaling pathway (Piek, Afrakhte et al. 1999). Mice with the null mutation for the ActRII receptor have a low prevalence of facial deformities, including macrognathia and cleft palates, which are distinct from the reproductive and vascular defects associated with Activin mutations (Matzuk,

Kumar et al. 1995). This suggests that ActRII receptors have a predominant in vivo role in the BMP signaling pathway.

Smurf-1 is a protein involved in the regulation of the BMP signaling pathway. Its function is carried out by binding Smads 1 and 5, BMP type I receptors, and Runx2 to promote their degradation in a ubiquitin- and proteasome- dependent mechanism (Murakami, Watabe et al. 2003; Chen, Zhao et al. 2004). Overexpression causes decreased osteoblast proliferation and multiple bone formation defects (Zhao, Qiao et al. 2004). Smurf-1 null mutant mice display increased trabecular bone volume and bone formation rates (Chen, Zhao et al. 2004). Consequently, this BMP antagonist plays a critical role in osteogenic signaling regulation.

Nephroblastoma overexpressed (Nov) is a protein expressed in bone and cartilage, and is involved in regulation of osteogenesis, chondrogenesis, and angiogenesis (Canalis 2007). It is a member of the CCN protein family and its transcription is controlled by TGF-β and BMP-2 (Parisi, Gazzerro et al. 2006). Nov overexpression impairs BMP2 signaling, bringing about decreased bone formation and osteopenia (Canalis 2007). Therefore, Nov is a BMP antagonist, inhibiting osteogenesis.

Clinical Use of BMPs

Traditionally, a common technique that was used to promote bone healing was the use of bone graft harvested from the iliac crest. For example, all spinal fusion surgeries in the past involved harvesting of bone graft from the iliac crest. This resulted in additional pain, risk of infection, blood loss, and morbidity. With the advent of recombinant human BMPs (rhBMPs), this method has been challenged and is becoming antiquated. Other osteoinductive factors such as platelet rich plasma (PRP) or demineralized bone matrix have also been used to enhance bone healing. Over the past 5 years, the use of recombinant human BMPs has increased tremendously. Currently, rhBMP2 and rhBMP7 are available for clinical use. Their applications include lumbar spinal fusion procedures, alveolar cleft reconstructions, fractures, and non-unions of fractures.

The clinical use of rhBMP2 and rhBMP7 can be attributed to their ability to induce osteoblast differentiation and facilitate osteoid matrix production. To deliver and integrate the BMPs in their specific locations, various carrier mechanisms are employed. These include direct injection, liposome mediated polymers, DNA polymers, and viral methods, primarily utilizing adenovirus strains (Luo, Sun et al. 2005). Although these applications have shown remarkable potential in a variety of orthopedic procedures, they have some limitations. In the current clinical practice, the carrier that is used for delivery of rhBMPs in patients is a collagen sponge that has high affinity for the BMP protein. The carriers used to deliver the rhBMPs are thought to deliver only at 50% capacity at 1 week and 5% at 3 weeks, which is often insufficient time for proper bone healing (Seeherman, Wozney et al. 2002). This suggests that the initial stimulus from the rhBMPs is sufficient to start the bone regeneration cascade since it is cleared relatively quickly in the human body. The current source for BMPs in clinical use is via recombinant proteins since it is impractical to purify the proteins from human sources. It takes 1 kg of bone to make 0.002mg of rhBMP2 (Takaoka, Nakahara et al. 1988).

The gold standard in achieving union of associated bones in spinal fusion is autologous bone graft from the iliac crest. However, failure rates in these procedures occur in 10%-40% of single level fusions (Boden, Kang et al. 2002). In the past, multiple animal studies in rabbits, canines, primates, and sheep have revealed that spinal fusion was achieved with the use of BMPs and was superior to autografts (Luo, Sun et al. 2005). Recently, clinical trials in humans demonstrated a similar result in several fusion applications, including intervertebral and posterolateral lumbar fusions (Sandhu 2003). These trials demonstrated that rhBMP2 was superior to autologous bone graft with lower failure rates, quicker improvement times, and fewer associated complications (Boden, Hair et al. 2000; Boden, Kang et al. 2002; Govender, Csimma et al. 2002). While the use of rhBMP2 in fusion of single level degenerative joint disease appears to be preferable, there is currently no evidence examining its efficacy in spinal disorders, such as spondylolisthesis (Garrison, Donell et al. 2007). While rhBMP7 has shown potential in animal studies, up to this point there has been no clinically significant improvement in human fusion surgeries.

The use of rhBMPs in spinal procedures has been mainly restricted to the lumbar spine. The Food and Drug Administration has approved the use of rhBMPs for lumbar interbody and posterolateral fusions. There are reports of its use in the cervical spine, but as an off-label use. BMPs can mount a significant inflammatory response and the use in the cervical spine has resulted in swelling in the neck (Perri, Cooper et al. 2007; Tumialan and Rodts 2007). When severe, the swelling has led to airway compromise and the need for administration of steroids to reverse the swelling (Perri, Cooper et al. 2007; Tumialan and Rodts 2007). Nevertheless, there have been studies demonstrating success of using lower doses of BMPs in the cervical spine to promote fusion (Buttermann 2008). Therefore, the use of rhBMPs for cervical spine fusions remains under intense investigation.

Recently, the Food and Drug Administration has approved the use of rhBMP2 for alveolar cleft reconstruction, which holds promise in the pediatric population, where the avoidance of iliac crest bone harvest may lead to reduced morbidity, growth disturbance, as well as reduction in pain and hospital stay. Although not fully studied prospectively, rhBMP2 has been demonstrated to improve bony healing when surgically administered to skeletally mature patients who had unilateral alveolar clefts (Dickinson, Ashley et al. 2008). Patients in this study who received this form of BMP had a greater percentage of bone volume and improved mineralization compared to conventionally grafted control subjects. Further animal data and randomized, prospective human trials are still required prior to complete acceptance of rhBMP use in the craniofacial skeleton.

The use of rhBMPs in therapeutic applications involving fractures has been shown to enhance the healing process. Segmental bone defects, also known as critical sized defects, pose a significant problem for orthopaedic surgeons, since they cannot heal spontaneously (Luo, Sun et al. 2005). While some surgeries for fracture non-union result in successful recoveries, the healing process is often delayed or prevented by underlying conditions, such as diabetes, infection, mechanical stress, or high energy injuries. Bone allografts and autografts, as well as vascularized bone grafts, have traditionally been used to improve the treatment of these injuries. However, it has recently been shown that the biologic augmentation of the fracture healing in animal studies by rhBMP2 and rhBMP7 have enabled quicker recovery times, lower rates of therapeutic failure, and fewer complications associated with the therapeutic interventions (Luo, Sun et al. 2005). In humans, rhBMP augmentation of conventional therapy reduced the number of secondary interventions needed in union of acute open tibial fractures (Garrison, Donell et al. 2007). Additional clinical trials have shown efficacy in long bone applications of rhBMP7 in cases of distal tibial fractures, tibial nonunions, scaphoid nonunions, and atrophic nonunion bone loss (White, Vaccaro et al. 2007). The BESTT trial and subsequent follow-ups demonstrated that treatment of open tibial fractures with rhBMP2 improved the healing time, reduced the incidence of infection, and decreased the number of hardware failures (Govender, Csimma et al. 2002; Bishop and Einhorn 2007). Many animal studies have shown that BMP4 can assist in the healing of critical sized defects, and BMP9 aids in spinal fusion success (Luo, Sun et al. 2005). Based on in vitro and in vivo animal data, it appears that BMP6 and BMP9 have tremendous potential for clinical use, since these two BMPs demonstrate robust bone generation capacity (Helm, Alden et al. 2000; Jane, Dunford et al. 2002; Luu, Song et al. 2007). However, currently, only rhBMP2 and rhBMP7 are available for clinical use.

BMPs in Bone Tumors

Disruption of the stem cell differentiation pathway has been associated with tumorigenesis (Reya, Morrison et al. 2001; Haydon, Luu et al. 2007). By preventing the differentiation of MSCs, the proliferative capability of these cells increases the risk for uncontrolled cell division, eventually becoming malignant. We believe that osteosarcoma arises from defects in osteoblastic differentiation (Thomas, Johnson et al. 2004; Luo, Chen et al. 2008). We have shown that differentiating agents, such as PPARγ and 9-cis-retinoic acid, are able to induce differentiation and inhibit osteosarcoma cell growth (Haydon, Zhou et al. 2002; Haydon, Luu et al. 2007).

The osteogenic BMPs 2, 6, and 9 are potent stimulators of the osteogenic pathway, and therefore, would appear to have the capability to serve as osteosarcoma differentiation agents. Recently, however, we showed that osteogenic BMPs promoted tumor growth and failed to induce terminal differentiation (Luo, Chen et al. 2008). In this same study, exogenous Runx2 expression successfully inhibited tumor proliferation. Runx2 is downstream of the BMPs in the signaling cascade and is able to bypass the differentiation defect in some of the osteosarcoma tumor lines. Therefore, it is likely that osteosarcoma results from early defects in the differentiation pathway. Signaling factors that are downstream of the defect are able to bypass these defects and can inhibit tumor growth through induction of terminal differentiation.

BMPs and Cartilage

Although the use of BMPs for cartilage repair in humans has not been studied, in vitro and animal studies demonstrate their ability to facilitate cartilage regeneration and repair. BMPs 2, 4, 6, 7, 13, and 14 are able to stimulate progenitor chrondrocyte differentiation (Majumdar, Wang et al. 2001; Jorgensen, Gordeladze et al. 2004; Sekiya, Larson et al. 2005). Overexpression of many of the BMPs lead to increased chrondrocyte proliferation, an increase in volume of cartilagenous elements, and, sometimes, fused skeletal elements (Duprez, Bell et al. 1996; Brunet, McMahon et al. 1998; Wijgerde, Karp et al. 2005). Mice lacking BMP type 1 receptors, BMP7, or BMP14 display severe chondrodysplasia and appendicular skeletal defects, with multiple abnormalities in early chondrogenesis (Chen, Zhao et al. 2004; Yoon, Ovchinnikov et al. 2005). Furthermore, mutations in BMP5 and BMP14 result in chondrodysplasia in humans. Therefore, it appears that BMPs play an essential role in chondrogenesis throughout skeletal development.

Cartilaginous defects or injuries are a major source of morbidity, without reliable therapeutic options. In the case of defects in articular cartilage, BMPs are able to stimulate MSCs to facilitate the healing process. In vivo studies established that recombinant BMP2, BMP4, and BMP7 improve the repair of full-thickness cartilage defects (Di Cesare, Frenkel et al. 2006; Kuo, Rodrigo et al. 2006; Kuroda, Usas et al. 2006). The underlying mechanism of endogenous cartilage repair is associated with BMP2 stimulation from the deeper hypoxic zones of cartilage tissue, facilitating HIF-1α (Hypoxia Inducible Factor 1 alpha) activity (Gelse, Muhle et al. 2008). To control the heterotopic ossification associated with BMP facilitated repair, antagonists, such as Noggin, are able to regulate site specific localization (Hannallah, Peng et al. 2004). Much like the successful rhBMP associated fracture repair and spinal fusions that we have discussed, BMPs will likely be used to facilitate cartilaginous defect procedures in the near future.

BMPs and Tendons

Tendon injury repair via surgical management is an effective method of achieving stabilization and facilitating the healing process. However, there are notable complications of these procedures, including re-rupture risk, scar tissue formation, infection, and muscular atrophy (Bolt, Clerk et al. 2007). Especially in tendon to bone healing procedures, such as rotator cuff repair, there is a high surgical failure rate (Kovacevic and Rodeo 2008). Recently, strategies involving the manipulation of molecular mechanisms associated with BMPs have demonstrated the potential for future therapeutic options. Initial obstacles to molecular tendon repair are based on tenocytes being highly differentiated cells, with limited replication potential, embedded in an extensive extracellular matrix. Tendon healing is a slow process that results in a mechanically inferior extracellular matrix, due in part to its low mitotic potential and poor vascularity

(Ahmed, Lutton et al. 1998; Woo, Hildebrand et al. 1999). Further complications involve the method of tissue localization and regulation of tenocyte differentiation without ectopic chondrogenesis or osteogenesis. Animal studies have investigated delivering BMPs through a carrier to facilitate localization and sustained release into the surrounding tissues. Although recombinant protein injection is associated with the induction of bone or cartilage in tendon repair, it appears that adenoviral models overcome this complication (Lou 2000; Lou, Tu et al. 2001; Forslund and Aspenberg 2003; Forslund, Rueger et al. 2003; Bolt, Clerk et al. 2007).

BMPs have been shown to regulate the tenocyte differentiation process in a fashion similar to the way in which they control chondrogenesis and osteogenesis. Specifically, BMP12 has the ability to induce tenogenic differentiation of bone marrow derived MSCs (Violini, Ramelli et al. 2009). Furthermore, BMPs are able to augment tendon repair in a variety of in vivo procedures. In such repairs, BMPs stimulate a more active cellular response, resulting in an increased number of tenocytes at the site of injury. In Achilles tendon repair, BMP12, BMP13, and BMP14 improve tensile strength and facilitate a quicker recovery in rat and rabbit models (Forslund and Aspenberg 2003; Forslund, Rueger et al. 2003; Bolt, Clerk et al. 2007; Majewski, Betz et al. 2008). In rabbit and chicken models, flexor tendon laceration repair is augmented through the use of BMP12 gene therapy (Lou 2000; Lou, Tu et al. 2001). Tendon to bone healing is an especially difficult process since tendon healing occurs through fibrous scar tissue formation rather than normal tissue regeneration. Therefore, BMPs could have a significant impact on treatment of these injuries, such as rotator cuff repair. An example of the tendon to bone healing model is seen in patellar sheep tendons, where injury is associated with increased BMP2 and BMP7 levels, along with their associated signaling factors (Yu, Bliss et al. 2007). BMP2 and BMP12 augment ACL (Anterior Cruciate Ligament) and rotator cuff repairs, demonstrating the potential for BMP regeneration of the tendon and ligament to bone interface (Hashimoto, Yoshida et al. 2007; Kovacevic and Rodeo 2008).

Human Disease Due to Disruptions in BMP Signaling

Fibrodysplasia ossificans progressive (FOP) is a rare inherited disorder characterized by disabling congenital malformations of the great toes and progressive heterotopic endochondral ossification in a predictable anatomic pattern. The episodic flares of heterotopic ossification result in immobility, morbidity, and mortality. Although there have been no mutations identified in the BMP4 genome, elevated expression of BMP4 in early fibromatous lesions suggest that a component of the BMP4 signaling pathway is responsible for the mutation in FOP (Gannon, Kaplan et al. 1997; Xu, Feldman et al. 2000). Mutations in FOP have been identified in the BMP antagonist Noggin and a BMP type 1 receptor involved in the BMP4 cascade, supporting the association of BMP4 pathway dysregulation with FOP (Fontaine, Semonin et al. 2005; Kaplan, Le Merrer et al. 2008). Further investigation into the pathway underlying this skeletal metamorphosis process is needed in order to target therapies according to the defects.

Sclerostosis and Van Buchen disease are autosomal recessive disorders associated with a progressive increase in bone mass, hyperostotic skeleton, and increased bone mineral density (Li, Ominsky et al. 2008). Homozygous mutations for the secreted protein sclerostin (SOST) are responsible for the osteosclerotic characteristics of sclerostosis (Balemans, Ebeling et al. 2001; Brunkow, Gardner et al. 2001). In Van Buchen disease, a deletion downstream from the SOST genome results in the absence of SOST expression and therefore precludes functional sclerostin production (Loots, Kneissel et al. 2005). Sclerostin is a member of the BMP antagonist family, which is expressed in osteocytes and binds BMPs 5, 6, and 7 with high affinity (Chen, Zhao et al. 2004). It also is directly involved in the regulation of osteoblastic differentiation through the Wnt signaling pathway (Li, Ominsky et al. 2008). The osteosclerosis in both diseases occurs as a direct result of the collective deficiency in the sclerostin protein. This deficiency likely enables a decrease in endogenous regulation of BMP associated bone formation and an increase in the degree of BMP activity.

Another member of the family of BMP antagonists associated with inherited disorders is the secreted protein Noggin. This polypeptide displays high affinity binding capabilities for BMPs 2, 4, and 7. It carries out its antagonism by blocking the binding epitopes for BMP type I and type II receptors (Chen, Zhao et al. 2004). Mutation of this protein is responsible for disorders of joint morphogenesis. Proximal symphalangism is an autosomal dominant disorder characterized by the ankylosis of proximal interphalangeal joints, carpal and tarsal bone fusion, and stapes ankylosis leading to conductive deafness (Gong, Krakow et al. 1999). Similar characteristics are noted in the disorder known as multiple synostoses syndrome, however, patients often experience a more severe phenotype with consistent, severe ankylosis of the cervical spine and hip joints

(Krakow, Reinker et al. 1998). Missense mutations in the Noggin coding regions have been identified in patients with this disorder (Gong, Krakow et al. 1999). Other disorders of joint morphogenesis have also been linked to Noggin mutations (Gong, Krakow et al. 1999; Chen, Zhao et al. 2004). Interestingly, despite the broad expression of Noggin in a variety of tissues during development, the characteristics associated with these mutations seem be joint tissue specific, without widespread deformities.

Chondrodysplasia is a condition associated with defects in the cartilage differentiation pathway. It is characterized by severe abnormalities of the limbs and joints, including proximal limb shortening and lack of articulation in the fingers and toes, though the axial skeleton remains relatively normal (Thomas, Lin et al. 1996; Thomas, Kilpatrick et al. 1997). Homozygous mutation in the cartilage-derived morphogenetic protein-1 (CDMP-1), which is also known as BMP14, leads to chondrodysplasia. Heterozygotes are of average stature, with minimal skeletal abnormalities, such as polydactyly, brachydactyly, valgus toe deviation, and flexion contractures of fingers (Thomas, Kilpatrick et al. 1997). It seems that CDMP-1 is co-expressed with many of the BMPs and its mutation interferes with BMP expression throughout skeletal development. This leads to the associated skeletal abnormalities in both homozygous and heterozygous individuals.

A non-osteogenic disorder related to BMP mutations is pulmonary hypertension. Although most cases of pulmonary hypertension are sporadic, familial primary pulmonary hypertension is a rare autosomal dominant disorder associated with excessive proliferation of endothelial cells in pulmonary arterioles. This results in increased pulmonary artery pressure and right ventricular failure, with a high associated mortality (Chen, Zhao et al. 2004). Over 25 different mutations in the type II BMP receptor, BMPR-II, have been linked to this disorder (Deng, Morse et al. 2000; Lane, Machado et al. 2000; Newman, Wheeler et al. 2001). However, only 10%-20% with a mutation in BMPR-II actually develop the disorder, which points to an underlying combination of genetic and environmental insults (Newman, Wheeler et al. 2001). When mutated, the anti-proliferative role of BMPR-II and its associated signaling pathway is inhibited, leading to the observed vascular abnormalities.

Conclusion

Bone Morphogenetic Proteins are important regulators of bone, cartilage, tendon, and ligament development, and repair in the musculoskeletal system. The BMPs also play an important role in embryogenesis as well as downstream development of multiple solid organs. Aberrations in BMP signaling have been identified in a number of human diseases which include multiple musculoskeletal abnormalities. In the last decade, we have begun to harness the power of the BMPs by using recombinant human BMP2 and BMP7 in patients. Both rhBMP2 and rhBMP7 have demonstrated efficacy in promoting fracture unions spinal fusions, as well as osseointegration of alveolar clefts. Other BMPs are currently being explored for tendon, ligament, and cartilage repair/regeneration. Thus, BMPs are important regulatory proteins in the musculoskeletal system and can be used in the treatment of a number of clinical scenarios.

References

Ahmed, T., Lutton, J. D., et al. (1998). "Gene transfer of alpha interferon into hematopoietic stem cells." *Leukemia research*, *22(2)*, 119-124.

Bahamonde, M. E. & Lyons, K. M. (2001). "BMP3: to be or not to be a BMP." *The Journal of bone and joint surgery.American volume*, *83-A Suppl 1 (Pt 1)*, S56-62.

Bailon-Plaza, A., Lee, A. O., et al. (1999). "BMP-5 deficiency alters chondrocytic activity in the mouse proximal tibial growth plate." *Bone*, *24(3)*, 211-216.

Balemans, W., Ebeling, M., et al. (2001). "Increased bone density in sclerosteosis is due to the deficiency of a novel secreted protein (SOST)." *Human molecular genetics*, *10(5)*, 537-543.

Bell, E., Munoz-Sanjuan, I., et al. (2003). "Cell fate specification and competence by Coco, a maternal BMP, TGFbeta and Wnt inhibitor." *Development (Cambridge, England)*, *130(7)*, 1381-1389.

Belo, J. A., Bachiller, D., et al. (2000). "Cerberus-like is a secreted BMP and nodal antagonist not essential for mouse development." *Genesis (New York, N.Y.: 2000)*, *26(4)*, 265-270.

Bishop, G. B. & Einhorn, T. A. (2007). "Current and future clinical applications of bone morphogenetic proteins in orthopaedic trauma surgery." *International orthopaedics*, *31(6)*, 721-727.

Boden, S. D., Hair, G. A., et al. (2000). "Gene therapy for spine fusion." *Clinical orthopaedics and related research*, *(379 Suppl)(379 Suppl)*, S225-33.

Boden, S. D., Kang, J., et al. (2002). "Use of recombinant human bone morphogenetic protein-2 to achieve posterolateral lumbar spine fusion in humans: a prospective, randomized clinical pilot trial: 2002 Volvo Award in clinical studies." *Spine*, *27(23)*, 2662-2673.

Bolt, P., Clerk, A. N., et al. (2007). "BMP-14 gene therapy increases tendon tensile strength in a rat model of Achilles tendon injury." *The Journal of bone and joint surgery.American volume*, *89(6)*, 1315-1320.

Brunet, L. J., McMahon, J. A., et al. (1998). "Noggin, cartilage morphogenesis, and joint formation in the mammalian skeleton." *Science (New York, N.Y.)*, *280(5368)*, 1455-1457.

Brunkow, M. E., Gardner, J. C., et al. (2001). "Bone dysplasia sclerosteosis results from loss of the SOST gene product, a novel cystine knot-containing protein." *American Journal of Human Genetics*, *68(3)*, 577-589.

Buttermann, G. R. (2008). "Prospective nonrandomized comparison of an allograft with bone morphogenic protein versus an iliac-crest autograft in anterior cervical discectomy and fusion." *The spine journal : official journal of the North American Spine Society*, *8(3)*, 426-435.

Canalis, E. (2007). "Nephroblastoma overexpressed (Nov) is a novel bone morphogenetic protein antagonist." *Annals of the New York Academy of Sciences*, *1116*, 50-58.

Canalis, E., Economides, A. N., et al. (2003). "Bone morphogenetic proteins, their antagonists, and the skeleton." *Endocrine reviews*, *24(2)*, 218-235.

Candia, A. F., Watabe, T., et al. (1997). "Cellular interpretation of multiple TGF-beta signals: intracellular antagonism between activin/BVg1 and BMP-2/4 signaling mediated by Smads." *Development (Cambridge, England)*, *124(22)*, 4467-4480.

Chang, C., Holtzman, D. A., et al. (2001). "Twisted gastrulation can function as a BMP antagonist." *Nature*, *410(6827)*, 483-487.

Chen, D., Zhao, M., et al. (2004). "Bone morphogenetic proteins." *Growth factors (Chur, Switzerland)*, *22(4)*, 233-241.

Chen, H., Shi, S., et al. (2004). "BMP10 is essential for maintaining cardiac growth during murine cardiogenesis." *Development (Cambridge, England)*, *131(9)*, 2219-2231.

Chen, H., Yong, W., et al. (2006). "Overexpression of bone morphogenetic protein 10 in myocardium disrupts cardiac postnatal hypertrophic growth." *The Journal of biological chemistry*, *281(37)*, 27481-27491.

Cheng, H., Jiang, W., et al. (2003). "Osteogenic activity of the fourteen types of human bone morphogenetic proteins (BMPs)." *The Journal of bone and joint surgery.American volume*, *85-A(8)*, 1544-1552.

Christiansen, J. H., Coles, E. G., et al. (2000). "Molecular control of neural crest formation, migration and differentiation." *Current opinion in cell biology*, *12(6)*, 719-724.

Coucouvanis, E. & Martin, G. R. (1999). "BMP signaling plays a role in visceral endoderm differentiation and cavitation in the early mouse embryo." *Development (Cambridge, England)*, *126(3)*, 535-546.

Daluiski, A., Engstrand, T., et al. (2001). "Bone morphogenetic protein-3 is a negative regulator of bone density." *Nature genetics*, *27(1)*, 84-88.

Deng, Z., Morse, J. H., et al. (2000). "Familial primary pulmonary hypertension (gene PPH1) is caused by mutations in the bone morphogenetic protein receptor-II gene." *American Journal of Human Genetics*, *67(3)*, 737-744.

Di Cesare, P. E., Frenkel, S. R., et al. (2006). "Regional gene therapy for full-thickness articular cartilage lesions using naked DNA with a collagen matrix." *Journal of orthopaedic research : official publication of the Orthopaedic Research Society*, *24(5)*, 1118-1127.

Dickinson, B. P., Ashley, R. K., et al. (2008). "Reduced morbidity and improved healing with bone morphogenic protein-2 in older patients with alveolar cleft defects." *Plast Reconstr Surg*, *121(1)*, 209-17.

Dionne, M. S., Skarnes, W. C., et al. (2001). "Mutation and analysis of Dan, the founding member of the Dan family of transforming growth factor beta antagonists." *Molecular and cellular biology*, *21(2)*, 636-643.

Dong, J., Albertini, D. F., et al. (1996). "Growth differentiation factor-9 is required during early ovarian folliculogenesis." *Nature*, *383(6600)*, 531-535.

Dudley, A. T., Lyons, K. M., et al. (1995). "A requirement for bone morphogenetic protein-7 during development of the mammalian kidney and eye." *Genes & development*, *9(22)*, 2795-2807.

Duprez, D., Bell, E. J., et al. (1996). "Overexpression of BMP-2 and BMP-4 alters the size and shape of developing skeletal elements in the chick limb." *Mechanisms of development*, *57(2)*, 145-157.

Fontaine, K., Semonin, O., et al. (2005). "A new mutation of the noggin gene in a French Fibrodysplasia ossificans progressiva (FOP) family." *Genetic counseling (Geneva, Switzerland)*, *16(2)*, 149-154.

Forslund, C. & Aspenberg, P. (2003). "Improved healing of transected rabbit Achilles tendon after a single injection of cartilage-derived morphogenetic protein-2." *The American Journal of Sports Medicine*, *31(4)*, 555-559.

Forslund, C., Rueger, D., et al. (2003). "A comparative dose-response study of cartilage-derived morphogenetic protein (CDMP)-1, -2 and -3 for tendon healing in rats." *Journal of orthopaedic research : official publication of the Orthopaedic Research Society*, *21(4)*, 617-621.

Frost, H. M. (2001). "From Wolff's law to the Utah paradigm: insights about bone physiology and its clinical applications." *The Anatomical Record*, *262(4)*, 398-419.

Galvin, K. M., Donovan, M. J., et al. (2000). "A role for smad6 in development and homeostasis of the cardiovascular system." *Nature genetics*, *24(2)*, 171-174.

Gannon, F. H., Kaplan, F. S., et al. (1997). "Bone morphogenetic protein 2/4 in early fibromatous lesions of fibrodysplasia ossificans progressiva." *Human pathology*, *28(3)*, 339-343.

Garrison, K. R., Donell, S., et al. (2007). "Clinical effectiveness and cost-effectiveness of bone morphogenetic proteins in the non-healing of fractures and spinal fusion: a systematic review." *Health technology assessment (Winchester, England)*, *11(30)*, 1-150, iii-iv.

Gazzerro, E. & Canalis, E. (2006). "Bone morphogenetic proteins and their antagonists." *Reviews in endocrine & metabolic disorders*, *7(1-2)*, 51-65.

Gazzerro, E., Smerdel-Ramoya, A., et al. (2007). "Conditional deletion of gremlin causes a transient increase in bone formation and bone mass." *The Journal of biological chemistry*, *282(43)*, 31549-31557.

Gazzerro, E., Pereira, R. C., et al. (2005). "Skeletal overexpression of gremlin impairs bone formation and causes osteopenia." *Endocrinology*, *146(2)*, 655-665.

Gelse, K., Muhle, C., et al. (2008). "Cell-based resurfacing of large cartilage defects: long-term evaluation of grafts from autologous transgene-activated periosteal cells in a porcine model of osteoarthritis." *Arthritis and Rheumatism*, *58(2)*, 475-488.

Gerlach-Bank, L. M., Ellis, A. D., et al. (2002). "Cloning and expression analysis of the chick DAN gene, an antagonist of the BMP family of growth factors." *Developmental dynamics : an official publication of the American Association of Anatomists*, *224(1)*, 109-115.

Gong, Y., Krakow, D., et al. (1999). "Heterozygous mutations in the gene encoding noggin affect human joint morphogenesis." *Nature genetics*, *21(3)*, 302-304.

Gosselet, F. P., Magnaldo, T., et al. (2007). "BMP2 and BMP6 control p57(Kip2) expression and cell growth arrest/terminal differentiation in normal primary human epidermal keratinocytes." *Cellular signaling, 19(4)*, 731-739.

Govender, S., Csimma, C., et al. (2002). "Recombinant human bone morphogenetic protein-2 for treatment of open tibial fractures: a prospective, controlled, randomized study of four hundred and fifty patients." *The Journal of bone and joint surgery.American volume,84-A(12)*, 2123-2134.

Groppe, J., Greenwald, J., et al. (2002). "Structural basis of BMP signalling inhibition by the cystine knot protein Noggin." *Nature, 420(6916)*, 636-642.

Gysin, R., Wergedal, J. E., et al. (2002). "Ex vivo gene therapy with stromal cells transduced with a retroviral vector containing the BMP4 gene completely heals critical size calvarial defect in rats." *Gene therapy, 9(15)*, 991-999.

Hannallah, D., Peng, H., et al. (2004). "Retroviral delivery of Noggin inhibits the formation of heterotopic ossification induced by BMP-4, demineralized bone matrix, and trauma in an animal model." *The Journal of bone and joint surgery.American volume, 86-A(1)*, 80-91.

Hanyu, A., Ishidou, Y., et al. (2001). "The N domain of Smad7 is essential for specific inhibition of transforming growth factor-beta signaling." *The Journal of cell biology, 155(6)*, 1017-1027.

Hashimoto, Y., Yoshida, G., et al. (2007). "Generation of tendon-to-bone interface "enthesis" with use of recombinant BMP-2 in a rabbit model." *Journal of orthopaedic research : official publication of the Orthopaedic Research Society, 25(11)*, 1415-1424.

Haydon, R. C., Luu, H. H., et al. (2007). "Osteosarcoma and osteoblastic differentiation: a new perspective on oncogenesis." *Clinical orthopaedics and related research, 454*, 237-246.

Haydon, R. C., Zhou, L., et al. (2002). "Nuclear receptor agonists as potential differentiation therapy agents for human osteosarcoma." *Clinical cancer research : an official journal of the American Association for Cancer Research, 8(5)*, 1288-1294.

Helm, G. A., Alden, T. D., et al. (2000). "Use of bone morphogenetic protein-9 gene therapy to induce spinal arthrodesis in the rodent." *Journal of neurosurgery, 92(2 Suppl)*, 191-196.

Ho, A. M., Marker, P. C., et al. (2008). "Dominant negative Bmp5 mutation reveals key role of BMPs in skeletal response to mechanical stimulation." *BMC developmental biology, 8*, 35.

Hoffmann, A. & Gross, G. (2001). "BMP signaling pathways in cartilage and bone formation." *Critical reviews in eukaryotic gene expression*, *11(1-3)*, 23-45.

Hogan, B. L. (1996). "Bone morphogenetic proteins: multifunctional regulators of vertebrate development." *Genes & development*, *10(13)*, 1580-1594.

Hu, J., Chen, Y. X., et al. (2004). "Developmental expression and function of Bmp4 in spermatogenesis and in maintaining epididymal integrity." *Developmental biology*, *276(1)*, 158-171.

Imamura, T., Takase, M., et al. (1997). "Smad6 inhibits signalling by the TGF-beta superfamily." *Nature*, *389(6651)*, 622-626.

Jane, J. A. Jr., Dunford, B. A., et al. (2002). "Ectopic osteogenesis using adenoviral bone morphogenetic protein (BMP)-4 and BMP-6 gene transfer." *Molecular therapy : the journal of the American Society of Gene Therapy*, *6(4)*, 464-470.

Jorgensen, C., Gordeladze, J., et al. (2004). "Tissue engineering through autologous mesenchymal stem cells." *Current opinion in biotechnology*, *15(5)*, 406-410.

Kaiser, S., Schirmacher, P., et al. (1998). "Induction of bone morphogenetic protein-6 in skin wounds. Delayed reepitheliazation and scar formation in BMP-6 overexpressing transgenic mice." *The Journal of investigative dermatology*, *111(6)*, 1145-1152.

Kalantry, S., Manning, S., et al. (2001). "The amnionless gene, essential for mouse gastrulation, encodes a visceral-endoderm-specific protein with an extracellular cysteine-rich domain." *Nature genetics*, *27(4)*, 412-416.

Kang, Q., Sun, M. H., et al. (2004). "Characterization of the distinct orthotopic bone-forming activity of 14 BMPs using recombinant adenovirus-mediated gene delivery." *Gene therapy*, *11(17)*, 1312-1320.

Kaplan, F. S., Le Merrer, M., et al. (2008). "Fibrodysplasia ossificans progressiva." *Best practice & research.Clinical rheumatology*, *22(1)*, 191-205.

Katoh, M. & Katoh, M. (2004). "Identification and characterization of human CKTSF1B2 and CKTSF1B3 genes in silico." *Oncology reports*, *12(2)*, 423-427.

Katoh, M. & Katoh, M. (2006). "CER1 is a common target of WNT and NODAL signaling pathways in human embryonic stem cells." *International journal of molecular medicine*, *17(5)*, 795-799.

Kawabata, M., Imamura, T., et al. (1998). "Signal transduction by bone morphogenetic proteins." *Cytokine & growth factor reviews*, *9(1)*, 49-61.

Kawano, Y. & Kypta, R. (2003). "Secreted antagonists of the Wnt signalling pathway." *Journal of cell science, 116(Pt 13)*, 2627-2634.

Khokha, M. K., Hsu, D., et al. (2003). "Gremlin is the BMP antagonist required for maintenance of Shh and Fgf signals during limb patterning." *Nature genetics, 34(3)*, 303-307.

Kingsley, D. M. (1994). "What do BMPs do in mammals? Clues from the mouse short-ear mutation." *Trends in genetics : TIG, 10(1)*, 16-21.

Kingsley, D. M., Bland, A. E., et al. (1992). "The mouse short ear skeletal morphogenesis locus is associated with defects in a bone morphogenetic member of the TGF beta superfamily." *Cell, 71(3)*, 399-410.

Kolle, G., Georgas, K., et al. (2000). "CRIM1, a novel gene encoding a cysteine-rich repeat protein, is developmentally regulated and implicated in vertebrate CNS development and organogenesis." *Mechanisms of development, 90(2)*, 181-193.

Kovacevic, D. & Rodeo, S. A. (2008). "Biological augmentation of rotator cuff tendon repair." *Clinical orthopaedics and related research, 466(3)*, 622-633.

Krakow, D., Reinker, K., et al. (1998). "Localization of a multiple synostoses-syndrome disease gene to chromosome 17q21-22." *American Journal of Human Genetics, 63(1)*, 120-124.

Kuo, A. C., Rodrigo, J. J., et al. (2006). "Microfracture and bone morphogenetic protein 7 (BMP-7) synergistically stimulate articular cartilage repair." *Osteoarthritis and cartilage / OARS, Osteoarthritis Research Society, 14(11)*, 1126-1135.

Kuroda, R., Usas, A., et al. (2006). "Cartilage repair using bone morphogenetic protein 4 and muscle-derived stem cells." *Arthritis and Rheumatism, 54(2)*, 433-442.

Lane, K. B., Machado, R. D., et al. (2000). "Heterozygous germline mutations in BMPR2, encoding a TGF-beta receptor, cause familial primary pulmonary hypertension. The International PPH Consortium." *Nature genetics, 26(1)*, 81-84.

Larrain, J., Bachiller, D., et al. (2000). "BMP-binding modules in chordin: a model for signalling regulation in the extracellular space." *Development (Cambridge, England), 127(4)*, 821-830.

Lee, K. J., Mendelsohn, M., et al. (1998). "Neuronal patterning by BMPs: a requirement for GDF7 in the generation of a discrete class of commissural interneurons in the mouse spinal cord." *Genes & development, 12(21)*, 3394-3407.

Li, X., Ominsky, M. S., et al. (2008). "Targeted deletion of the sclerostin gene in mice results in increased bone formation and bone strength." *Journal of bone and mineral research : the official journal of the American Society for Bone and Mineral Research*, *23(6)*, 860-869.

Lintern, K. B., Guidato, S., et al. (2009). "Characterization of wise protein and its molecular mechanism to interact with both WNT and BMP signals." *The Journal of biological chemistry*.

Loots, G. G., Kneissel, M., et al. (2005). "Genomic deletion of a long-range bone enhancer misregulates sclerostin in Van Buchem disease." *Genome research*, *15(7)*, 928-935.

Lou, J. (2000). "In vivo gene transfer into tendon by recombinant adenovirus." *Clinical orthopaedics and related research (379 Suppl)*, (379 Suppl), S252-5.

Lou, J., Tu, Y., et al. (2001). "BMP-12 gene transfer augmentation of lacerated tendon repair." *Journal of orthopaedic research : official publication of the Orthopaedic Research Society*, *19(6)*, 1199-1202.

Luo, G., Hofmann, C., et al. (1995). "BMP-7 is an inducer of nephrogenesis, and is also required for eye development and skeletal patterning." *Genes & development*, *9(22)*, 2808-2820.

Luo, J., Sun, M. H., et al. (2005). "Gene therapy for bone regeneration." *Current gene therapy*, *5(2)*, 167-179.

Luo, X., Chen, J., et al. (2008). "Osteogenic BMPs promote tumor growth of human osteosarcomas that harbor differentiation defects." *Laboratory investigation; a journal of technical methods and pathology*, *88(12)*, 1264-1277.

Luu, H. H., Song, W. X., et al. (2007). "Distinct roles of bone morphogenetic proteins in osteogenic differentiation of mesenchymal stem cells." *Journal of orthopaedic research : official publication of the Orthopaedic Research Society*, *25(5)*, 665-677.

Mahoney, D. J., Mikecz, K., et al. (2008). "TSG-6 regulates bone remodeling through inhibition of osteoblastogenesis and osteoclast activation." *The Journal of biological chemistry*, *283(38)*, 25952-25962.

Mailhot, G., Yang, M., et al. (2008). "BMP-5 expression increases during chondrocyte differentiation in vivo and in vitro and promotes proliferation and cartilage matrix synthesis in primary chondrocyte cultures." *Journal of cellular physiology*, *214(1)*, 56-64.

Majewski, M., Betz, O., et al. (2008). "Ex vivo adenoviral transfer of bone morphogenetic protein 12 (BMP-12) cDNA improves Achilles tendon healing in a rat model." *Gene therapy*, *15(16)*, 1139-1146.

Majumdar, M. K., Wang, E., et al. (2001). "BMP-2 and BMP-9 promotes chondrogenic differentiation of human multipotential mesenchymal cells and overcomes the inhibitory effect of IL-1." *Journal of cellular physiology, 189(3)*, 275-284.

Massague, J., Seoane, J., et al. (2005). "Smad transcription factors." *Genes & development, 19(23)*, 2783-2810.

Matzuk, M. M., Kumar, T. R., et al. (1995). "Different phenotypes for mice deficient in either activins or activin receptor type II." *Nature, 374*(6520), 356-360.

McPherron, A. C., Lawler, A. M., et al. (1999). "Regulation of anterior/posterior patterning of the axial skeleton by growth/differentiation factor 11." *Nature genetics, 22(3)*, 260-264.

Michos, O., Panman, L., et al. (2004). "Gremlin-mediated BMP antagonism induces the epithelial-mesenchymal feedback signaling controlling metanephric kidney and limb organogenesis." *Development (Cambridge, England), 131(14)*, 3401-3410.

Mishina, Y., Suzuki, A., et al. (1995). "Bmpr encodes a type I bone morphogenetic protein receptor that is essential for gastrulation during mouse embryogenesis." *Genes & development, 9(24)*, 3027-3037.

Mitu, G. & Hirschberg, R. (2008). "Bone morphogenetic protein-7 (BMP7) in chronic kidney disease." *Frontiers in bioscience : a journal and virtual library, 13*, 4726-4739.

Moser, M., Binder, O., et al. (2003). "BMPER, a novel endothelial cell precursor-derived protein, antagonizes bone morphogenetic protein signaling and endothelial cell differentiation." *Molecular and cellular biology, 23(16)*, 5664-5679.

Murakami, G., Watabe, T., et al. (2003). "Cooperative inhibition of bone morphogenetic protein signaling by Smurfl and inhibitory Smads." *Molecular biology of the cell, 14(7)*, 2809-2817.

Nacamuli, R. P., Fong, K. D., et al. (2005). "Expression and possible mechanisms of regulation of BMP3 in rat cranial sutures." *Plast Reconstr Surg, 116(5)*, 1353-62.

Newman, J. H., Wheeler, L., et al. (2001). "Mutation in the gene for bone morphogenetic protein receptor II as a cause of primary pulmonary hypertension in a large kindred." *The New England journal of medicine, 345(5)*, 319-324.

Nosaka, T., Morita, S., et al. (2003). "Mammalian twisted gastrulation is essential for skeleto-lymphogenesis." *Molecular and cellular biology*, *23(8)*, 2969-2980.

Oelgeschlager, M., Larrain, J., et al. (2000). "The evolutionarily conserved BMP-binding protein Twisted gastrulation promotes BMP signalling." *Nature*, *405(6788)*, 757-763.

Parisi, M. S., Gazzerro, E., et al. (2006). "Expression and regulation of CCN genes in murine osteoblasts." *Bone*, *38(5)*, 671-677.

Perri, B., Cooper, M., et al. (2007). "Adverse swelling associated with use of rh-BMP-2 in anterior cervical discectomy and fusion: a case study." *The spine journal : official journal of the North American Spine Society*, *7(2)*, 235-239.

Piccolo, S., Agius, E., et al. (1999). "The head inducer Cerberus is a multifunctional antagonist of Nodal, BMP and Wnt signals." *Nature*, *397(6721)*, 707-710.

Piek, E., Afrakhte, M., et al. (1999). "Functional antagonism between activin and osteogenic protein-1 in human embryonal carcinoma cells." *Journal of cellular physiology*, *180(2)*, 141-149.

Pierre, A., Pisselet, C., et al. (2005). "Bone morphogenetic protein 5 expression in the rat ovary: biological effects on granulosa cell proliferation and steroidogenesis." *Biology of reproduction*, *73(6)*, 1102-1108.

Reddi, A. H. (2001). "Bone morphogenetic proteins: from basic science to clinical applications." *The Journal of bone and joint surgery.American volume*, *83-A Suppl 1(Pt 1)*, S1-6.

Retting, K. N., Song, B., et al. (2009). "BMP canonical Smad signaling through Smad1 and Smad5 is required for endochondral bone formation." *Development (Cambridge, England)*, *136(7)*, 1093-1104.

Reya, T., Morrison, S. J., et al. (2001). "Stem cells, cancer, and cancer stem cells." *Nature*, *414(6859)*, 105-111.

Rodriguez Esteban, C., Capdevila, J., et al. (1999). "The novel Cer-like protein Caronte mediates the establishment of embryonic left-right asymmetry." *Nature*, *401(6750)*, 243-251.

Rosen, V. & Thies, R. S. (1992). "The BMP proteins in bone formation and repair." *Trends in genetics : TIG*, *8(3)*, 97-102.

Ross, J. J., Shimmi, O., et al. (2001). "Twisted gastrulation is a conserved extracellular BMP antagonist." *Nature*, *410(6827)*, 479-483.

Sandhu, H. S. (2003). "Bone morphogenetic proteins and spinal surgery." *Spine*, *28(15 Suppl)*, S64-73.

Seeherman, H., Wozney, J., et al. (2002). "Bone morphogenetic protein delivery systems." *Spine, 27(16 Suppl 1)*, S16-23.

Sekiya, I., Larson, B. L., et al. (2005). "Comparison of effect of BMP-2, -4, and -6 on in vitro cartilage formation of human adult stem cells from bone marrow stroma." *Cell and tissue research, 320(2)*, 269-276.

Settle, S. H. Jr., Rountree, R. B., et al. (2003). "Multiple joint and skeletal patterning defects caused by single and double mutations in the mouse Gdf6 and Gdf5 genes." *Developmental biology, 254(1)*, 116-130.

Solloway, M. J., Dudley, A. T., et al. (1998). "Mice lacking Bmp6 function." *Developmental genetics, 22(4)*, 321-339.

Storm, E. E., Huynh, T. V., et al. (1994). "Limb alterations in brachypodism mice due to mutations in a new member of the TGF beta-superfamily." *Nature, 368(6472)*, 639-643.

Takao, M., Hino, J., et al. (1996). "Identification of rat bone morphogenetic protein-3b (BMP-3b), a new member of BMP-3." *Biochemical and biophysical research communications, 219(2)*, 656-662.

Takaoka, K., Nakahara, H., et al. (1988). "Ectopic bone induction on and in porous hydroxyapatite combined with collagen and bone morphogenetic protein." *Clinical orthopaedics and related research (234)*, (234), 250-254.

Thomas, D. M., Johnson, S. A., et al. (2004). "Terminal osteoblast differentiation, mediated by runx2 and p27KIP1, is disrupted in osteosarcoma." *The Journal of cell biology, 167(5)*, 925-934.

Thomas, J. T., Lin, K., et al. (1996). "A human chondrodysplasia due to a mutation in a TGF-beta superfamily member." *Nature genetics, 12(3)*, 315-317.

Thomas, J. T., Kilpatrick, M. W., et al. (1997). "Disruption of human limb morphogenesis by a dominant negative mutation in CDMP1." *Nature genetics, 17(1)*, 58-64.

Tremblay, K. D., Dunn, N. R., et al. (2001). "Mouse embryos lacking Smad1 signals display defects in extra-embryonic tissues and germ cell formation." *Development (Cambridge, England), 128(18)*, 3609-3621.

Tumialan, L. M. & Rodts, G. E. (2007). "Adverse swelling associated with use of rh-BMP-2 in anterior cervical discectomy and fusion." *The spine journal : official journal of the North American Spine Society, 7(4)*, 509-510.

Urist, M. R. (1965). "Bone: formation by autoinduction." *Science (New York, N.Y.), 150(698)*, 893-899.

Violini, S., Ramelli, P., et al. (2009). "Horse bone marrow mesenchymal stem cells express embryo stem cell markers and show the ability for tenogenic differentiation by in vitro exposure to BMP-12." *BMC cell biology*, *10*, 29.

Westendorf, J. J., Kahler, R. A., et al. (2004). "Wnt signaling in osteoblasts and bone diseases." *Gene*, *341*, 19-39.

White, A. P., Vaccaro, A. R., et al. (2007). "Clinical applications of BMP-7/OP-1 in fractures, nonunions and spinal fusion." *International orthopaedics*, *31(6)*, 735-741.

Wijgerde, M., Karp, S., et al. (2005). "Noggin antagonism of BMP4 signaling controls development of the axial skeleton in the mouse." *Developmental biology*, *286(1)*, 149-157.

Wilkinson, L., Kolle, G., et al. (2003). "CRIM1 regulates the rate of processing and delivery of bone morphogenetic proteins to the cell surface." *The Journal of biological chemistry*, *278(36)*, 34181-34188.

Winkler, D. G., Sutherland, M. K., et al. (2003). "Osteocyte control of bone formation via sclerostin, a novel BMP antagonist." *The EMBO journal*, *22(23)*, 6267-6276.

Winnier, G., Blessing, M., et al. (1995). "Bone morphogenetic protein-4 is required for mesoderm formation and patterning in the mouse." *Genes & development*, *9(17)*, 2105-2116.

Wolff, J. (1892). "Das Gesetz dez Transformation der Knochen (The Law of Bone Remodeling)." *Berlin*, *Springer*.

Woo, S. L., Hildebrand, K., et al. (1999). "Tissue engineering of ligament and tendon healing." *Clinical orthopaedics and related research (367 Suppl)*, (367 Suppl), S312-23.

Wright, V., Peng, H., et al. (2002). "BMP4-expressing muscle-derived stem cells differentiate into osteogenic lineage and improve bone healing in immunocompetent mice." *Molecular therapy : the journal of the American Society of Gene Therapy*, *6(2)*, 169-178.

Xu, M. Q., Feldman, G., et al. (2000). "Linkage exclusion and mutational analysis of the noggin gene in patients with fibrodysplasia ossificans progressiva (FOP)." *Clinical genetics*, *58(4)*, 291-298.

Yamaguchi, A., Ishizuya, T., et al. (1996). "Effects of BMP-2, BMP-4, and BMP-6 on osteoblastic differentiation of bone marrow-derived stromal cell lines, ST2 and MC3T3-G2/PA6." *Biochemical and biophysical research communications*, *220(2)*, 366-371.

Yang, X., Castilla, L. H., et al. (1999). "Angiogenesis defects and mesenchymal apoptosis in mice lacking SMAD5." *Development (Cambridge, England), 126(8)*, 1571-1580.

Yi, S. E., Daluiski, A., et al. (2000). "The type I BMP receptor BMPRIB is required for chondrogenesis in the mouse limb." *Development (Cambridge, England), 127(3)*, 621-630.

Yokouchi, Y., Vogan, K. J., et al. (1999). "Antagonistic signaling by Caronte, a novel Cerberus-related gene, establishes left-right asymmetric gene expression." *Cell, 98(5)*, 573-583.

Yoon, B. S., Ovchinnikov, D. A., et al. (2005). "Bmpr1a and Bmpr1b have overlapping functions and are essential for chondrogenesis in vivo." *Proceedings of the National Academy of Sciences of the United States of America, 102(14)*, 5062-5067.

Yu, Y., Bliss, J. P., et al. (2007). "Bone morphogenetic proteins and Smad expression in ovine tendon-bone healing." *Arthroscopy : The Journal of Arthroscopic & Related Surgery : Official Publication of the Arthroscopy Association of North America and the International Arthroscopy Association, 23(2)*, 205-210.

Zhang, H. & Bradley, A. (1996). "Mice deficient for BMP2 are nonviable and have defects in amnion/chorion and cardiac development." *Development (Cambridge, England), 122(10)*, 2977-2986.

Zhao, G. Q., Liaw, L., et al. (1998). "Bone morphogenetic protein 8A plays a role in the maintenance of spermatogenesis and the integrity of the epididymis." *Development (Cambridge, England), 125(6)*, 1103-1112.

Zhao, G. Q., Chen, Y. X., et al. (2001). "Mutation in Bmp7 exacerbates the phenotype of Bmp8a mutants in spermatogenesis and epididymis." *Developmental biology, 240(1)*, 212-222.

Zhao, M., Qiao, M., et al. (2003). "E3 ubiquitin ligase Smurf1 mediates core-binding factor alpha1/Runx2 degradation and plays a specific role in osteoblast differentiation." *The Journal of biological chemistry, 278(30)*, 27939-27944.

Zhao, M., Qiao, M., et al. (2004). "Smurf1 inhibits osteoblast differentiation and bone formation in vitro and in vivo." *The Journal of biological chemistry, 279(13)*, 12854-12859.

Zhu, H., Kavsak, P., et al. (1999). "A SMAD ubiquitin ligase targets the BMP pathway and affects embryonic pattern formation." *Nature, 400*(6745), 687-693.

In: Bone Regeneration: Growth Factors... ISBN: 978-1-60876-621-5
Editors: V. Legard et al., pp. 93-118

Chapter III

Bone Regeneration: Growth Factors, Augmentation Procedures and Tissue Engineering Application

Ana Claudia Muniz Renno[a], Mariza Matsumoto[b] and Daniel Araki Ribeiro[a]

[a]Department of Biosciences, Federal University of São Paulo (UNIFESP).
[b]Department of Oral and Maxillofacial Surgery and daggerOral Biology Program, University of Sacred Heart (USC), Bauru, Sp, Brazil.

Bone tissue plays a vital role in human system for its biological and mechanical activities, resulting from its regenerative capacity and permanent remodeling process (Gautier et al., 2005). Bone damage, either due to pathologies or traumas, is a very common occurrence and represents a major problem in orthopedics (Wiesmann et al, 2004). Most bone lesions, such as fractures or small size defects, heal well with conventional therapy due to the high regeneration potential of bone (Wiesmann et al, 2004). However, in large traumatic or post-surgical bone defects, fractures result in non-unions, others exhibit delayed healing, indicating that, in some cases, bone cannot heal through its normal wound healing response (Zimmermann & Wentzensen, 2007). In this context, there is a critical need for technologies that are able to enhance bone healing (Gautier et al., 2005). The development of effective therapies for bone regeneration is one of the most clinically important long-term goals of research in the mineralized tissue field. Gene therapy, tissue engineering and biomaterials have emerged as promising approaches for bone repair and reconstruction (Humatcher et al, 2000).

The Role of Growth Factors for Bone Regeneration

Tissue development is regulated by several signals, including soluble signaling molecules, insoluble ligands, mechanical cues and cell–cell interactions (Bell, 2000). Numerous peptides and proteins are involved in this signaling possess as a result of biological activity that marks them potential therapeutics (Boontheekul and Mooney, 2003). Soluble growth factors and immobilized ligands can regulate the adhesion, migration, proliferation and differentiation of various cell types. These factors may be a key to success in tissue engineering, an emerging field combining engineering and life sciences with the goal of repairing damaged or lost tissues (Bottaro et al., 2002). Tissue regeneration is affected by either transplanted or host cells, in concert with local environmental signals (Boontheekul and Mooney, 2003). One approach to direct the process of tissue formation is the incorporation of bioactive components, including growth factors and peptides mimicking the function of extracellular matrix (ECM) molecules, into biomaterials (Hubbel, 1999). Advances in material engineering have led to new modes of presenting these molecules to control the cell response and new tissue development (Boontheekul and Mooney, 2003). Transplantation of cells that secrete the desired growth factors is also a promising strategy to deliver these bioactive proteins, as are approaches based on gene therapy (Raffi and Lyden, 2003).

Growth factors act through autocrine and paracrine signaling mechanisms to induce the migration, proliferation and differentiation of osteoprogenitor cells and/or type I collagen synthesis and matrix apposition by mature osteoblasts (Boontheekul and Mooney, 2003). These factors are stored in the extracellular matrix during bone formation. Upon matrix destruction, either through the natural bone remodeling process or bone fracture, these factors are released to initiate osseous healing and to maintain the cyclic anabolic and catabolic processes that continually remodel bone (Boontheekul and Mooney, 2003). Therefore, a primary goal of growth factor delivery for bone tissue engineering is to accelerate the regeneration process and engineer bone in situations that normally lead to incomplete regeneration.

Nowadays, a plethora of mediators associated with bone regeneration has been purposed. These include BMP-2 and 4 (Barnes et al., 1999; Kloen et al., 2003), bFGF (Rodan et al., 1989), TGF-β (Bostrom and Asnis, 1998; Erlebacher et al., 1998), PDGF (Mandrachia et al., 2001) and VEGF (Deckers et al., 2000; Street et al., 2002; Ferrara et al., 2003), and others.

Accumulating evidence suggests that neovascularization plays a crucial role in treating many diseases (e.g. coronary artery disease) and in virtually all approaches to tissue engineering (Boontheekul and Mooney, 2003). A lack of vascularization leads to insufficient nutrient delivery and waste removal, cell death and limited tissue development and tissue loss. Angiogenesis, new blood vessel formation by sprouting from the sides and ends of pre-existing microvascular vessels, has been widely studied to determine the rules guiding blood vessel formation in adult tissues (Boontheekul and Mooney, 2003). Numerous growth factors have been identified as regulators of this process so far. For example, vascular endothelial growth factor (VEGF) is a key mediator of angiogenesis, as it is a potent mitogen for endothelial cells (ECs) and induces EC migration by upregulation of several endothelial integrin receptors (Ferrara et al., 2003). There is a family of VEGF growth factors, VEGF-A through VEGF-E, and two isoforms of VEGF-A (VEGF121 and VEGF165) have progressed to clinical trials in the treatment of ischemic diseases (Ennett and Mooney, 2002). Furthermore, several biomaterials have been used for the controlled release of VEGF, including PGA, alginate and fibrin (Lee et al., 2001; Wong et al., 2003). It has recently been noted that the VEGF release rate from polymers can be responsive to the local mechanical environment and this may be crucial in applications involving a mechanically dynamic environment (Lee et al., 2001). The VEGFs and their corresponding receptors are key regulators in a cascade of molecular and cellular events that ultimately lead to the development of the vascular system, either by vasculogenesis, angiogenesis or in the formation of the lymphatic vascular system (Tammela et al., 2005; Rossant and Howard, 2002). Although VEGF is a critical regulator in physiological angiogenesis, it also plays a significant role in skeletal growth (Zelzer et al., 2002; Gerber et al., 1999) and repair (Street et al., 2002). Sojo et al. (2005) have assumed that angiogenesis occurs predominantly before the onset of osteogenesis in bone lengthening in an osteodistraction model. Studies in the rat femoral drill-hole injury have shown differential expression of VEGF splicing isoforms along with its receptors indicating an important role in the bone healing process (Uchida et al., 2003).

Additionally, it is important to stress that osteoclasts release heparinase that induces the release of the active form of VEGF from heparin, activating not only angiogenesis but also osteoclast recruitment, differentiation and activity leading to the remodelling of the fracture callus during endochondral ossification (Saijo et al., 2003; Nakagawa et al., 2000). In the bone remodelling sequence, it has been postulated that VEGF can act with sRANKL to promote osteoclastogenesis and that a single injection of recombinant human VEGF can similarly induce osteoclast recruitment in op/op mice, suggesting that VEGF can act as a substitute for M-CSF in osteoclastogenesis(Niida et al., 1999). Others have yet revealed that the inhibition of VEGF in fractured mouse femurs results in a decrease in blood vessel invasion with a reduction in osteoclastic bone remodelling, impaired callus mineralization and reduced trabecular bone healing (Street et al., 2002; Gerber et al., 1999). Taken as a whole, it seems that osteogenesis and fracture repair maybe enhanced by the synchronized endogenous production of angiogenic and osteogenic mediators (Glowacki, 1998).

Fibroblast growth factor 2 (FGF-2), another well-studied angiogenic factor, is able to exert diverse biological effects on numerous cell types, including ECs, and has been used in clinical trials to induce angiogenesis (Montero et al., 2000; Collin-Osdoby et al., 2002; Khurana and Simons, 2002). It also has the capability to accelerate fracture repair when added exogenously to the early healing stage of a bone (Kawaguchi et al., 2001, Chen et al., 2004), since it stimulates the proliferation and differentiation of osteoblasts (Globus et al., 1988; Nakamura et al., 1995; Walsh et al., 2000; Pun et al., 2001). In light of these important features, FGF-2 has been incorporated into various polymers, such as gelatin (Iwakura et al., 2003), collagen (Pieper et al., 2002), chitosan (Ishihara et al., 2003) and PLA (Lee et al., 2002). FGF-2 reversibly binds to heparin-like molecules and heparin sulfate proteoglycans. This can be exploited to increase the FGF-2 binding capacity to polymeric delivery vehicles, resulting in a more gradual and sustained release of FGF-2 and an enhancement of angiogenesis (Pieper et al., 2002).

Transforming growth factor (TGF) family of proteins has an essential role in bone formation through the regulation of osteoprogenitor and osteoblast proliferation and differentiation (Forslund, 2003). A variety of materials have been used for TGF-b1 delivery, including PLGA-PEG and coral and combined TGF-b1 and insulin-like growth factor (IGF) release from PLA carriers accelerates osteotomy healing (Forslund, 2003). However, the optimal conditions for TGF-b1 release during bone regeneration have yet to be defined.

Bone morphogenetic proteins (BMP), members of the TGF superfamily appear to act as differentiation factors, causing mesenchymal cells to differentiate into bone-forming cells (Sellers et al., 2000). Therefore, BMPs are widely recognized to possess the greatest in vivo bone stimulatory capacity and are the only growth factors known to stimulate mesenchymal stem cells (MSC) to differentiate along osteoblastic lineages (Kofron et al., 2004). This has allowed the use of BMP in combination with mesodermal-derived stem cells from bone marrow (Blum et al., 2004), muscle (Rose et al., 2003; Lee et al., 2002a) or adipose tissue (Dragoo et al., 2003) in so far as to be explored for use in bone tissue engineering applications. In particular, recombinant human BMP-2 and BMP-7 are currently in clinical use as osteoinductive agents (Kofron et al., 2003). Several carriers for BMP have been developed, including collagen (Akamaru et al., 2003), PLGA (Bessho et al., 2002), PLA–p-dioxanone– PEG block copolymers (Saito et al., 2001), PLA-PEG and PEG hydrogels (Burdick et al., 2002). These systems allow control over the BMP release rate and promote pre-osteoblast differentiation and mineralization in vitro and ectopic bone formation in vivo (Burdick et al., 2002).

Bone Augmentation

Bone fractures and bone loss caused by ablative surgeries and severe inflammatory conditions were the main causes for the development of therapies and surgical techniques. However, in various situations, not only function is required, but also aesthetics, considering the increasing demand for perfection by modern society, being more evident when the face and head are affected. Undoubtedly, the phenomenon of osseointegration presented by Branemark et al. in late 1970's, attesting a 10-year result of endosseous integration for dental rehabilitation, has emerged an uncountable number of clinical and scientific researches on bone biology and reconstruction, especially in the field of oral and maxillofacial surgery due to the high impact of oral and craniofacial rehabilitation in both function and aesthetics, and the particular influence of load, mastication, and oral habits. Although edentulous caused by poor oral hygiene still represents a health problem in developing countries, it is currently the main focus on bone augmentation worldwide, including trauma and congenital disorders. In this way, much of our discussion in this section will be based on dental implantology.

Autogenous Bone Grafts

Autogenous bone grafts are still considered the "gold standard" material for bone replacement (Friedrich et al., 2008; Sbordone et al. 2009). The lack of immunogenicity reaction and the presence of osteogenic cells releasing growth factors contributing to bone repair are the main advantages of autogenous bones. The choice of the donor area and type of bone graft, free or vascularized, depends on the extension and site of the defect, the quality of the donor bone, as well as the function desired after reconstruction (Ghert et al., 2007; Gerressen et al. 2009). Iliac crest is one of the most used bones, especially for reconstruction of larger defects (Barone & Covani, 2007; van Gemert et al., 2009; Wong et al., 2009), along with rib and tibia (O'Keefe et al., 1991; Catone et al., 1992). When considering rehabilitation using osseointegrated implants, craniofacial bones like the mandible and calvaria are preferred, although limited amount of graft is a disadvantage (Misch, 1997; Garg et al., 1998; Cordaro et al., 2002; Brugnami et al., 2009).

In past decades, discussion on the embryological origin of bone grafts intrigued some researchers (Zins & Whitaker, 1983; Kusiak et al., 1985; Hardesty & Marsh, 1990), once different bones were put in close contact with each other in grafting procedures, resulting in their integration. Despite the results favoring intramembranous bone grafts in relation to revascularization and minimal resorption in comparison to endochondral grafts, little was known about the biological mechanisms involved. Nowadays, molecular biology has been helping in the elucidation of these mechanisms, as demonstrated by recent publications. Lu & Rabie (2006) observed the expression of the metalloproteinase-9 (MMP-9), a key factor involved in bone resorption, in endochondral (EC) and intramembranous (IM) bone grafts of rabbits by immunohistochemistry and in situ hybridization. Results showed a significantly higher expression of MMP-9 in EC grafts, associated with percentage of resorption of 138% higher than IM grafts. Identification of specific genes involved in EC ossification are being revealed (Vaes et al., 2006), as well as the expression of different BMPs (Suttapreyasri et al., 2006) in EC and IM bones.

Bone Substitutes

In order to minimize costs, a second surgical site, and additional morbidity in donor graft areas, biomaterials serving as bone substitutes are being extensively studied (Boccaccini & Blaker, 2005; El-Ghannam, 2005; Kanczler & Oreffo, 2008) and presenting satisfactory results (Del Fabbro et al., 2004; Aghaloo & Moy, 2007). They aggregate a number of materials from natural to synthetic origin, in which the so-called bioactive materials are preferred for their capacity of chemically adhere to bone tissue, due to the presence of calcium and phosphate in their composition (Hubbell, 1999; Hench & Best 2004).

Bioceramics represent a broad range of inorganic/nonmetallic compositions (Hench & Best, 2004), including hydroxyapatite, bioactive glass, calcium phosphate, and others and they have been extensively used in particulate form, acting as fillers and scaffold for bone defects reconstruction promoting a tri-dimensional matrix in order to stabilize and maintain the shape of the filled area. It also permits and supports cell migration and angiogenesis, resulting in new bone formation during repair (Spector, 2008; Walsh et al., 2008). Each alloplastic biomaterial presents specific chemical and physical characteristics, which directly influence rate and time of bone formation. Although they are osteoconductive in the majority, bone substitute biomaterial should not only substitute bone, but act in a way that its complete absorption occurs, permitting the formation and its replacement by a new bone tissue. This response is mainly dependent on its origin, shape, and porosity characteristics (Spector, 2008).

Synthetic hydroxyapatite (HA) has been one of the most frequently used ceramic because of its chemical composition similarity to human bone, non toxicity, high chemical stability, and absence of inflammatory or antigenic reactions. One of the advantages of hydroxyapatite is that its microstructure can be controlled to promote the formation of pores that can allow the migration of blood vessels and bone tissue into the material (Silva et al, 2005). This material is known as a slowly biodegradable material with high osteocompatibility and bone binding capability, and its resorption rate is relatively slow compared with the rate of new bone formation (Ricci et al., 1992). Artiz et al. (2004) stated that the low-lasting presence of bovine HA incorporated to bone, creates a dense cancellous network, by the strengthening of bone tissue mass.

Many studies have investigated the effects of hydroxyapatite on bone metabolism and as a substitute for autogenous bone grafts in the reconstruction of bone defects. Silva et al. (2005) evaluated the repair of calvarial bone defects in rats comparing autogenous bone grafts and porous bioceramic discs of hydroxyapatite/phosphate cement mixture. They found that by week 24, the defects filled with bioceramic material were almost completely closed as a result of the joining of ceramic fragments and the neoformed bone tissue. Moreover, Orr et al. (2001) found an increase in the modulus of elasticity of the bone defects in rabbits implanted with a synthetic hydroxyapatite material 26 weeks post surgery.

In oral and maxillofacial reconstruction, bovine HA represents one of the most important biomaterial for bone reconstruction aiming rehabilitation with endosseous implants (Olson et al., 2000; Schlegel et al., 2003; Hallman & Nordin, 2004; Dahlin et al., 2009). An interesting result was presented by Schelegel et al. (2003) comparing maxillary sinus augmentation using autogenous bone grafts and bovine HA in beagle dogs. After 90 days, a slight improve in bone-implant contact and volume in the bone grafts group; however, after 180 days, these results inverted, with a decrease in bone volume of sinuses treated with bone grafts.

Chemically pure tricalcium phosphate (TCP) is another material currently used for bone grafts due to their similarity to the mineral phase of bone and excellent osteoconductivity. It is available in two distinct forms: α-TCP, modified by high temperatures, over 1.125°C, and the β-TCP, modified by low temperatures, under 1.125°C, being α-phase more soluble than β-phase. When α-TCP is used as bone substitute, its dissolution occurs before new bone formation, resulting in a fibrous tissue. Modifying material phase and increasing porosity, degradation and resorption rates can be adapted to follow bone regeneration (Wiltfang et al., 2002). Former types of TCP, due to their heterogeneous compounds, were often contaminated with foreign phases, leading to different degradation times (Merten et al., 2001).

β-TCP has biodegradable properties and it is also considered to have high osteocompatibility and to be resorbed at a higher rate than HA. Dias et al., (2004) showed that 12 weeks post surgery, β-TCP implants were surrounded by new bone tissue in tibia defects of rabbits. Study comparing the effects of HA and β-TCP implants in parietal bone in rats showed that, at 2 weeks, osteogenesis from the parietal bone was observed around both materials, and new bone had attached directly to the surfaces of HA and TCP. However, the volume of new bone around HA was larger than that around TCP. Also, the authors found that there was no remarkable change in the amount of remaining HA, but that of TCP was decreased (Fujita et al., 2003). Orztuck et al. (2006) evaluated the effects of hydroxyapatite/tri-calcium phosphate (HA/TCP) implants and demineralised bone matrix (DBM) on bone consolidation in rats. The authors found that 10 weeks after implantation, the use of DBM alone demonstrated improved healing on radiographic and histological studies compared to other groups and the control group. Interestingly, no difference was found between the HA/TCP treated group and the control group.

As previously mentioned, bone substitutes represent a welcome alternative to autogenous bone grafts considering oral rehabilitation using dental implants, especially in bone cavities as maxillary sinuses, since they avoid an extra surgical site for graft retrieval, but mainly because they sustain bone defects while healing, permitting the insertion of the implants before complete bone formation (Schelegel, 2003). The necessity for understanding the effects of these materials on cells and tissue has leading researches to search for relevant information from molecular biology techniques. The identification of growth factors involved in osteoblasts differentiation and synthesis, as well as in vascularization, contributes to the comprehension of different healing processes. Zerbo et al. (2005), in biopsies retrieved from human maxillary sinuses after 6 months reconstruction with β-TCP, observed bone growth into material porous, as well as soft connective tissue. Positive immunostaining for Runx-2/Cbfa-1 (core-binding factor-1), a transcriptor factor required for activation of osteoblast differentiation, osteopontin and bone sialoprotein in (pre) osteoblast, young osteocytes and in the connective tissue cells around or in the particles of β-TCP suggested that these cells were following osteogenic differentiation pathway. Our group also analyzed the presence of Runx2/Cbfa-1 and VEGF (vascular endothelial growth factor) in maxillary sinuses augmentation of rabbits using HA (Bio-Oss®) and β-TCP (Cerasorb®). Our results showed a stronger Runx2/Cbfa-1 immunostaining in HA specimens than β-TCP group, but similar rates of VEGF in both groups, in an attempt to contribute with the elucidation of biomaterials behavior (unpublished data). Inert implants, like some metals, make bone formation possible permitting a mechanical adhesion to the material, called biological fixation. However, the considered bioactive materials, such as the cited biomaterials in this section, induce a biological response in the interface area, creating a chemical adhesion with bone, called bioactive fixation, due to the presence of calcium and phosphate. The surface of these materials induces a biologically active carbonated layer of hidroxyapatite that creates an adhesive interface with the tissues. In addition, researches have shown the stimulation of osteoblastic differentiation *in vitro* by local rise in calcium and phosphate ions (Sugimoto et al., 1993).

Another promising treatment is the use of bioactive glasses as bone graft substitutes due to their ability to bond and integrate with living bone by forming a biologically active bonelike apatite layer on their surfaces (Hench & Polak, 2002; Hench, 2003). Bioactive glasses are resorbable and the dissolution products (soluble silicon and calcium) have been found to up-regulate seven families of genes in osteoblasts (Hench, 2003). It was found that both Ca and Si ions had to be present for the up-regulation to occur and the effect was found to be dose-dependant. It is therefore vital that the ion release from any bioactive scaffold can be tailored.

The original bioactive glass discovered by Hench and named Bioglass® 45S5 is a melt-derived glass with four components (46.1% SiO2, 24.4% Na2O, 26.9% CaO and 2.6% P2O5, in mol) (Hench, 2003)). Pores have been introduced into melt-derived bioactive glasses but the pores were few in number and were in the form of orientated channels of irregular diameter running through the glass so interconnectivity was poor.

Bioglass® 45S5 has been known for many years as the most bioactive composition among numerous bone-bonding glasses (Hench, 2003). It has been reached many clinical applications, including the repair of periodontal bone defects, maxillo-facial defects reconstruction, spinal surgery and bone replacement (Hench, 2003). Many works have proved that Bioglass® 45S5 has good bone bonding properties and a stimulatory effect on osteogenesis (Reilly at al., 2007). Human bone cell culture over the surface of Bioglass® 45S5 discs demonstrated enhanced formation of bone nodules, which is a three-dimensional structure composed of mineralized extracellular matrix and cell aggregates with organizational complexity similar to natural bone grown *in vivo* (Xynus et al., 2000). In addition, the ionic products of Bioglass® 45S5 induced an increase in both human osteoblast proliferation and expression of insulin-like growth factor II (IGF-II), a potent osteoblast mitogenic growth factor (Xynus et al., 2000). Indeed, the mechanism for *in-situ* tissue regeneration involves upregulation of seven families of genes that control the osteoblast cell cycle, mitosis and differentiation (Hench et al., 2004).

Additionally, Day et al (2005) demonstrated the angiogenic properties of Bioglass® 45S5-particles incorporated into biodegradable polymer composites. Based on i*n vitro* studies, which showed that fibroblasts cultured on discs consisting of specific quantities of Bioglass-particles mixed into poly (d,l-lactide-*co*-glycolide) secreted significantly increased quantities of vascular endothelial growth factor, the authors incorporated an optimal quantity of Bioglass-particles into a three-dimensional macroporous poly (d,l-lactide-*co*-glycolide) foam scaffolds. The foams were implanted subcutaneously into mice for periods of up to 6 weeks. Histological assessment was used to determine the area of granulation tissue around the foams, and the number of blood vessels within the granulation tissue was counted. The presence of Bioglass-particles in the foams produced a sustained increase in the area of granulation tissue surrounding the foams. The number of blood vessels surrounding the neat foams was reduced after 2 weeks of implantation; however, compression-molded foams containing Bioglass after 4 and 6 weeks of implantation, had a significant higher number of blood vessels surrounding the foams compared with foams containing no Bioglass These results indicate that composite polymer foam scaffolds containing Bioglass-particles retain granulation tissue and blood vessels surrounding the implanted foams.

Tissue Engineering Applications and Bone Repair

Tissue engineering is a relatively modern and an expanding field of medical research working to develop appropriate models and technologies to promote the bone regeneration. It attempts to regenerate and replace skeletal tissue lost through trauma or disease (Dickson et al, 2007). Strategies to engineer new bone are many and varied but aim ultimately at facilitating the regeneration of dynamic bony tissue with its various cell types (including osteoblasts, osteocytes and osteoclasts) and extracellular matrix constituents (fibrous type 1 collagen and numerous non-collagenous proteins) (Dickson et al., 2007). One of the most promising methods used in tissue engineering is the use of 3-D cellular scaffolds (Griffith and Naughton, 2002).

Scaffolds have been largely used as a way to promote growth of cells into three-dimensional tissues and serve as a cell delivery system (Palsson & Bhatia, 2004). It is a porous three-dimensional structure, which serves for initial cell attachment and subsequent tissue formation both *in vitro* and *in vivo*. Cells can be expanded in culture and seeded onto a scaffold that will slowly degrade and resorb, as the tissue structures grow *in vitro* and/ or *in vivo* and it seems to be more effective in bone repair or regeneration than pure biomaterials (Hassna et al, 2004).

Ideally, a scaffold should have the following characteristics:

(1) It needs to be three dimensional and highly porous with an interconnected pore network for cell / tissue growth and transport of nutrients and metabolic waste. Porous biomaterials are necessary for the efficient diffusion of nutrients, gases, growth factors, and other essential components to all cells within the scaffold, as well as diffusion of metabolic waste products away from the cells (Hutmacher et al, 2001). Moreover, appropriate pore distribution also allows for angiogenesis and

vascularisation, which are essential for long-term tissue integrity after implantation.

(2) biodegradable or bioresorbable with a controllable degradation and resorption rate to match cell / tissue growth *in vitro* and/or *in vivo*.

(3) appropriate surface chemistry and topography for cell attachment, proliferation, and differentiation;

(4) mechanical properties to match those of the tissues at the site of implantation. It seems that scaffolds are more biocompatible when their mechanical properties correspond with those of the intended *in vivo* site of implantation (Hutmacher 2001).

(5) Lack of toxicity: it should not induce any adverse responses within the recipient (Palsson and Bhatia, 2004).

A number of different materials, as well as scaffold design, have been experimentally and/or clinically studied. These materials may consist of naturally-derived molecules and/or synthetic substances, such as metals (as titanium and its alloys), ceramics, polymers and composites (Yang *et al.,* 2001; Palsson and Bhatia, 2004). For example, bioceramics, such as hydroxyapatite, bioactive glass, calcium phosphate ceramics and others, have been used as filler materials for bone defect repair (Yang et al., 2001). Substances such as these are currently being explored as potential biomaterials for the construction of synthetic cellular scaffolds for use in a range of clinical applications (Ang et al., 2006; Ryan et al., 2006).

As mentioned above, synthetic hydroxyapatite (HA) has been one of the most frequently used biomaterials. In the area of trabecular bone tissue engineering, many forms of porous HA have been explored as potential scaffolds (Kazakia et al., 2006).

Kazakia et al (2006) studied the effects of *in vitro* marrow stromal cell-mediated tissue deposition upon a trabeculated hydroxyapatite scaffold on the strength and toughness of the resulting bone substitute. The authors showed that the mechanical properties of the trabecular bone substitutes were significantly enhanced by *in vitro* tissue formation when compared to control. The strength increased by 85% and energy to fracture increased by 130% in the cellullarized scaffold. Accompanying the enhancement of mechanical properties was the deposition of significant amounts of bone matrix and mineral. This new mineralization was found to exist within a newly deposited parallel-fibered matrix both encasing and bridging between scaffold trabeculae.

Moreover, Zang et al (2003) studied a biodegradable nanocomposite porous scaffold comprising a b-tricalcium phosphate (b-TCP) matrix and HA nanofibers for bone tissue engineering. In this study, the role of HA nanofibers in enhancing the mechanical properties of the scaffold was investigated. Compression tests were performed to measure the compressive strength, modulus and toughness of the porous scaffolds. The results showed that the mechanical properties of the scaffold were significantly enhanced by the inclusion of HA nanofibers. The same authors also have tested the effect of a new scaffold of calcium phosphate cemente (CPC)/HA and a biodegradable polymer- chitosan (Zang et al, 2006). They used the Chitosan to make the material stronger, fastsetting and anti-washout; and mannitol to create macropores. The resulting porosity in CPC ranged from 34.4% to 83.3% volume fraction. At 70.2% porosity, the hydroxyapatite scaffold possessed flexural strength and elastic modulus. These values were within the range for sintered porous hydroxyapatite and cancellous bone. In conclusion, this new graft composition could be delivered during surgery in the form of a paste to harden *in situ* in the bone site to form macroporous hydroxyapatite.

Another class of biomaterials used for scaffolds fabrication is termed 'biodegradable' polymers and it included the application of homopolymers as polylactic acid (PLA), polyglycolic acid (PGA) and polycaprolactone (PCL) and co-polymers such as polylactic-polyglycolic acid (PLG) and L- and DL- lactide (PLDLA) (Palsson and Bhatia, 2004). Polymer chains interact with one another to confer unique characteristics as compared to other solids, and are often tailored for a particular clinical application (Palsson and Bhatia, 2004). An advantage of these polyesters is that they can biodegrade within weeks or months, enabling natural tissue to fill the space occupied previously by the implant. However polymers in current use lose strength well before they lose mass and while early ingrowth of natural tissue is inhibited, the subsequent rapid loss of mass can produce inflammation resulting from the production of acidic degradation products (Dickson et al, 2007). In this context, it is common to combine polymers with bioceramics (Yang *et al.*, 2001). These composites have more appropriate mechanical properties for stimulating osteoblastic cell expression and bone tissue (Dickson et al, 2007).

Many works investigated the effects of Bioglass® 45S5 scaffolds. Chen et al (2005) tested a novel highly porous Bioglass® 45S5 scaffolds treated with 3-AminoPropyl-TriethoxySilane (APTS) and glutaraldehyde (GA). It was found that the aqueous heat-treatment applied at 80◦C for 4 hrs during the surface functionalization procedure accelerated the structural transition of the crystalline

Na2Ca2Si3O9 phase, present in the original scaffold structure as a result of the sintering process used for fabrication, to an amorphous phase during SBF immersion. The surface functionalized scaffolds exhibited an accelerated crystalline hydroxyapatite layer formation upon immersion in SBF caused by ion leaching and the increased surface roughness induced during the heat treatment step.

Moreover, Bretcanu et al (2009) investigated the in vitro biocompatibility of Bioglass® 45S5 scaffolds, using a human osteosarcoma cell line (HOS-TE85). Corroborating with other works, these authors showed that the scaffolds are highly bioactive, as after a few days of immersion in a simulated body fluid, a hydroxyapatite-like layer was formed on the scaffold's surface. Moreover, HOS cells seeded and cultured onto the scaffolds for up to 2 weeks, demonstrated an increase in cell proliferation, confirming that Bioglass/P(3HB) scaffolds have potential as osteoconductive tissue engineering substrates for maintenance and normal functioning of bone tissue.

Some studies demonstrated that rat-derived mesenchymal stem cells (MSCs) show elevated levels of alkaline phosphatase activity when grown on 45S5 bioactive glass scaffolds as compared to standard tissue culture plastic. Similarly, exposure to the dissolution products of 45S5 elevates alkaline phosphatase activity and other osteogenic markers in these cells (Reilly et al (2007).

Despite its well-known stimulatory effects on osteogenesis, the use of monolithic Bioglass® 45S5 for bone engineering applications has been limited mainly due to its relatively poor mechanical properties (Vallet-Regi, 2001). Considering this important matter, our research group has developed nucleation and growth thermal treatments to obtain a novel fully-crystallized bioactive glass-ceramic of the quaternary P_2O_5-Na_2O-CaO-SiO_2 system (Biosilicate®, patent application WO 2004/074199). Crystallinity significantly changes the fracture characteristics of glasses leading to glass-ceramics with improved mechanical properties (Hench, 2004; Vallet-Regi, 2001).

In an *in vitro* experiment, Moura et al (2007) demonstrated that Biosilicate® produced enhanced bone-like matrix formation in comparison to its parent glass and to Bioglass® 45S5 in an osteogenic cell culture system and Biosilicate® supported significantly larger areas of calcified matrix at day 17 post-seeding (Moura et al. 2007). The authors concluded that the full crystallization of bioactive glasses in a range of compositions of the system P2O5-Na2O-CaO-SiO2 may promote enhancement of *in vitro* bone-like tissue formation in an osteogenic cell culture system.

Another study performed by our research group investigated the osteoblastic cell proliferation on Biosilicate® scaffolds and determined whether the application of laser phototherapy at 830 nm (continuous; 10 J/cm^2) would result in an increasing of osteoblast proliferation on the glass-ceramic scaffold. As *in vitro* osteoblast growth is a slow process, laser phototerapy can be used to stimulate osteoblast proliferation. Thus, the combination of cell seeding and laser phototerapy in scaffolds may lead to potential advantages in the tissue healing process. In this study, seeded scaffolds were irradiated with a single exposure of laser at 10 J/cm^2 (λ=830 nm). Cell proliferation was assessed 24 hours after irradiation. As a result, we found that osteoblastic MC3T3 cells were successfully grown on discs composed of a glass-ceramic composite (Figure 1). However, laser irradiation at 830 nm produced a 13% decrease in MC3T3 cell proliferation on glass-ceramic discs compared to control (non-irradiated) discs. In conclusion, successful seeding of bioscaffolds with osteoblasts, laser phototherapy (830nm at 10 J/cm^2) resulted in a reduction in cell proliferation compared to non-irradiated controls. However, future research combining laser phototherapy and glass-ceramic scaffolds should take in to account possible interactions of the laser with matrix compounds (Renno et al, 2009 in press).

Despite the positive effects of Biosilicate® on bone cell proliferation, studies investigating its effects on bone healing are fairly limited in the literature. In order to progress our understanding of the physiological processes of the Biosilicate® on fracture consolidation, we investigated the *in vivo* biological performance of the Biosilicate® as a potential biomaterial for bone augmentation restoration. Granito et al (2009) investigated the mechanical and histological characteristics of bone defects filled with two different particle sizes of Biosilicate materials (180-212 and 300-355 micron mean size) and compared these characteristics to those obtained with a Bioglass® material of similar particle sizes. In this study, a bilateral 2.4 mm non-critical size bone defects were surgically created at the upper third of the tibia of rats. The bone cavities were completely filled with the corresponding biomaterial of the treated groups and the rats were euthanized 15 days post-surgery. Biomechanical analyses showed that groups treated with Bioglass® and Biosilicate® had an increase of 10% and 20% of maximal load, respectively, when compared to the control group. In the histological analysis, the specimens treated with Biosilicate®, at the higher granulometry (300 to 355 μm), had the bone defect filled with biomaterial but woven bone was in apposition to the surface of the biomaterial. Therefore, it seems that the total amount of fully-formed bone was higher (10%) when compared to the control group. The same picture occurred regarding Biosilicate®

with granulometry 180 to 200 μm, i.e., the amount of fully-formed bone was found higher (10%) for this group as well. In the group exposed to Bioglass 45S5 ® either 300-355 (G4) or 180-200 granulometry, bone defects were also filled with biomaterial containing granulation tissue throughout the bone defect. In this group, there was also the presence of woven bone circumjacent to the biomaterial. Such findings suggest that Biosilicate® exerts more osteogenic activity when compared to Bioglass® under subjective histopathological analysis.

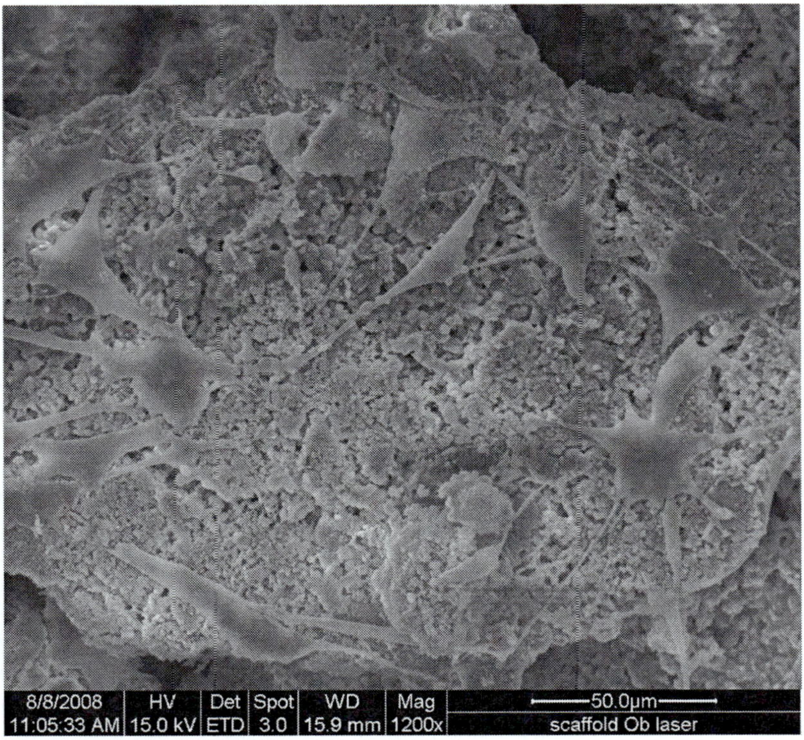

Figure 1. Scanning electron microscope (SEM) image of osteoblast cells successfully seeded on to a biosilicate disc.

In conclusion, we can assume that the development of effective therapies for bone regeneration is one of the most clinically important long-term goals of research in the mineralized tissue field. Bone tissue engineering is a very promising alternative approach to repair bone defects and in this context, a wide range of growth factors, biomaterials and synthetic scaffolds have been proposed to improve and to accelerate bone consolidation.

References

Aghaloo, T. L. & Moy, P. K. (2007). Which hard tissue augmentation techniques are the most successful in furnishing bony support for implant placement? *Int J Oral Maxillofac Implants*, *22 Suppl*, 49-70.

Akamaru, T., Suh, D., Boden, S. D., Kim, H. S., Minamide, A. & Louis-Ugbo, J. (2003). Simple carrier matrix modifications can enhance delivery of recombinant human bone morphogenetic protein-2 for posterolateral spine fusion. *Spine*, *28*, 429-434.

Ang, K. C., Leong, K. F., Chua, C. K. & Chandrasekaran, M. (2007). Compressive properties and degradability of poly(epsilon-caprolatone)/hydroxyapatite composites under accelerated hydrolytic degradation. *Journal of Biomedical Materials Research. Part A*, *80*, 655-60.

Artzi, Z., Weinreb, M.; Givol, N., Rohrer, M. D., Nemcovsky, C. E., Prasad, H. S. & Tal, H. (2004). Biomaterial resorption rate and healing site morphology of inorganic bovine bone and β-tricalcium phosphate in the canine: a 24-month longitudinal histologic study and morphometric analysis International *Journal of Maxillofacial Implants*, *19*, 357-368

Barnes, G. L., Kostenuik, P. J., Gerstenfeld, L. C. & Einhorn, T. A. (1999). Growth factor regulation of fracture repair. *J Bone Miner Res*, *14*, 1805-1815.

Barone, A. & Covani, U. (2007). Maxillary alveolar ridge reconstruction with nonvascularized autogenous block bone: clinical results. *J Oral Maxillofac Surg*, *65*, 2039-46.

Bell, E. (2000). Tissue engineering in perspective. In Principles of Tissue Engineering, edn 2. In: R. P. Lanza, R. S. Langer, & J. P. Vacanti (Eds.), San Diego: Academic Press, xxxv-xli.

Bessho, K., Carnes, D. L., Cavin, R. & Ong, J. L. (2002). Experimental studies on bone induction using low-molecular-weight poly (DL-lactideco- glycolide) as a carrier for recombinant human bone morphogenetic protein-2. *J Biomed Mater Res*, *61*, 61-65.

Boccaccini AR, Blaker JJ (2005). Bioactive composite materials for tissue engineering scaffolds. Expert Rev Med Devices 2, 303-17.

Blum, J. S., Parrott, M. B., Mikos, A. G. & Barry, M. A. (2004). Early osteoblastic differentiation induced by dexamethasone enhances adenoviral gene delivery to marrow stromal cells. *J Orthop Res*, *22*,411-416.

Boontheekul, T. & Moone,y,. .DJ (2003). Protein-based signaling systems in tissue engineering. *Current Opinion in Biotechnology*, 2003, *14*, 559-565.

Bostrom, M. P. & Asnis, P. (1998). Transforming growth factor beta in fracture repair. *Clin Orthop Relat Res*, 355S, S124-S131.

Bottaro, D. P., Liebmann-Vinson, A. & Heidaran, M. A. (2002). Molecular signaling in bioengineered tissue microenvironments. *Ann NY Acad Sci*, *961*, 143-153.

Branemark, P. I., Brånemark, P. I., Hansson, B. O., Adell, R., Breine, U., Lindström, J., Hallén, O. & Ohman, A. (1977). Osseointegrated implants in the treatment of edentulous jaw Experience from a 10-year period Scandinavian *Journal of Plastic Surgery Supplement*, *16*, 1-132.

Bretcanu, O., Misra, S., Roy, I., Renghini, C., Fiori, F., Boccaccini, A. R. & Salih, V. (2009). In vitro biocompatibility of 45S5 Bioglass-derived glass-ceramic scaffolds coated with poly(3-hydroxybutyrate).*J Tissue Eng Regen Med.*, *3*, 139-48.

Brugnami, F., Caiazzo, A. & Leone, C. (2009). Local intraoral autologous bone harvesting for dental implant treatment: alternative sources and criteria of choice. *Keio J Med*, *58*, 24-28.

Burdick, J. A., Mason, M. N., Hinman, A. D., Thorne, K. & Anseth, K. S. (2002). Delivery of osteoinductive growth factor from degradable PEG hydrogels influences osteoblast differentiation and mineralization. *J Control Release*, *83*, 53-63.

Catone, G. A., Reimer, B. L., McNeir, D. & Ray, R. (1992). Tibial autogenous cancellous bone as an alternative donor site in maxillofacial surgery: A preliminary report. *J Oral Maxillofac Surg*, *50*, 1258-1263.

Chen, W. J., Jingushi, S., Aoyama, I., Anzai, J., Hirata, G., Tamura, M. & Iwamoto, Y. (2004). Effects of FGF-2 on metaphyseal fracture repair in rabbit tibiae. *J Bone Miner Metab*, *22*, 303-309.

Collin-Osdoby, P., Rothe, L., Bekker, S., Anderson, F., Huang, Y. & Osdoby, P. (2002). Basic fibroblast growth factor stimulates osteoclast recruitment, development, and bone pit resorption in association with angiogenesis in vivo on the chick chorioallantoic membrane and activates isolated avian osteoclast resorption in vitro. *J Bone Miner Res*, *17*, 1859-1871.

Cordaro, L., Amadé, D. S. & Cordaro, M. (2002). Clinical results of alveolar ridge augmentation with mandibular block bone grafts in partially edentulous patients prior to implant placement. *Clin Oral Implants Res*, *13*, 103-11.

Dahlin, C., Simion, M. & Hatano, N. (2009). Long-Term Follow-Up on Soft and Hard Tissue Levels Following Guided Bone Regeneration Treatment in

Combination with a Xenogeneic Filling Material: A 5-Year Prospective Clinical Study Clinical Implant Dentistry and Related Research, in press.

Day, R. M., Maquet, V., Boccaccini, A. R. & Jerome, R. (2005). In vitro nad in vivo analysis of macroporous biodegardable poly (d, L-lactide-co-glycolide) scaffolds cointaining bioactive galss. *J Biomed Mater Res A*, *75*, 778-87.

Deckers, M. M., Karperien, M., van der Bent, C., Yamashita, T., Papapoulos, S. E. & Lowik, C. W. (2000). Expression of vascular endothelial growth factors and their receptors during osteoblast differentiation. *Endocrinology*, *141*, 1667-1674.

Del Fabbro, M., Testori, T., Francetti, L. & Weinstein, R. (2004). Sistematic review of survival rates for implants placed in the grafted maxillary sinus. *Int J Periodontics Restorative Dent*, *24*, 565-577.

Dias, A. G., Lopes, M. A., Santos, J. D., Afonso, A., et al. (2006). In vivo performance of biodegradable calcium phosphate glass ceramics using the rabbit model: Histological and SEM Observation *Journal of Biomaterials Application*, *20*, 253-256.

Dickson, G., Buchananb, F., Marsha, D. & Harkin-Jonesb, E. (2007). Orthopaedic tissue engineering and bone regeneration. *Technology and Health Care*, *15*, 57-67.

Dragoo, J. L., Samimi, B., Zhu, M., Hame, S. L., Thomas, B. J., Lieberman, J. R., Hedrick, M. H. & Benhaim, P. (2003). Tissue-engineered cartilage and bone using stem cells from human infrapatellar fat pads. *J Bone Joint Surg Br*, *85*, 740-747.

El-Ghannam, A. (2005). Bone reconstruction: from bioceramics to tissue engineering. *Expert Rev Med Devices*, *2*, 87-101.

Ennett, A. B. & Mooney, D. J. (2002). Tissue engineering strategies for in vivo neovascularisation. *Expert Opin Biol Ther*, *2*, 805-818.

Erlebacher, A., Filvaroff, E. H., Ye, J. Q. & Derynck, R. (1998). Osteoblastic responses to TGF-beta during bone remodeling. *Mol Biol Cell*, *9*, 1903-1918.

Ferrara, N., Gerber, H. P. & LeCouter, J. (2003). The biology of VEGF and its receptors. *Nat Med*, *9*, 669-676.

Forslund, C. (2003). BMP treatment for improving tendon repair. Studies on rat and rabbit Achilles tendons. *Acta Orthop Scand Suppl*, *74(308)*, I1-I30.

Friedrich, J. B., Moran, S. L., Bishop, A. T., Wood, C. M. & Shin, A. Y. (2008). Free vascularized fibular graft salvage of complications of long-bone allograft after tumor reconstruction. *J Bone Joint Surg Am*, *90*, 93-100.

Fujita, R., Yokoyama, A., Kawasaki, T. & Kohgo, T. (2003). Bone augmentation osteogenesis using hydroxyapatite and β-tricalcium phosphate blocks *Jounal of Oral and Maxillofacial Surgery*, *61*, 1045-1053.

Garg, A. K., Morales, M. J., Navarro, I. & Duarte, F. (1998). Autogenous mandibular bone grafts in the treatment of the resorbed maxillary anterior alveolar ridge: rationale and approach. *Implant Dent*, *7*, 169-76.

Gautier, E. (2003). Sommer Guidelines for the clinical application of the LCP. C. *Injury.*, 34 B63-76. Review.

Gerber, H. P., Vu, T. H., Ryan, A. M., Kowalski, J., Werb, Z. & Ferrara, N. (1999). VEGF couples hypertrophic cartilage remodeling, ossification and angiogenesis during endochondral bone formation. *Nat Med*, *5*, 623-628.

Gerressen, M., Hermanns-Sachweh, B., Riediger, D., Hilgers, R. D., Spiekermann, H. & Ghassemi, A. (2009). Purely cancellous vs. corticocancellous bone in sinus floor augmentation with autogenous iliac crest: a prospective clinical trial. *Clin Oral Implants Res*, *20*, 109-15.

Ghert, M., Colterjohn, N. & Manfrini, M. (2007). The use of free vascularized fibular grafts in skeletal reconstruction for bone tumors in children. *J Am Acad Orthop Surg*, *15*, 577-87.

Granito, R. Effects of a highly bioactive glass-ceramic (Biosilicate®). on the bone wound healing response in rats: a biomechanical and histomorphometric study. PhD thesis, Federal University of Sao Carlos, Sao Carlos, Brazil.

Griffith, L. & Naughton, G. (2002). Tissue Engineering - Current Challenges and Expanding Opportunities. *Science*, *295*, 1009-13.

Hallman, M. & Nordin, T. (2004). Sinus floor augmentation with bovine hydroxyapatite mixed with fibrin glue and later placement of nonsubmerged implants: a retrospective study in 50 patients International *Journal of Oral and Maxillofacial Implants*, *19*, 222-227.

Hardesty, R. A. & Marsh, J. L. (1990). Craniofacial onlay bone grafting: a prospective evaluation of graft morphology, orientation, and embryonic origin. *Plast Reconstr Surg*, *85*, 5-14.

Hassna, R. R. & Ramay, M. Zhang (2004). Biphasic calcium phosphate nanocomposite porous scaffolds for load-bearing bone tissue engineering. *Biomaterials*, *25*, 5171-5180.

Hench, L. L. Glass and Genes (2003). The 2001 W E S Turner Memorial *Lecture Glass Technology*, *44*, 235-4.

Hench, L. L. & Best, S. (2004). Ceramic, glasses, and glass-ceramic. In: B. D. Ratner, A. S. Hoffman, F. J. Schoen, & J. E. Lemons (Eds.), Biomaterials

Science: an introduction to materials in medicine, 1st edition, p.153. San Diego: Elsevier Academic Press.

Hench, L. L. & Polak, J. M. T. (2002). *Third-Generation Biomedical Materials Science*, *295*, 1014-1017.

Hench, L. L., Xynos, I. D. & Polak, J. M. (2004). Bioactive glasses for in situ tissue regeneration *Journal of Biomaterials Science and Polymeration Edition*, *15*, 543-62

Hubbell, J. A. (1999). Bioactive biomaterials. *Curr Opin Biotechnol*, *10*, 123-129.

Hutmatcher, D. W. (2001). Scaffold design and fabrication technologies for engineering tissues - state of the art and future perspectives.Journal of Biomaterials Science. *Polymer Edition*, *12*, 107-124.

Ishihara, M., Obara, K., Ishizuka, T., Fujita, M., Sato, M., Masuoka, K., Saito, Y., Yura, H., Matsui, T., Hattori, H., et al. (2003). Controlled release of fibroblast growth factors and heparin from photocrosslinked chitosan hydrogels and subsequent effect on in vivo vascularization. *J Biomed Mater Res*, *64A*, 551-559.

Iwakura, A., Fujita, M., Kataoka, K., Tambara, K., Sakakibara, Y., Komeda, M. & Tabata, Y. (2003). Intramyocardial sustained delivery of basic fibroblast growth factor improves angiogenesis and ventricular function in a rat infarct model. *Heart Vessels*, *18*, 93-99.

Kanczler, J. M. & Oreffo, R. O. (2008). Osteogenesis and angiogenesis: the potential for engineering bone. *Eur Cell Mater*, *15*, 100-14.

Kawaguchi, H., Nakamura, K., Tabata, Y., Ikada, Y., Aoyama, I., Anzai, J., Nakamura, T., Hiyama, Y. & Tamura, M. (2001). Acceleration of fracture healing in nonhuman primates by fibroblast growth factor-2. *J Clin Endocrinol Metab*, *86*, 875-880.

Kazakia, G. A. Nauman, Donna, Bernard, H. & Keaveny, T. (2006). Effects of *in vitro* bone formation on the mechanical properties of a trabeculated hydroxyapatite bone substitute. *J Biomed Mater Res*, *77A*, 688-699.

Khurana, R. & Simons, M. (2003). Insights from angiogenesis trials using fibroblast growth factor for advanced arteriosclerotic disease. *Trends Cardiovasc Med*, *13*, 116-122.

Kloen, P., Di, P. M., Borens, O., Richmond, J., Perino, G., Helfet, D L. & Goumans, M. J. (2003). BMP signaling components are expressed in human fracture callus. *Bone*, *33*, 362-371.

Kofron, M. D., Zhang, J., Lieberman, J. R. & Laurencin, C. T (2003). Genetically modified mesodermal-derived cells for bone. tissue engineering. *IEEE Eng Med Biol Mag*, *22*, 57-64.

Kumegawa, M. & Hakeda, Y. (2000). Vascular endothelial growth factor (VEGF) directly enhances osteoclastic bone resorption and survival of mature osteoclasts. *FEBS Lett*, *473*, 161-164.

Kusiak, J. F., Zins, J. E. & Whitaker, L. A. (1985). The early revascularization of membranous bone. *Plast Reconstr Surg*, *76*, 510-16.

Lee, H., Cusick, R. A., Browne, F., Kim, T. H., Ma, P. X., Utsunomiya, H., Langer, R. & Vacanti, J. P. (2002). Local delivery of basic fibroblast growth factor increases both angiogenesis and engraftment of hepatocytes in tissue-engineered polymer devices. *Transplantation*, *73*, 1589-1593.

Lee, J. Y., Peng, H., Usas, A., Musgrave, D., Cummins, J., Pelinkovic, D., Jankowski, R., Ziran, B., Robbins, P. & Huard, J. (2002). Enhancement of bone healing based on ex vivo gene therapy using human musclederived cells expressing bone morphogenetic protein 2. *Hum Gene Ther*, *13*, 1201-1211a.

Lee, K. Y., Peters, M. C. & Mooney, D. J. (2001). Controlled drug delivery from polymers by mechanical signals. *Adv Mater*, *13*, 837-839.

Lu, M. & Rabie, A. B. M. (2006). Matrix metalloproteinase-9 regulates graft bone resorption. *Angl Orthod*, *76*, 598-604.

Mandracchia, V. J., Nelson, S. C. & Barp, E. A. (2001). Current concepts of bone healing. *Clin Podiatr Med Surg*, *18*, 55- 77.

Merten, H. A., Wiltfang, J., Grohmann, U. & Hoenig, J. F. (2001). Intraindividual comparative animal study of α and β-Tricalcium phosphate degradation in conjunction with simultaneous insertion of dental implants *Journal of Craniofacial Surgery*, *12*, 59-67.

Michelle D. Kofron, Xudong, Li. & Cato T Laurencin (2004). Protein- and gene-based tissue engineering in bone repair. *Current Opinion in Biotechnology*, *15*, 399-405.

Misch, C. M. (1997). Comparison of intraoral donor sites for onlay grafting prior to implant placement. *Int J Oral Maxillofac Implants*, *6*, 767-76.

Montero, A., Okada, Y., Tomita, M., Ito, M., Tsurukami, H., Nakamura, T., Doetschman, T., Coffin, J. D. & Hurley, M. M. (2000). Disruption of the fibroblast growth factor-2 gene results in decreased bone mass and bone formation. *J Clin Invest*, *105*, 1085-1093.

Moura, J., Teixeira, L. N., Ravagnani, C., Peitl, O., Zanotto, E. D., Beloti, M. M., Panzeri, H., Rosa, A. L. & de Oliveira, P. T. (2007). In vitro osteogenesis on a highly bioactive glass-ceramic (Biosilicate). *J Biomed Mater Res A.*, *82*, 545-57.

Nakagawa, M., Kaneda, T., Arakawa, T., Morita, S., Sato, T., Yomada, T., Hanada, K. morphogenetic protein-2 (rhBMP-2). *J Bone Joint Surg Am*, *82*, 151-60.

Nandi, S. K., Kundu, B., Datta, S., De, D. K. & Basu, D. (2009). The repair of segmental bone defects with porous bioglass: An experimental study in goat. *Res Vet Sci.*, 2. (Epub ahead of print).

Niida, S., Kaku, M., Amano, H., Yoshida, H., Kataoka, H., Nishikawa, S., Tanne, K., Maeda, N., Nishikawa, S. & Kodama, H. (1999). Vascular endothelial growth factor can substitute for macrophage colony-stimulating factor in the support of osteoclastic bone resorption. *J Exp Med*, *190*, 293-298.

O'Keefe, R. M., Reimer, B. L. & Botterfield, S. L. (1991). Harvesting of autogenous bone graft from the proximal tibial metaphysis. A review of 230 cases. *J. Orthop Trauma*, *5*, 469.

Olson, J. W., Dent, C. D., Morris, H. F. & Ochi, S. (2000). Long-term assessment (5 to 71 months) of endosseous dental implants placed in the augmented maxillary sinus *Annals of Periodontology*, *5*, 152-156.

Orr, T. E., Villars, P. & Mitchell, S. (2001). Compressive properties of cancellous bone defects in a rabbit model treated with particles of natural bone mineral and synthetic hydroxyapatite *Biomaterials*, *22*, 1953-9.

Oztürk, A., Yetkin, H., Memis, L., Cila, E., Bolukbasi, S. & Gemalmaz, C. (2006). Demineralized bone matrix and hydroxyapatite/tri-calcium phosphate mixture for bone healing in rats International *Orthopedics*, *30*, 147-52.

Palsson, B. & Bhatia, S. N. (2004). Tissue Engineering, New Jersey, Pearson Prentice Hall.

Pieper, J. S., Hafmans, T., Wachem, P. B., Luyn, M. J., Brouwer, L. A., Veerkamp, J. H. & Kuppevelt, T. H. (2002). Loading of collagen-heparan sulfate matrices with bFGF promotes angiogenesis and tissue generation in rats. *J Biomed Mater Res*, *62*, 185-194.

Rafii, S. & Lyden, D. (2003). Therapeutic stem and progenitor cell transplantation for organ vascularization and regeneration. *Nat Med*, *9*, 702-712.

Reilly, G. C., Radin, S., Chen, A. T. & Ducheyne, P. (2007). Differential alkaline phosphatase responses of rat and human bone marrow derived mesenchymal stem cells to 45S5 bioactive glass. *Biomaterials.*, *28(28)*, 4091-7.

Renno, A. C. M., McDonnell, A. & Laakso, L. (2009). Effect of 830nm laser phototherapy on osteoblasts grown *in vitro* on Biosilicate® scaffolds. *Photomedicine and laser in surgery* (Epub ahead of print).

Ricci, J. L., Blummenthal, N. C., Spivak, J. M., et al. (1992). Evaluation of a low temperature calcium phosphate particulate implant material: Physical chemical properties and in vivo bone response *Journal of Oral Maxillofacial Surgery*, *50*, 969.

Rodan, S. B., Wesolowski, G., Thomas, K. A., Yoon, K. & Rodan, G. A. (1989). Effects of acidic and basic fibroblast growth factors on osteoblastic cells. *Connect Tissue Res*, *20*, 283- 288.

Rose, T., Peng, H., Shen, H. C, Usas, A., Kuroda, R., Lill, H., Fu, F. H. & Huard, J. (2003). The role of cell type in bone healing mediated by ex vivo gene therapy. *Langenbecks Arch Surg*, *388*, 347-355.

Rossant, J. & Howard, L. (2002). Signaling pathways in vascular development. *Annu Rev Cell Dev Biol*, *18*, 541- 573.

Ryan, G., Pandit, A. & Apatsidis, D. P. (2006). Fabrication methods of porous metals for use in orthopaedic applications. *Biomaterials*, *27*, 2651-2670.

Saijo, M., Kitazawa, R., Nakajima, M., Kurosaka, M., Maeda, S. & Kitazawa, S. (2003). Heparanase mRNA expression during fracture repair in mice. *Histochem Cell Biol*, *120*, 493-503.

Saito, N., Okada, T., Horiuchi, H., Murakami, N., Takahashi, J., Nawata, M., Ota, H., Nozaki, K. & Takaoka, K. (2001). A biodegradable polymer as a cytokine delivery system for inducing bone formation. *Nat Biotechnol*, *19*, 332-335.

Sbordone, L., Toti, P., Menchini-Fabris, G., Sbordone, C. & Guidetti, F. (2009). Implant success in sinus-lifted maxillae and native bone: a 3-year clinical and computerized tomographic follow-up. *Int J Oral Maxillofac Implants*, *24*, 316-24.

Schlegel, K. A., Fichtner, G., Schultze-Mosgau, S. & Wiltfang, J. (2003). Histologic findings in sinus augmentation with autogenous bone chips versus a bovine bone substitute International *Journal of Oral and Maxillofacial Implants*, *18*, 53-58.

Sellers, R. S., Zhang, R., Glasson, S. S., et al. Repair of articular cartilage defects one year after treatment with recombinant human bone. *J Bone Joint Am Surg*, *82*, 151-160.

Silva, R. V., Camilli, J. A., Bertran, C. A. & Moreira, N. H. (2005). The use of hydroxyapatite and autogenous cancellous bone grafts to repair bone defects in rats International *Journal of Oral Maxillofacial Surgery*, *34*, 178-184.

Sojo, K., Sawaki, Y., Hattori, H., Mizutani, H. & Ueda, M. (2005). Immunohistochemical study of vascular endothelial growth factor (VEGF) and bone morphogenetic protein- 2,-4(BMP-2,-4) on lengthened rat femurs. *J Craniomaxillofac Surg*, *33*, 238-245.

Spector, M. (2008). Basic principles of scaffolds in tissue engineering. In: S. E. Lynch, R. E. Marx, M. Nevins, & L. A. Wisner-Lynch (Eds.), Tissue Engineering: applications in oral and maxillofacial surgery and periodontics Hanover Park: Quintessence Publishing Co, Chapter 2, 26-36

Street, J., Bao, M., deGuzman, L., Bunting, S., Peale, F. V. Jr., Ferrara, N., Steinmetz, H., Hoeffel, J., Cleland, J. L., Daugherty, A., van, B. N., Redmond, H. P., Carano, R. A. & Filvaroff, E. H. (2002). Vascular endothelial growth factor stimulates bone repair by promoting angiogenesis and bone turnover. *Proc Natl Acad Sci*, U.S.A., *99*, 9656-9661.

Suguimoto, T., Kanatani, M., Kano, J., Kaiji, H., Tsukamoto, T., Yamaguchi, T., Fukase, M. & Chihara, K. (1993). Effects of high calcium concentration on the functions and interactions of osteoblastic cells and monocytes and on the formation of osteoclast-like cells *Journal of Bone Mineral Research*, 8, 1445-1452.

Suttapreyasri, S., Koontongkaew, S., Phongdara, A. & Leggat, U. (2006). Expression of bone morphogenetic proteins in normal human intramembranous and endochondral bones. *Int J Oral Maxillofac Surg*, *35*, 444-452.

Tammela, T., Enholm, B., Alitalo, K. & Paavonen, K. (2005). The biology of vascular endothelial growth factors. *Cardiovasc Res*, *65*, 550-563.

Uchida, S., Sakai, A., Kudo, H., Otomo, H., Watanuki, M., Tanaka, M., Nagashima, M. & Nakamura, T. (2003). Vascular endothelial growth factor is expressed along with its receptors during the healing process of bone and bone marrow after drill-hole injury in rats. *Bone*, *32*, 491-501.

Vaes, B. L., Ducy, P., Sijbers, A. M., Hendriks, J. M., van Someren, E. P., de Jong, N. G., van den Heuvel, E. R., Olijve, W., van Zoelen, E. J. & Dechering, K. J. (2006). Microarray analysis on Runx2-deficient mouse embryos reveals novel Runx2 functions and target genes during intramembranous and endochondral bone formation. *Bone*, *39*, 724-738.

Vallet-Regi, M. (2001). Ceramics for medical applications. *J Chem Soc. Dalton Trans*, *78*, 97-108.

Van Gemert, J. T., van Es, R. J., Van Cann, E. M. & Koole, R. (2009). Nonvascularized bone grafts for segmental reconstruction of the mandible – a reappraisal. *J Oral Maxillofac Surg*, *67*, 1446-1452.

Vargas, G. E., Mesones, R. V., Bretcanu, O., López, J. M., Boccaccini, A. R. & Gorustovich, A. (2009). Biocompatibility and bone mineralization potential of 45S5 Bioglass-derived glass-ceramic scaffolds in chick embryos. *Acta Biomater.*, *5*, 374-80.

Walsh, R. W., Vizesi, F., Michael, D., Auld, J., Langdown, A., Oliver, R., Yu, Y., Irie, H. & Bruce, W. (2008). B-TCP bone graft substitutes in a bilateral rabbit tibial defect model *Biomaterials*, *29*, 266-271.

Wiltfang, J., Merten, H. A., Schlegel, K. A., Schultze-Mosgau, S., Kloss, F. R., Rupprecht, S. & Kessler, P. (2002). Degradation characteristics of α and β Tri-calcium-phosphate (TCP) in minipigs *Journal of Biomedical Materials Research (Appl Biomater)*, *63*, 115-121.

Wong, C., Inman, E., Spaethe, R. & Helgerson, S. (2003). Fibrin-based biomaterials to deliver human growth factors. *Thromb Haemost*, *89*, 573-582.

Wong, T. C., Ip, F. K. & Wu, W. C. (2009). Bone-periosteum-bone graft reconstruction for chronic ulnar instability of the metacarpophalangeal joint of the thumb--minimum 5-year follow-up evaluation. *J Hand Surg*, *34*, 304-308.

Xynos, I. D., Edgar, A. J., Buttery, L. D., Hench, L. L. & Polak, J. M. (2000). Ionic products of bioactive glass dissolution increase proliferation of human osteoblasts and induce insulin-like growth factor II mRNA expression and protein synthesis. *Biochem Biophys Res Commun.*, *24*, *276*, 461-5.

Yang, X. B., Webb, D., Blaker, J., Boccaccini, A. R., Maquet, V., Cooper, C. & Oreffo, R. O. (2006). Evaluation of human bone marrow stromal cell growth on biodegradable polymer/bioglass composites. *Biochem Biophys Res Commun.*, *342*, 1098-107.

Zelzer, E., McLean, W., Ng, Y. S., Fukai, N., Reginato, A. M., Lovejoy, S., D'Amore, P. A. & Olsen, B. R. (2002). Skeletal defects in VEGF(120/120) mice reveal multiple roles for VEGF in skeletogenesis. *Development*, *129*, 1893-1904.

Zerbo, I. R., Bronckers, A. L. J. J., de Lange, G. & Burger, E. H. (2005). Localisation of osteogenic and osteoclastic cells in porous β-tricalcium phosphate particles used for human maxillary sinus floor elevation *Biomaterials*, *26*, 1445-1451.

Zhang, Y., Ni, M., Zhang, M. & Ratner, B. (2003). Calcium Phosphate-Chitosan Composite Scaffolds for Bone Tissue Engineering. *Tissue Engineering*, 9, 337-345.

Zimmermann, G., Muller, U. & Wentzensen, A. (2007). The value of laboratory and imaging studies in the evaluation of long-bone non-unions. Injury, *Int. J. Care Injured*, *38S*, S33-S37.

Zins, J. E. & Whitaker, L. A. (1983). Membranous versus endochondral bone: implications for craniofacial reconstruction. *Plast Reconstr Surg*, *72*, 778-85.

In: Bone Regeneration: Growth Factors... ISBN: 978-1-60876-621-5
Editors: V. Legard et al., pp. 119-146

Chapter IV

Bone Tissue Engineering: Synthetic Materials and Stem Cells

E.I. Kulneva, S.V. Korzhikova and A.S. Teplyashin
Center of cells technologies (Beauty Plaza Science)

Abstract

There are different approaches for treatment of bone fractures; one of them is bone tissue engineering. There are great varieties of clinical cases when the bone grafts are needed. All of them implies to fill a relevant big gap of bone with bone tissue or to supply the bone with a good bioconductive material. For this purpose synthetic materials are used. These materials are bioactive — they produce a specific reaction on the boundary of the bone and material — they also provides osteoconductive and osteoinductive properties of the transplant. These material serve not only for mechanical stability of bone, but also as cells and/or factors delivery system for bone growth. There are different approaches in creation of bone tissue transplant. All of them means to use different scaffolds (bioceramics, bioglass), with different properties, different cells (differentiated or stem cells), and different methods of stem cell placement and retention in those materials — it implies different number of cells, different methods of loading those cells, usage of some factors that helps for bone formation and blood vessels invasion. Here we would like to make a review of different approaches in bone tissue engineering and to tell what was done in our laboratory in this area.

Introduction

There are numerous clinical cases when the treatment of large bone defects is needed – tumor excision, bone substitution after revision prosthesis surgery, treating bone fractures after different traumas, nonunions, alveoli filling after extraction of teeth, correction of nose defects, legs lengthening in aesthetic surgery. All these cases require a relatively large bone graft to fill such defect.

There are several approaches to treat large bone defects – Ilizarov method (bone transport) and bone graft transplant. Ilizarov technique implies osteotomy and the following bone distraction, and based on the bone's potential to regeneration. This method is successful in treating of large bone defects up to ten centimeters [1]. But this method has several considerable disadvantages, it is very inconvenient for the patient, requires a long recovery period, and numerous complications are encountered, most commonly wire-site sepsis and fixation instability [2, 3]. Another method is the usage of bone graft, it can be autologous, allogenic or xenogenic origin, or different biological implants of natural or synthetic origin. All of them have certain advantages and disadvantages. But the main purpose of these grafts is to fill the defect and to provide bone's mechanical integrity. Their osteoconductive (they support cell and nutrition infiltration through the three-dimensional porous structure) properties provide ingrowth of the bone, but some of them are also osteoinductive (contain special proteins, facilitating bone ingrowth and differentiation of osteogenic precursors) and osteogenic (supply the graft with bone-forming cells) materials. Here are the main grafts used for bone defects treatment.

Autografts and Allografts

The "gold" standard in bone graft transplantation is autograft, it doesn't cause any immunologic responses in the patient, no host rejection, there is no any risk of disease transmission and they also provide osteoinductive and osteogenic properties for bone defects treatment, because autograft can supply the bone growth with some proteins and osteogenic cells (but there are several indications, that these cells often don't survive the transplant) [4, 5]. But this graft is not convenient for treatment of significantly large bone defects, because it can be obtained in limited quantity. The most frequently used donor regions for autograft obtainment are iliac crest and ribs [6, 7, 8]. It is also not feasible to take the autologous bone for treatment large bone defects in pediatric patients and elderly people, or in people with congenital disorders of bone tissue or metastatic disease. Even though, if it is possible to overcome all of these difficulties, the gap are large enough to fill it with the autologous bone, the patient is an adult without any diseases, there are still some inconveniences concerning autograft obtainment. As the obtainment of autologous bone is an extra surgery operation, so the patient has postoperative donor site morbidity and complications at the donor site. Often, in several weeks, or even days, such pain or discomfort is over, but some times patients have such complications during a long period of time [8]. Except pain, there are some other significant risks of such operation, including hernia formation, blood loss, postoperative infection, and hematoma formation [8, 9]. But Bunwart et.al showed that some of these complications can be avoided [10].

The alternative to autograft in treatment osseous defects is allograft. There are a lot of banks with preserved allogenic bones, so these grafts are available in reasonable quantities and can be versatile in shape, contour, and mineral density. So this graft can be altered to a particular clinical indication [11, 12, 13]

Allograft as well as autograft has osteoconductive properties, can provide structural support, but its osteoinductive properties are rather low. This is because of preservation, storage and sterilization procedures done with allografts to bank the obtained bone and to reduce the risk of disease transmittion and rejection [14]. But all these procedures decrease the number of osteogenic cells and osteoinductive potential, besides, irradiation and freeze-drying affect the structural integrity of the allograft [15, 16, 17]. Still, even such processing is not able to totally eliminate the risk of disease transfer. Cases of superficial and late infection are documented [18]. There was a study made by Palmer, who showed that 8% of osteoarthritic femoral heads, which would have been considered suitable bone-bank donation, showed evidence of disease, other than osteoarthritis, that were undiagnosed before surgery [19].

Natural Materials

Demineralized bone matrix (DBM) is a product of banked allograft, which is prepared by a standardized process. The bone is treated with acid, but first, it is choped, crushed into particles, then the residual acid is eliminated by rinsing in sterile water, ethanol, and ethyl ether. These are the basic steps, but variables may be in time, acid application, or temperature [20, 21]. DBM has greater osteoinductive potential than allograft, because it still contains growth factors such as bone morphogenic proteins and BMP-7 is most abundant [21].

But it should be noted, that osteoinductive capacity differs from donor to donor and, also, in case of safety, each batch of DBM should be obtained from a single human donor. The form of demineralized bone matrix vary a lot, it is available as a freeze-dried powder, crushed granules, chips, gel or paste [22]. These different forms allow perfect matrix application depending on the injury site shape.

It was shown that osteoinductive potential increases while the size of demineralized particles decreases, perhaps, due to increased contact surface between growth factors and mesenchymal cells, at the same time if the particles are less than 0.1mm, the inductive potential decreases [23]. There was a study performed on rabbits, were DBM was compared with autogenous cortical bone. In this study DBM showed satisfactory healing, comparable to autogenous grafting [20]. DBM is often mixed with such grafts like bone marrow allogenic or autologous origin [24]. Bone marrow itself serves a perfect source of osteogenic cells, but it doesn't provide any structural support. Instead of bone marrow, DBM is also grafted with autogenous cortical bone [25]. The results of this study showed that DBM is effective in decreasing the resorption of autogenous cortical bone graft.

Another natural source of material for graft is natural coral exoskeleton. The structure of the commonly used coral, Porites, is similar to that of cancellous bone, and is used as a bone graft substitute [26, 27, 28]. Natural coral exhibited progressive resorption, it's biocompatible and osteoinductive properties were evaluated in different animals and in human studies. Before using, the coral exoskeleton is just cut into the blocks of necessary size and shape and sterilized [29].

But the most common usage of coral exoskeleton is obtainment of natural hydroxyapatite (HA) by the hydrothermal conversion of the calcium carbonate skeleton in the presence of ammonium phosphate [30, 31]. Commonly used species of coral for coralline HA developing are Porites and Goniopora, their exoskeleton provide the bone with physical and chemical characteristics to stimulate bone ingrowth [32]. But the heating makes the coral HA very brittle, which becomes powder even under the slightest pressure.

There is another source of HA of natural origin – bovine-derived hydroxyapatite is also widely used in treatment of bone defects [33, 34].

Natural polymers are also common for bone loss treatment. But more often they are used for growth factors delivery or for coating of bioinert materials. The most widely used is collagen, the most prevalent protein in human body. There are at least 19 types of collagen, and collagen I is the most abundant [35]. Derived collagen is almost often xenogenic origin, it is obtained from skin of swine or cow. But different proceeding techniques allow to eliminate the immunologic response. But the risk of microbial transmission from animal to human is still should bear in mind [35]. As a natural material, collagen provides excellent cell attachment, but is not able to mechanically support the bone. Collagen can be presented in shape of absorbable sponges, membranes, for growth factors delivery, for example for bone morphogenic protein-2 delivery. [36].

Collagen is often used in composite scaffolds, as fibrin, hyaluronic acid, or chitosan [37, 38]. Fibrin is presented in medicine as glue [39]. The usage of fibrin presented contradictory data in literature. Several studies revealed no significant bone healing, and some of them report that fibrin impair bone healing process. But there are also data about perfect bone healing with fibrin treatment [40, 41, 42, 43, 44]. Chitosan is prepared from exoskeleton of insects and crustaceans. Well characterized scaffolds are made out of chitosan. It gives toughness to the composite scaffold, containing deglued bone and gelatin. For example, as it was reported by Sarswathy et.al, in vivo defects, filled with chitosan bone graft paste, demonstrated a high degree of bone ingrowth [45, 46].

Synthetic Materials

Above-listed are the most frequently used natural scaffolds for bone tissue engineering, but many of them have significant disadvantages. And the main inconvenience is based on the inability to change their parameters for example – to control the rate of their resorption, which is sufficient for new bone growth and growth factors delivery.

So, the alternatives to natural materials are synthetic ones. Based on their chemical composition, synthetic bone grafts can be devided into several groups: metallic implants, ceramics, glass ceramics, and polymers.

Metallic materials are always used in orthopedic or dental surgery for implants fabrication (plates, screws, nails to fix up all the bone particle). The main materials are titan and its alloys, and stainless steel. The main disadvantage of metal scaffolds is the lack of biological recognition on the material surface. To overcome this problem different approaches on surface modification are presented. These include chemical and electrochemical treatment, ion implantation, mechanical treatment, thermal spraying, covalent collagen immobilization, calcium-phosphate ceramic coating, collagen-mimetic peptide (GFOGER – glycine-phenylalanine-hydroxyproline-glycine-glutamatearginine) or proteoglycans coating [47, 48, 49, 50, 51]. Besides these implants which contact only on its surface with tissue, there are porous metallic scaffolds with interconnected pores for easy bone ingrowth and vascularization, study of which showed good osteointegration [52, 53, 54]. Metallic porous scaffold also can be used as growth factors delivery device. These scaffolds where good enough to release TGF-β1 (tissue growth factor) and the calcium phosphate doesn't further enhanced its bone-conducting property [55].

Another limitation of metal implant application is their possible release of toxic metallic ions. This is due to their corrosion or wear and can cause inflammation and rejection, infection and impair bone growth [56, 57, 58]. But these difficulties can be overcome by coating with biocompatible materials.

And, at last, the final inconvenience, with metal implants is the so called stress shielding, when the most amount of load is directed through the load-bearing implant and this can result in fragile bone or nonunion. This occurs in fact of discrepancy between the much higher stiffness of metal implant compared with bone [59, 60].

Biodegradable synthetic materials present a great challenge in orthopedics and bone tissue substitution. Synthetic materials can be modified accordingly to the case of injury. And when the damage occurs, fixation helps to set the structures and promotes effective bone healing.

When the healing is achieved, all fixative materials should be removed (for example in case of metal implants); this requires a second surgical event. Consequently, degradable materials are increasingly developed and used in bone tissue repair that helps to avoid a second surgical procedure.

Synthetic polymer materials are commonly used for implant manufacture and represent an effective alternative to standard metallic fixation. The use of resorbable plates and screw for fixation of pediatric scaffolds is very effective and numerously documented [61, 62]. But there are several complications, concerning late degradation tissue response to polymer particles. Both poly-L-lactide acid and polyglycolic acid causes these tissue responses, which occurs in 2-4 month after implantation to 5 years. Local inflammatory and osteolytic reactions to bioabsorbable implants are documented. Microscopic evaluation of poly-L-lactic acid polymer showed a dense fibrous capsule surrounding the remnants of degradable material. It is thought that poly-L-lactic acid slowly degrades into particles with a high crystallinity, and the degradation rate is very low. These particles are still not fully resorbed even after 5, 7 years of implantation [63, 64].

The most commonly used biodegradable polymers for bone tissue engineering are – polyesters: poly(glycolic acid) (PGA), poli-L-lactic acid (PLLA) and their copolymers (PLGA), poly(caprolactone) (PCL) and poly(propylene fumarat) (PPF); and others: polyurethanes (PU), polyethylene glycol (PEG).

Besides implants manufacturing, biodegradable polymers are used in bone tissue engineering. Synthetic polymer scaffolds commonly have porous structure with interconnected pores. Biodegradable, with porous structure that can be modified for appropriate bone ingrowth, these polymers represent a promising material for scaffold fabrication [65, 66, 67, 68, 69].

Poly(glycolic acid) as a biodegradable polymer and its degradation product – glycolic acid – is a natural metabolite. It was investigated whether new vascularized bone could be engineered by transplantation of osteoblasts using biodegradable poly(glycolic acid) as a cell delivery vehicle, and the results showed an organized trabecular bone with a vesicular pedicle [66]. Poly-L-lactic acid is more hydrophobic than PGA and, its degradation product is naturally produced by the body. It is used for controlled growth factors delivery, for example – rhBMP-2 [70]. PGA and PLLA can be copolymerized, and properties of this copolymer are very different from the individual polymers. PLGA degrades slower and demonstrate lower mechanical strength compared to the individual polymers [71]. These copolymers are used for drug and growth factors delivery. Poly(caprolactone) degrades at a much lower rate than PLA and is a useful base polymer for developing long-term, implantable drug delivery systems. PPF based degradable polymer compositions are also known as a good growth factor delivery system. In the study of model drug (Texas red dextran or FITC-dextran) delivery the injectable PPF-based microsphere/scaffold composites are promising to serve as vehicles for controlled drug delivery for bone tissue engineering [72, 73, 74], PPF-scaffolds can be incorporated with microparticle carriers (PLGA and PEG) for localized delivery of bioactive molecules without altering scaffold mechanical or structural properties up to 26 weeks in vitro [75]. Polyurethanes (PU) represent excellent mechanical properties and good biocompatability, and often are used as scaffold for tissue engineering [76].

Bioceramics are well-studied materials for hard tissue, such as bone, teeth, joints, regeneration. According to a type of reaction with bone tissue, bioceramics can be devided into bioinert, bioactive and resorbable.

Nearly inert bioceramic, such as Al_2O_3, ZrO_2 , is nontoxic, non-allergenic, but is less reactive compared with other synthetic materials. These implants bind to the bone due to bone's attachment to the roughness of the material [77, 78].

Bioactive materials consist of two groups – slowly and rapidly resorbing ceramics. They allow forming a chemical bond with living tissue. It is considered, that on the boundary of bioglass and bone, as a result of ion change, Si-OH groups are formed on the surface of the glass, than amorphous calcium phosphate formation occurs which results in hydroxyapatite crystallization and formation of apatite layer. Hydroxyapatite is also bioactive, but there is no silicon in HA implants. It is considered that during incubation biological silicon could be concentrated at the alkaline pH HA interface, this fact produce a surface similar to that of bioactive glasses [79]. Bioactive materials are used as coating on metal implants, providing appropriate biocompatable properties [80].

Rapidly resorbable ceramics – tricalcium phosphate constructs – are used for bone replacement, which especially effective if the velocity of composites degradation is similar to bone growth. Slowly resorbable ceramic – hydroxyapatite – has a rather low rate of degrability. It was evaluated that 100% synthetic HA ceramics have not been resorbed after more than five years from the implantation [81]. The rate of degradation of resorbable bioactive ceramics can be controlled by the creation of composite material with different proportion of HA/TCP content. The higher concentration of HA – less resorbing composit is, on the contrary, more TCP – more resorbable scaffold is [82].

Bioceramics, of course, have sufficient advantages – it is resorbable and this property can be manipulated, this scaffold can be highly porous, and their size, degree of porosity can be modified. They can be performed in different state and shape. But bioceramic has several disadvantages – it is very brittle, and it does not provide any osteogenic properties.

There are enormous kinds of materials, which are used for bone defects treatment. They all have advantages and disadvantages, so there was made an approach to try to avoid some disadvantages of one material by making a composite scaffold with another. Such composite materials are - composites of ceramics with polymer materials, ceramics with metals, natural polymers with bioceramic. It is often bone marrow or autograft/allograft used with DBM, and other materials to provede the new graft with osteoinductive properties.Traditional scaffolds based on biodegradable polymers such as poly(lactic acid) and poly(lactic acid-co-glycolic acid) are weak and non-osteoconductive. For bone tissue engineering, polymer-based composite scaffolds containing bioceramics such as hydroxyapatite can be produced and used. The bioceramics can be either incorporated in the scaffolds as a dispersed secondary phase or form a thin coating on the pore surface of polymer scaffolds. This bioceramic phase makes scaffolds bioactive and also strengthens them [83, 84, 85, 86].

Synthetic scaffold alone does not provide any osteogenic properties. So, another approach for bone tissue regeneration was performed. It implies that scaffold is filled with osteogenic cells. The most promising and effective considered to be mesenchymal stem cells.

Mesenchymal Stem Cells

Mesenchymal stem cells (MSCs) are nonhematopoietic cells, plastic adherent when they are obtained, with fibroblast-like morphology, which reside in the bone marrow (Fig 1). They were first obtained and characterized by Fridenstein in 60s years of 20th century [87]. The number of these cells in bone marrow – the main source of MSCs – is rather low – 0.01-3% [87, 89, 90, 91]. But they can be cultivated in vitro to increase their number, after 2-3 passages their number is 50-300 mln cells.

Appart from bone marrow, MSCs are also located in other tissues of human body. The number of reports of new tissue sources increases intensively. These cells can be easily obtained from adipose tissue (Fig 2) [92], skin (Fig 3) [93], muscles, heart, liver [94], umbilical cord blood [95, 96], placenta (Fig 4) [97, 98], and peripheral blood [99].

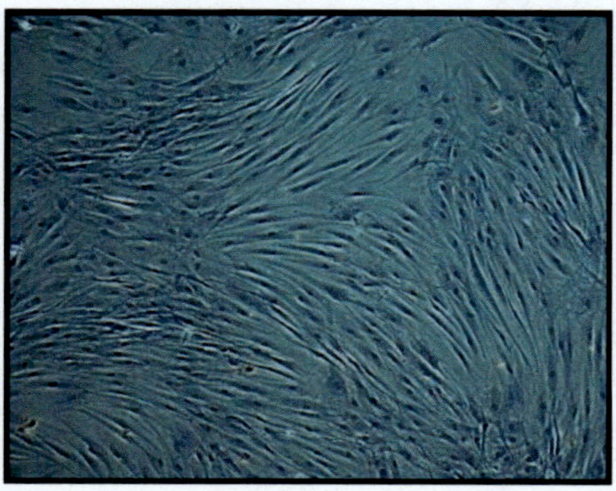

Fig 1. Monolayer of MSC-like cells (x 200) (from bone marrow).

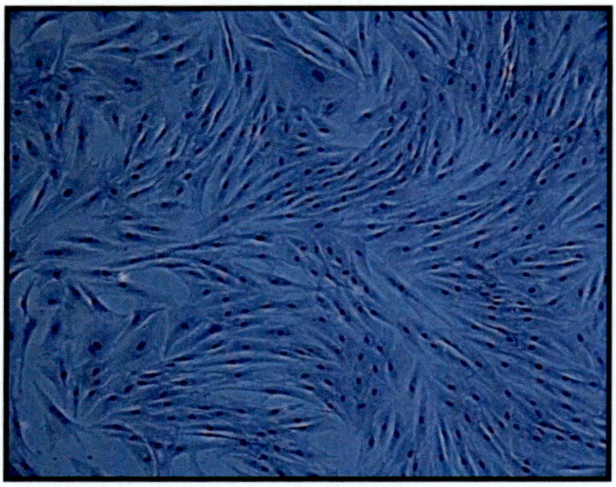

Fig 2. Monolayer of MSC-like cells (x 200) (from adipose tissue).

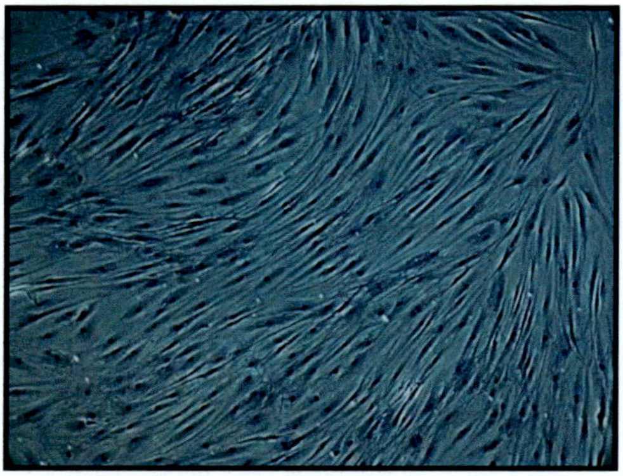

Fig 3. Monolayer of MSC-like cells (x 200) (from derma).

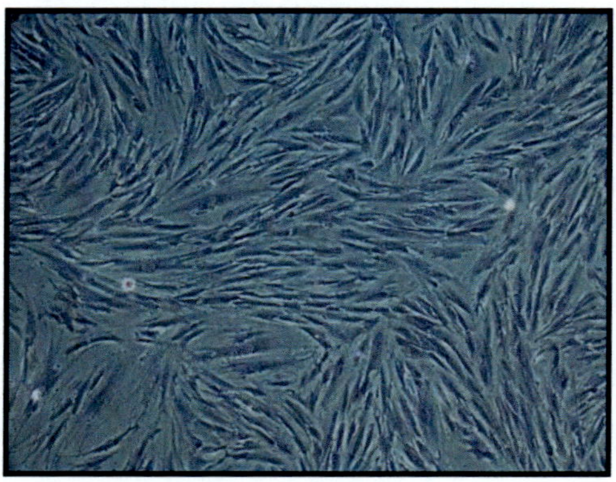

Fig 4. Monolayer of MSC-like cells (x 200) (from placenta).

The amount of obtained MSCs vary due to method of isolation and cultivation MSCs. There are also several data, that the number of MSCs decreases with age [100].

MSCs is a heterogeneous population of cells – they differ in morphology, physiology and expression of surface antigens. Still, there is no single specific marker to identify MCSs. They express a complex of markers, characteristic for mesenchymal, epithelial and muscle cells, and lack hemopoietic surface antigens [101]. MSCs isolated from bone marrow express: CD44, CD105 (SH2; endoglin), CD106 (vascular cell adhesion molecule; VCAM-1), CD166, CD29, CD73 (SH3 and SH4), CD90 (Thy-1), CD117, and STRO-1 [102, 103]. At the same time, MSCs do not possess markers typical for hematopoietic and endothelial cell lineages: CD11b, CD14, CD31, CD33, CD34, CD133 and CD45 [103].

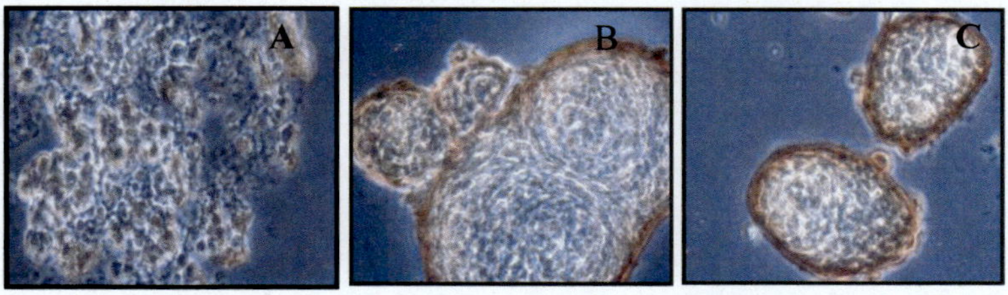

Fig 5. A - Collagen II, DAB-kit stain 21 days after transfer to chondrogenic medium (x200) (MSCs from adipose tissue). B - Aggrecan, DAB-kit stain 21 days after transfer to chondrogenic medium (x200) (MSCs from adipose tissue). C - Aggrecan, DAB-kit stain 21 days after transfer to chondrogenic medium (x200) (MSCs from bone marrow).

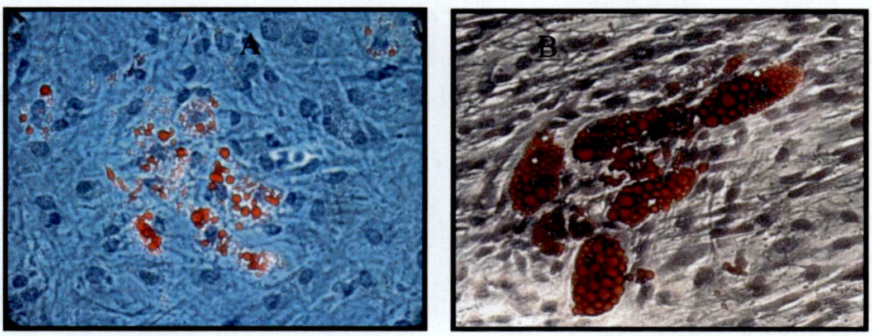

Fig 6. A - Oil Red 21 days after transfer to adipogenic medium (x 400) (MSCs from bone marrow). B - Oil Red 21 days after transfer to adipogenic medium (x 400) (MSCs from adipose tissue).

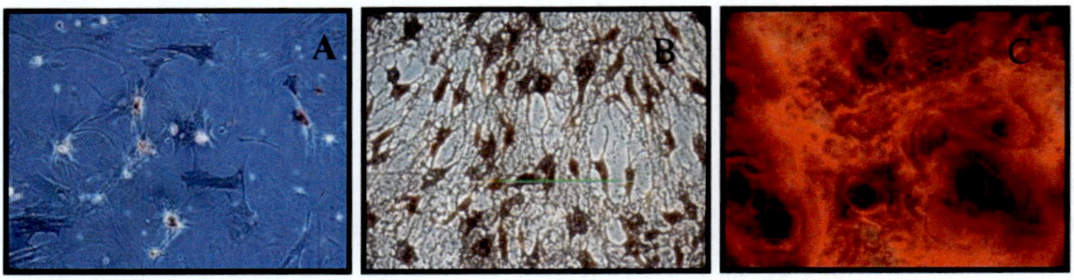

Fig 7. A - Alkaline Phosphatase 14 days after transfer to osteogenic medium (x 200) (MSCs from bone marrow). B - Von Kossa 21 days after transfer to osteogenic medium (x 200) (MSCs from bone marrow). C - Alizarin Red 21 days after transfer to osteogenic medium (**x** 200) (MSCs from bone marrow).

MSCs are shown to posess a potential to multilineage differentiation. Under appropriate conditions MSC can be differentiated into chondocytes (Fig 5) [104], adipocytes and osteoblasts (Fig 6, 7) [105, 106, 101]. But it is still not clear if there is one MSC that gives rise to each cell of mesenchymal origin, or a mixture of progenitor cells commited to different cell lineages. Thera are also controversal data about the ability of MSCs to differentiate into cells of three germ layers [107, 108, 109].

One of the main MSCs advantages is their immunological properties. The immunosupressive effects of MSCs were shown [110]. Injection of MSCs could cure severe graft versus host disease (GVHD) [111]. Furthermore, MSCs injection in immunocompetent baboons prolonged skin allograft survival [112].

MSCs lack MHCI and II expression, so they could not be recognized with immune system, these antigenes are absent on the surface, but are detected in the cell. [113]. So these cells can be applied as allogenic transplant [114].

Taking into account all above mentioned characteristics of mesenchymal stem cells it becomes clear, that they will be a convenient and promising tool for bone tissue engineering.

These cells have already been shown to be effective in bone tissue formation. MSC showed good osteoinductive properties while they where systematically transplanted into patients with osteogenesys imperfecta [115, 116]. Goshima et.al. were the first to show new bone formation in porous bioceramic scaffold, filled with mesenchymal stem cells; new bone deposition was evaluated in mouse subcutaneous model [117]. Different studies were obtained ever since. And numerous studies were undertaken for developing a transplant with mesenchymal stem cells seeded. Though mesenchymal stem cells can be obtained from different tissues, bone marrow MSCs are more commonly used for bone tissue engineering. Still, it is provided, that MSCs from different resources have almost common characteristics [118]. In our research we compared bone marrow derived mesenchymal stem cells with MSCs from adipose tissue and showed, that that although the multilineage potential of bone marrow and adipose tissue derived MSCs was similar according to cell morphology and histology, some minor differences in marker gene expression occurred before and after induction of diverse differentiation pathways. So, BMMSCs are more preferable for bone tissue engineering [119, 120].

So, these cells, with an effective potential to bone formation, are widely used in development of bone tissue transplant.

To deliver cells into the defect, the suitable vehicle should be chosen. There are a great variety of scaffolds to take, one should choose the appropriate one for the study, there are different approaches to fill these scaffolds with cells, different manipulations, which in fact, can effect the result of bone healing.

The scaffold is porous and it should be interconnected for cell contact, good nutrients delivery and vascularization. Concerning on literature data, the mean pore size should be about 400μm, and the poristosy of the scaffold should be about 60% - these parameters are the most appropriate for cell loading and bone formation [114, 122, 81, 123].

These cells should be appropriately seeded into the matrix. Their number should be sufficient for uniform filling of the whole scaffold's volume. The cell number varies a lot, not only because of the size of scaffold; it varies from $4x10^6$ cells/cm^3 to $185x10^6/cm^3$ [124, 114, 121, 122]. Our laboratory results show that seeding cell concentration influences not only on settling scaffolds with cells, but also on a differentiation of BM SSCs (bone marrow derived stromal stem cells) in osteogenic pathway. We compared 2.5×10^5, 1×10^6, 2×10^6 concentrations to assess the efficiency of seeding concentration of cells on bone formation with BM SSCs (bone marrow stromal stem cells) in calcium phosphate scaffolds (Table 2) [125]. The seeding cell concentration $2 \times 10^6/cm^3$ was optimal for engineering of bone tissue in vitro (Fig 8). For uniformly cell loading we used orbital shaker for 2 hours. And this approach showed good results in cell distribution throughout the scaffold. This method was found to be effective not only *in vitro*, but also in *in vivo* studies in sheep's critical size defect model. This study revealed that MSC affect active process of bone reparation, influence protein metabolism state, and determine the way of biochemical reaction during osteogenesis [126].

But it is easy to load cells onto a small scaffold, about 3x3x3 or 4x4x4 or 7x7x7 mm, which are commonly used for subcutaneous implantation to study ectopic bone formation. They are just poured onto the scaffold and are allowed to attach for several hours, this method allows uniformly spread cells only when the scaffold is relatively small [67]. And it is more difficult to achieve uniform loading when the defect is big. In this case, the scaffold should be big enough to fill the defect in critical size-defect animal model, and pouring the cell suspension is noneffective. Such big scaffolds are filled with cells by the method of vacuum cell loading. The cell suspension is added to a syringe containing the implant. All loaded carriers are subjected to a vacuum to allow penetration of the mesenchymal stem cell suspension. They also can be seeded with the help of orbital shaker [127, 114]. MSC cells are sometimes suspended in Tissucol (Baxter) or fibrin and after loading, they are treated with trombin to make a fibrin clot around and with ceramic, to entrap the cells [124, 128]. For the best adhesion of cells, the scaffold can be treated with patient's serum [121], fibronectin, laminin and other extracellular proteins, which help cells to attach.

Table 2.

Number of cells	$2.5x10^5$	$1x10^6$	$2x10^6$
Effectiveness of load, %	82,5	85,0	67,0
Number of cells out of matrix after 24 hour	43750	150000	738000

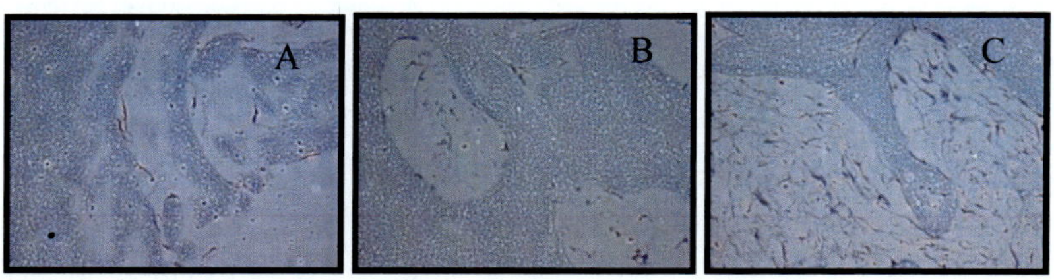

Fig 8. A - $2.5x10^5$ cells. B - $1x10^6$ cells. C - $2x10^6$ cells.

The effectiveness of regeneration can be increased when activators of osteogenesis are supplemented. It can be bone morphogenic proteins (BMP: BMP-2, BMP-3, BMP-4, BMP-6 or 7) which in their turn are members of TGF-β superfamily proteins. These inductive molecules trigger endochondrial cascade in non differentiated mesenchymal cells to bone formation. It was shown, that BMP-2 directly stimulates differentiation into osteoblasts, and this process can be stimulated with dexamethasone. [129]. Therefore, protein containing implants, which stimulate bone regeneration, are more effective in compare with mineral implant [130, 131, 132].

It is still not decided what will be preferable, to use not differentiated mesenchymal stem cells, or to trigger their differentiational pathway. Mostly MSCs are used without differentiation [128, 81]. But there are enough literatura data, evidencing the differentiation of MSCs, loaded into the scaffold before implantation [133]. An interesting approach was undertaken by Kim et.al., who used porous synthetic polymer scaffolds, that released biologically active dexamethasone and ascorbate-2-phosphate. These scaffolds were seeded with human MSC and implanted subsutaneously into athymic mice. The successful formation of mineralized bone tissue in vivo was achieved [67].

Bone is a well vascularized organ, so it is important if a bone transplant will be also vascularized, to supply the cells in the inner part of scaffold with nutrients, otherwise, they will be eliminated.

Itself scaffold should be porous, and the size of these pores should be enough to allow vessels to grow. But it is not sufficient when the scaffold is rather big. One of the most difficult problems of bone tissue engineered scaffolds is vascularization of bone implants. One of the approaches is - arteriovenous loop fabrication. But for large defects this method could be applied only with the use of bioreactor [134]. The most common is VEGF (vascular endothelial growth factor) treatment. Biodegradable polymer scaffold are loaded with VEGF, which is released in vivo, attracting endothelial cells [135]. There alternative method is to use endothelial differentiated cells. These cells are mesenchymal stem cells, obtained from different tissues, for example Wharton's jelly, bone marrow, or umbilical cord blood [136, 137]. Fabrication of such tissue engineered constructs is one of the most promising in tissue engineering. It would be more effective to transplant such full-grown bio-construct, to provide the defect with an excellent bioactive, osteoinductive and osteogenic implant, to achieve the most rapid and accurate bone tissue regeneration.

Conclusion

Here we reviewed one of the bone fracture treatments – bone tissue engineering. Different materials for scaffold fabrication were examined – natural, synthetic, composite materials, all of them have their advantages and disadvantages for bone healing, when they are used alone. But synthetic materials are the most commonly used and the most convenient for scaffold fabrication. They can be modified for the specific case of fracture/trauma. But these materials do not have any osteogenic properties, on themselves. So they are used for bone tissue engineering as scaffold for cells and factors delivery. Different approaches to bone tissue engineered transplant were reviewed, and we consider that correctly chosen material for scaffold fabrication, the use of mesenchymal stem cells in combination with growth factors, that promote bone and vessel formation, is the most effective and promising method for healing large bone defect, nonunions and fractures. Still, there are many questions to be solved and the transplant vascularization is one of them.

References

[1] Paley, D; Maar, DC. Ilizarov bone transport treatment for tibial defects. *J Orthop Trauma*, 2000 Feb, 14(2), 76-85.

[2] Green, SA; Jackson, JM; Wall, DM; Marinow, H; Ishkanian, J. Management of segmental defects by the Ilizarov intercalary bone transport method. *Clin Orthop Relat Res*, 1992 Jul, 280, 136-142.

[3] Cirpar, M; Cetik O; Uslu M; Eksioglu F. Common complications of segmental bone transport with Ilizarov technique in defective tibia pseudoarthrosis: a review. *European Journal of Orthopaedic Surgery & Traumatology*, 2006 Dec, 16(4), 380-385.

[4] Arrington, ED; Smith, WJ; Chambers, HG; Bucknell, AL; Davino, NA. Complications of iliac crest bone graft harvesting. *Clin Orthop Relat Res*, 1996 Aug, 329, 300-309.

[5] Sandhu, HS; Grewal, HS; Parvataneni, H. Bone grafting for spinal fusion. *Orthop Clin North Am*, 1999 Oct, 30(4), 685-98.

[6] Vaughan, LC. The use of bone autografts in canine orthopaedic surgery. *J Small Anim Pract*, 1972 Aug, 13(8), 455-477.

[7] Cotler, HB; Cotler, JM; Stoloff, A; Cohn, HE; Jerrell, BE; Martinez, L; Northrup, BE; Osterholm, JL; Rosato, FE. The use of autografts for vertebral body replacement of the thoracic and lumbar spine. *Spine*, 1985 Oct, 10(8), 748-756.

[8] Cricchio, G; Lundgren, S. Donor site morbidity in two different approaches to anterior iliac crest bone harvesting. *Clin Implant Dent Relat Res*, 2003, 5(3), 161-169.

[9] Yadav, SS. Dual-fibular grafting for massive bone gaps in the lower extremity. *J Bone Joint Surg Am*, 1990 Apr, 72(4), 486-494.

[10] Banwart, JC; Asher, MA; Hassanein, RS. Iliac crest bone graft harvest donor site morbidity. A statistical evaluation. *Spine*, 1995 May, 20(9), 1055-1060.

[11] Khan, MT; Stockley, I; Ibbotson, C. Allograft bone transplantation: a Sheffield experience. *Ann R Coll Surg Engl*, 1998 Mar, 80(2), 150-153.

[12] Malinin, TI; Carpenter, EM; Temple, HT. Particulate bone allograft incorporation in regeneration of osseous defects; importance of particle sizes. *Open Orthop J.*, 2007 Dec, 1, 19-24.

[13] Aho, AJ; Hirn, M; Aro, HT; Heikkilä, JT; Meurman, O. Bone bank service in Finland. Experience of bacteriologic, serologic and clinical results of the Turku Bone Bank 1972-1995. *Acta Orthop Scand*, 1998 Dec, 69(6), 559-565.

[14] Dunsmuir, RA; Gallacher, G. Microwave sterilization of femoral head allograft. *J Clin Microbiol.*, 2003 Oct, 41(10), 4755-7.

[15] Marmor, L. BONE BANKS—A New Concept in Procurement and Storage for Homologous Grafts. *Calif Med*, 1960 June, 92(6), 407–408.

[16] Pelker, RR; Friedlaender, GE. Biomechanical aspects of bone autografts and allografts.*Orthop Clin North Am.*, 1987 Apr, 18(2), 235-239.

[17] Pelker, RR; Friedlaender, GE; Markham, TC. Biomechanical properties of bone allografts. *Clin Orthop Relat Res.*, 1983 Apr, (174), 54-57.

[18] Francés, A; Moro, E; Cebrian, JL; Marco, F; García-López, A; Serfaty, D; López-Durán, L. Reconstruction of bone defects with impacted allograft in femoral stem revision surgery. *Int Orthop.*, 2007 Aug, 31(4), 457-464.

[19] Palmer, SH; Gibbons, CL; Athanasou, NA. The pathology of bone allograft. *J Bone Joint Surg Br.*, 1999 Mar, 81(2), 333-335.

[20] Bigham, AS; Dehghani, SN; Shafiei, Z; Torabi Nezhad, S. Xenogenic demineralized bone matrix and fresh autogenous cortical bone effects on experimental bone healing: radiological, histopathological and biomechanical evaluation. *J Orthop Traumatol.*, 2008 Jun, 9(2), 73-80.

[21] Pietrzak, WS; Perns, SV; Keyes, J; Woodell, JM; McDonald, NM. Demineralized Bone Matrix Graft: A Scientific and Clinical Case Study Assessment. *Journal of Foot and Ankle Surgery*, 2005 Sept., 44(5), 345-353
[22] Wang, JC; Alanay, A; Mark, D; Kanim, LE; Campbell, PA; Dawson, EG; Lieberman, JR. A comparison of commercially available demineralized bone matrix for spinal fusion. *Eur Spine J.*, 2007 Aug, 16(8), 1233-1240.
[23] Sampath, TK; Reddi, AH. Importance of geometry of the extracellular matrix in endochondral bone differentiation. *J Cell Biol.*, 1984 Jun; 98(6), 2192-7.
[24] Rougraff, BT; Kling, TJ. Treatment of active unicameral bone cysts with percutaneous injection of demineralized bone matrix and autogenous bone marrow. *J Bone Joint Surg Am*, 2002 Jun, 84-A(6), 921-9.
[25] Hatice, AL; Hakan, SA; Cagry, D. Effect of Demineralized Bone Matrix on Resorption of Autogenous Cortical Bone Graft in Rats. *Turk J Med Sci*, 2005, 35 , 209-216.
[26] Guillemin, G; Patat, JL; Fournie, J; Chetailm, M. The use of coral as a bone graft substitute. *J Biomed Mater Res*, 1987 May, 21(5), 557-67.
[27] Demers, C; Hamdym CR; Corsi, K; Chellatm, F; Tabrizianm, M; Yahia, L. Natural coral exoskeleton as a bone graft substitute: a review. *Biomed Mater Eng.*, 2002; 12(1), 15-35.
[28] Alvarez, K; Camero, S; Alarcón, ME; Rivas, A; González, G. Physical and mechanical properties evaluation of Acropora palmata coralline species for bone substitution applications. *J Mater Sci Mater Med*, 2002 May; 13(5), 509-15.
[29] Zuki, A; Fadilah A; Zamri-Saad M; Loqman M; Norimah, Y; Asnah H. Post-implantation Evaluation of the Natural Coral (Porites sp.) and Calcium Phosphate Cement in Sheep Femur: A Comparative Study. *Asian Journal of Animal and Veterinary Advances*, 2006, 1(1), 1-12.
[30] Damien, E; Revel, P. Coralline hydroxyapatite bone graft substitute: Areview of experimental studies and biomedicalapplications. *Journal of Applied Biomaterials & Biomechanics*, 2004, 2, 65-7.
[31] Chattopadhyay, P; Pal, S;Wahi, A; Singh, L; Verma, A. Synthesis of Crystalline Hydroxyapatite from Coral (Gergonacea sp) and Cytotoxicity Evaluation. *Trends Biomater. Artif. Organs*, 2007, 20(2), 142-144.
[32] Sartoris, DJ; Gershuni, DH; Akeson, WH; Holmes, RE; Resnick, D. Coralline hydroxyapatite bone graft substitutes: preliminary report of radiographic evaluation. *Radiology*, 1986 Apr; 159(1), 133-7.

[33] Eto, AL; Joly, JC; Jeffcoat, M; de Araújo, NS; de Araújo, VC; Cury, PR. Use of anorganic bovine-derived hydroxyapatite matrix/cell-binding peptide (P-15) in the treatment of class II furcation defects: a clinical and radiographic study in humans. *J Periodontol*, 2007 Dec; 78(12), 2277-83.

[34] Mangano, C; Scarano, A; Perrotti, V; Iezzi, G; Piattelli, A. Maxillary sinus augmentation with a porous synthetic hydroxyapatite and bovine-derived hydroxyapatite: a comparative clinical and histologic study. *Int J Oral Maxillofac Implants.*, 2007 Nov-Dec, 22(6), 980-6.

[35] Lee, CH; Singla, A; Lee, Y. Biomedical applications of collagen. *Int J Pharm*, 2001 Jun 19; 221(1-2), 1-22.

[36] Welch, RD; Jones, AL; Bucholz, RW; Reinert, CM; Tjia, JS; Pierce, WA; Wozney, JM; Li, XJ. Effect of recombinant human bone morphogenetic protein-2 on fracture healing in a goat tibial fracture model. *J Bone Miner Res*, 1998 Sep, 13(9), 1483-90.

[37] Harley, BA; Leung, JH; Silva, EC; Gibson, LJ. Mechanical characterization of collagen-glycosaminoglycan scaffolds. *Acta Biomater.*, 2007, Jul, 3(4), 463-74. Epub 2007 Mar 8.

[38] Aslan, M; Simsek, G; Dayi, E. The effect of hyaluronic acid-supplemented bone graft in bone healing: experimental study in rabbits. *J Biomater Appl*, 2006, Jan, 20(3), 209-20.

[39] Thompson, DF; Letassy, NA; Thompson, GD. Fibrin glue: a review of its preparation, efficacy, and adverse effects as a topical hemostat. *Drug Intell Clin Pharm*, 1988, Dec, 22(12), 946-52.

[40] Zárate-Kalfópulos, B; Estrada-Villaseñor, E; Lecona-Buitrón, H; Arenas-Soto, ML; Garza-Hernández, AC; Reyes-Sánchez, A. Liofilizado de fibrina combinado con injerto óseo autólogo en artrodesis posterolateral. Estudio en conejos de Nueva Zelanda. *Cir Ciruj*, 2007, 75 (3), 201-205.

[41] Lucht, U; Bünger, C; Møller, JT; Joyce, F; Plenk, H Jr. Fibrin sealant in bone transplantation. No effects on blood flow and bone formation in dogs. *Acta Orthop Scand*, 1986 Feb, 57(1), 19-24.

[42] Lendeckel, S; Jödicke, A; Christophis, P; Heidinger, K; Wolff, J; Fraser, JK; Hedrick, MH; Berthold, L; Howaldt, HP. Autologous stem cells (adipose) and fibrin glue used to treat widespread traumatic calvarial defects: case report. *J Craniomaxillofac Surg*, 2004 Dec, 32(6), 370-3.

[43] Le Guéhennec, L; Layrolle, P; Daculsi, G. A review of bioceramics and fibrin sealant. *Eur Cell Mater*, 2004 Sep 13, 8, 1-10; discussion 10-1.

[44] Lucht, U; Bünger, C; Møller, JT; Joyce, F; Plenk, H Jr. Fibrin sealant in bone transplantation. No effects on blood flow and bone formation in dogs. *Acta Orthop Scand*, 1986 Feb, 57(1), 19-24.

[45] Saltarrelli, J Jr, Mukherjee, D. In vivo testing of a bone graft containing chitosan, calcium sulfate and osteoblasts in a paste form in a critical size defect model in rats. *J. Biomedical Science and Engineering*, 2009, 2, 24-29

[46] Saraswathy, G; Pal, S; Rose, C; Sastry, T. A novel bio-inorganic bone implant containing deglued bone, chitosan and gelatin. *Bull. Mater. Sci.*, 24(4), August 2001, 415-420.

[47] Klinger, MM; Rahemtulla, F; Prince, CW; Lucas, LC; Lemons, JE. Proteoglycans at the bone-implant interface. *Crit Rev Oral Biol Med*, 1998, 9(4), 449-63.

[48] Catherine, R; Timothy, P; Kellie, L; Zvi, S; Andrés, G. Biomolecular surface coating to enhance orthopaedic tissue healing and integration. *Biomaterials*, 2007 July, 28(21), 3228–3235.

[49] Müller, R; Abke, J; Schnell, E; Macionczyk, F; Gbureck, U; Mehrl, R; Ruszczak, Z; Kujat, R; Englert, C; Nerlich, M; Angele, P. Surface engineering of stainless steel materials by covalent collagen immobilization to improve implant biocompatibility. *Biomaterials*, 2005 Dec, 26(34), 6962-72.

[50] Liu, X; Chu, P; Ding, C. Surface modification of titanium, titanium alloys, and related materials for biomedical applications. *Materials science & engineering*. R, Reports, 2004, 47(3-4), 49-121

[51] Moroni, A; Pegreffi, F; Hoang-Kim, A; Tesei, F; Giannini, S; Wippermann, B. Fixation of HA-coated unicortical locking screws in a sheep gap model: a comparative biomechanical study. *J Orthop Trauma*, 2008 Jan, 22(1), 37-42.

[52] Linder, L; Carlsson, A; Marsal, L; Bjursten, LM; Brånemark, PI. Clinical aspects of osseointegration in joint replacement. A histological study of titanium implants. *J Bone Joint Surg Br*, 1988 Aug, 70(4), 550-5.

[53] Wen, C; Yamada, Y; Hodgson, P. Fabrication of novel metal alloy foams for biomedical applications. *Materials forum*, 2005, 29, 274-278.

[54] Vasconcellosa, LMR; Oliveirab, MV; Graçac, A; Vasconcellosa, LGO; Carvalhoa, YR; Cairoc, AA. Porous Titanium Scaffolds Produced by Powder Metallurgy for Biomedical Applications. *Materials Research*, 2008, 11(3), 275-280.

[55] Vehof, JW; Haus, MT; de Ruijter, AE; Spauwen, PH; Jansen, JA. Bone formation in transforming growth factor beta-I-loaded titanium fiber mesh implants. *Clin Oral Implants Res*, 2002 Feb, 13(1), 94-102.

[56] Jacobs, JJ; Skipor, AK; Patterson, LM; Hallab, NJ; Paprosky, WG; Black, J; Galante, JO. Metal release in patients who have had a primary total hip arthroplasty. A prospective, controlled, longitudinal study. *J Bone Joint Surg Am*, 1998 Oct, 80(10), 1447-58.

[57] Mudali, K; Sridhar, T; Baldev, R; Corrosion of bio implants. *Sadhana*, 28 (3, 4), June/August 2003, 601-637.

[58] Uo, M; Watari, F; Yokoyama, A; Matsuno, H; Kawasaki, T. Tissue reaction around metal implants observed by X-ray scanning analytical microscopy. *Biomaterials*, 2001 Apr, 22(7), 677-85.

[59] Uhthoff, HK; Boisvert, D; Finnegan, M. Cortical porosis under plates. Reaction to unloading or to necrosis? *J Bone Joint Surg Am*, 1994, 76, 1507-1512.

[60] Brodke, DS; Gollogly, S; Alexander Mohr, R; Nguyen, BK; Dailey, AT; Bachus, aK. Dynamic cervical plates: biomechanical evaluation of load sharing and stiffness. *Spine*, 2001 Jun 15, 26(12), 1324-9.

[61] Eppley, BL. Use of Resorbable Plates and Screws in Pediatric Facial Fractures. *J Oral Maxillofac Surg*, 2005 Mar, 63(3), 385-91.

[62] Edwards, RC; Kiely, KD; Eppley, BL. Fixation of bimaxillary osteotomies with resorbable plates and screws: experience in 20 consecutive cases. *J Oral Maxillofac Surg*, 2001 Mar, 59(3), 271-6.

[63] Bergsma, JE; de Bruijn, WC; Rozema, FR; Bos, RR; Boering, G. Late degradation tissue response to poly(L-lactide) bone plates and screws. *Biomaterials*, 1995 Jan, 16(1), 25-31.

[64] Böstman, OM; Pihlajamäki, HK. Late foreign-body reaction to an intraosseous bioabsorbable polylactic acid screw. A case report. *J Bone Joint Surg Am*, 1998 Dec, 80(12), 1791-4.

[65] Coraça, DC; Duek, EA; Padovani, CA; Camilli, JA. Osteointegration of poly(L: -lactic acid)PLLA and poly(L: -lactic acid)PLLA/poly(ethylene oxide)PEO implants in rat tibiae. *J Mater Sci Mater Med.*, 2008 Jul, 19(7), 2699-2704.

[66] Kim, WS; Kim, HK. Tissue engineered vascularized bone formation using in vivo implanted osteoblast-polyglycolic acid scaffold. *J Korean Med Sci*, 2005 Jun, 20(3), 479-82.

[67] Kim, H; Suh, H; Jo, SA; Kim, HW; Lee, JM; Kim, EH; Reinwald, Y; Park, SH; Min, BH; Jo, I. In vivo bone formation by human marrow stromal cells

in biodegradable scaffolds that release dexamethasone and ascorbate-2-phosphate. *Biochem Biophys Res Commun*, 2005 Jul 15, 332(4), 1053-60.

[68] Young, CS; Terada, S; Vacanti, JP; Honda, M; Bartlett, JD; Yelick, PC. Tissue engineering of complex tooth structures on biodegradable polymer scaffolds. *J Dent Res*, 2002 Oct, 81(10), 695-700.

[69] Gunatillake, PA; Adhikari, R. Biodegradable synthetic polymers for tissue engineering. *Eur Cell Mater*, 2003 May 20, 5, 1-16; discussion 16.

[70] Yang, XB; Whitaker, MJ; Sebald, W; Clarke, N; Howdle, SM; Shakesheff, KM; Oreffo, RO. Human osteoprogenitor bone formation using encapsulated bone morphogenetic protein 2 in porous polymer scaffolds. *Tissue Eng*, 2004 Jul-Aug, 10(7-8), 1037-45.

[71] Yoon, DM; Fisher, JP. Natural and synthetic polymeric scaffolds. In: Springer. *Biomedical Materials*; US; Roger Narayan; 2009; 415-442.

[72] Kempen DH. Controlled drug release from a novel injectable biodegradable microsphere/scaffold composite based on poly(propylene fimarate). 2006

[73] Hedberg, EL; Tang, A; Crowther, RS; Carney, DH; Mikos, AG. Controlled release of an osteogenic peptide from injectable biodegradable polymeric composites. *J Control Release*, 2002 Dec 5, 84(3), 137-50.

[74] Kacey, M; Campbell, P; DiMilla, P; Kumta, P; Mooney, M; Szem, J; Weiss L. Novel Three Dimensional Biodegradable Scaffolds for Bone Tissue Engineering. Materials Research Society Fall Meeting, December, 1998.

[75] Hedberg, EL; Kroese-Deutman, HC; Shih, CK; Crowther, Rs; Carney, DH; Mikos, AG; Jansen, JA. In vivo degradation of porous poly(propylene fumarate)/poly(DL-lactic-co-glycolic acid) composite scaffolds. *Biomaterials*, 2005 Aug, 26(22), 4616-23.

[76] Huang, M; Wang, Y; Luo, Y. Biodegradable and bioactive porous polyurethanes scaffolds for bone tissue engineering. *Journal of Biomedical Science and Engineering (JbiSE)*, 2009 Feb, 1.

[77] Marti, A. Inert bioceramics (Al2O3, ZrO2) for medical application. *Injury*, 2000 Dec, 31 Suppl 4, 33-6.

[78] Hulbert, SF; Young, FA; Mathews, RS; Klawitter, JJ; Talbert, CD; Stelling, FH. Potential of ceramic materials as permanently implantable skeletal prostheses. *J Biomed Mater Res*, 1970 Sep, 4(3), 433-56.

[79] Hench, LL. Bioactive ceramics. *Ann N Y Acad Sci.*, 1988, 523, 54-71.

[80] Moroni, A; Pegreffi, F; Hoang-Kim, A; Tesei, F; Giannini, S; Wippermann, B. Fixation of HA-coated unicortical locking screws in a sheep gap model: a comparative biomechanical study. *J Orthop Trauma*, 2008 Jan, 22(1), 37-42.

[81] Mastrogiacomo, M; Muraglia, A; Komlev, V; Peyrin, F; Rustichelli, F; Crovace, A; Cancedda, R. Tissue engineering of bone: search for a better scaffold. *Orthod Craniofac Res*, 2005 Nov, 8(4), 277-84.

[82] Betz, RR. Limitations of autograft and allograft: new synthetic solutions. *Orthopedics*, 2002 May, 25(5 Suppl), s561-70.

[83] Wang, M. Composite scaffolds for bone tissue engineering. *American Journal of Biochemistry and Biotechnology*, 2006, 2 (2): 80-84.

[84] Wang, H; Li, Y; Zuo, Y; Li, J; Ma, S; Cheng, L. Biocompatibility and osteogenesis of biomimetic nano-hydroxyapatite/polyamide composite scaffolds for bone tissue engineering. *Biomaterials*, 2007 Aug, 28(22), 3338-48.

[85] Li, J; Habibovic, P; Yuan, H; van den Doel, M; Wilson, CE; de Wijn, JR; van Blitterswijk, CA; de Groot K. Biological performance in goats of a porous titanium alloy-biphasic calcium phosphate composite. *Biomaterials*, 2007 Oct, 28(29), 4209-18.

[86] Reyes, CD; Petrie, TA; Burns, KL; Schwartz, Z; García, AJ. Biomolecular surface coating to enhance orthopaedic tissue healing and integration. *Biomaterials*, 2007 Jul, 28(21), 3228-35. Epub 2007 Apr 5.

[87] Kuznetsov, SA; Krebsbach, PH; Satomura, K; Kerr, J; Riminucci, M; Benayahu, D; Robey, PG. Single-colony derived strains of human marrow stromal fibroblasts form bone after transplantation in vivo. *J Bone Miner Res.*, 1997 Sep, 12(9), 1335-47.

[88] Friedenstein, AJ; Gorskaja, JF; Kulagina, NN. Fibroblast precursors in normal and irradiated mouse hematopoietic organs. *Exp Hematol.*, 1976 Sep, 4(5), 267-74.

[89] Haynesworth, SE; Goshima, J; Goldberg, VM; Caplan, AI. Characterization of cells with osteogenic potential from human marrow. *Bone*, 1992, 13(1), 81-8.

[90] Majumdar, MK; Banks, V; Peluso, DP; Morris, EA. Isolation, characterization, and chondrogenic potential of human bone marrow-derived multipotential stromal cells. *J Cell Physiol.*, 2000 Oct, 185(1), 98-106.

[91] Pittenger, MF; Mackay, AM; Beck, SC; Jaiswal, RK; Douglas, R; Mosca, JD;, Moorman, MA; Simonetti, DW; Craig, S; Marshak, DR. Multilineage potential of adult humn mesenchymal stem cells. *Science*, 1999 Apr, 284(5411), 143-7.

[92] Zuk, PA; Zhu, M; Ashjian, P; De Ugarte, DA; Huang, JI; Mizuno, H; Alfonso, ZC; Fraser, JK; Benhaim, P; Hedrick, MH. Human adipose tissue

is a source of multipotent stem cells. *Mol Biol Cell.*, 2002 Dec, 13(12), 4279-95.

[93] Toma, JG; Akhavan, M; Fernandes, KJ; Barnabé-Heider, F; Sadikot, A; Kaplan, DR; Miller, FD. Isolation of multipotent adult stem cells from the dermis of mammalian skin. *Nat Cell Biol.*, 2001 Sep, 3(9), 778-84.

[94] Young, HE; Mancini, ML; Wright, RP; Smith, JC; Black, AC Jr; Reagan, CR; Lucas, PA. Mesenchymal stem cells reside within the connective tissues of many organs. *Dev Dyn.*, 1995 Feb, 202(2), 137-44.

[95] Mihu, CM; Rus Ciucă, D; Soritău, O; Suşman, S; Mihu, D. Isolation and characterization of mesenchymal stem cells from the amniotic membrane. *Rom J Morphol Embryol*, 2009, 50(1), 73-7.

[96] Romanov, YA; Svintsitskaya, VA; Smirnov, VN. Searching for alternative sources of postnatal human mesenchymal stem cells: candidate MSC-like cells from umbilical cord. *Stem Cells*, 2003, 21(1), 105-10.

[97] Igura, K; Zhang, X; Takahashi, K; Mitsuru, A; Yamaguchi, S; Takashi, TA. Isolation and characterization of mesenchymal progenitor cells from chorionic villi of human placenta. *Cytotherapy*, 2004, 6(6), 543-53.

[98] Miao, Z; Jin, J; Chen, L; Zhu, J; Huang, W; Zhao, J; Qian, H; Zhang, X. Isolation of mesenchymal stem cells from human placenta: comparison with human bone marrow mesenchymal stem cells. *Cell Biol Int*, 2006 Sep, 30(9), 681-7. Epub 2006 Apr 22.

[99] Zvaifler, NJ; Marinova-Mutafchieva, L; Adams, G; Edwards, CJ; Moss, J; Burger, JA; Maini, RN. Mesenchymal precursor cells in the blood of normal individuals. *Arthritis Res*, 2000, 2(6), 477-88. Epub 2000 Aug 31.

[100] Stolzing, A; Jones, E; McGonagle, D; Scutt, A. Age-related changes in human bone marrow-derived mesenchymal stem cells: consequences for cell therapies. *Mech Ageing Dev*, 2008 Mar, 129(3), 163-73. Epub 2007 Dec 17.

[101] Pittenger, MF; Mackay, AM; Beck, SC; Jaiswal, RK; Douglas, R; Mosca, JD;, Moorman, MA; Simonetti, DW; Craig, S; Marshak, DR. Multilineage potential of adult humn mesenchymal stem cells. *Science*, 1999 Apr, 284(5411), 143-7.

[102] Gronthos, S; Zannettino, AC; Hay, DJ; Shi, S; Graves, SE; Kortesidis, A; Simmonos, PJ. Molecular and cellular characterization of highly purified stromal stems derived from human bone marrow. *J Cell Sci*, 2003, 116, 1827-1835.

[103] Dennis, JE; Carbillet, JP; Caplan, AI; Charbord, P. The STRO-1+ marrow cell population is multipotential. *Cells Tissues Organs*, 2002, 170, 73-82.

[104] Teplyashin, AS; Korjikova, SV; Sharifullina, SZ; Rostovskaya, MS; Chupikova, NI; Vasyunina, NYu; Andronova, NV; Treshalina, EM; Savchenkova, IP. Differentiation of multipotent mesenchymal stromal cells of human bone marrow into cells of cartilage tissue by culturing in three-dimensional OPLA scaffolds. *Cell and Tissue Biology*, V.1 (2), 125-132. 2007

[105] Awad, HA; Butler, DL; Boivin, GP; Smith, FN; Malaviya, P; Huibregtse, B; Caplan, AI. Autologous mesenchymal stem cell-mediated repair of tendon. *Tissue Eng.*, 1999 Jun, 5(3), 267-77.

[106] De Ugarte, DA; Morizono, K; Elbarbary, A; Alfonso, Z; Zuk, PA; Zhu, M; Dragoo, JL; Ashjian, P; Thomas, B; Benhaim, P; Chen, I; Fraser, J; Hedrick, MH. Comparison of multi-lineage cells from human adipose tissue and bone marrow. *Cells Tissues Organs*, 2003, 174(3), 101-9.

[107] Jiang, Y; Jahagirdar, BN; Reinhardt, RL; Schwartz, RE; Keene, CD; Ortiz-Gonzalez, XR; Reyes, M; Lenvik, T; Lund, T; Blackstad, M; Du, J; Aldrich, S; Lisberg, A; Low, WC; Largaespada, DA; Verfaillie, CM. Pluripotency of mesenchymal stem cells derived from adult marrow. *Nature*, 2002 Jul, 418(6893), 41-9.

[108] Schwartz, RE; Reyes, M; Koodie, L; Jiang, Y; Blackstad, M; Lund, T; Lenvik, T; Johnson, S; Hu, WS; Verfaillie, CM. Multipotent adult progenitor cells from bone marrow differentiate into functional hepatocyte-like cells. *J Clin Invest.*, 2002 May, 109(10), 1291-302.

[109] Kørbling, M; Estrov, Z. Adult stem cells for tissue repair - a new therapeutic concept? *N Engl J Med.*, 2003 Aug 7, 349(6), 570-82.

[110] Nasef, A; Mathieu, N; Chapel, A; Frick, J; François, S; Mazurier, C; Boutarfa, A; Bouchet, S; Gorin, NC; Thierry, D; Fouillard, L. Immunosuppressive effects of mesenchymal stem cells: involvement of HLA-G. *Transplantation*, 2007 Jul6 27, 84(2), 231-7.

[111] Le Blanc, K; Frassoni, F; Ball, L; Locatelli, F; Roelofs, H; Lewis, I; Lanino, E; Sundberg, B; Bernardo, ME; Remberger, M; Dini, G; Egeler, RM; Bacigalupo, A; Fibbe, W; Ringdén, O. Mesenchymal stem cells for treatment of steroid-resistant, severe, acute graft-versus-host disease: a phase II study. *Lancet.*, 2008 May, 371(9624), 1579-86.

[112] Bartholomew, A; Sturgeon, C; Siatskas, M; Ferrer, K; McIntosh, K; Patil, S; Hardy, W; Devine, S; Ucker, D; Deans, R; Moseley, A; Hoffman, R. Mesenchymal stem cells suppress lymphocyte proliferation in vitro and prolong skin graft survival in vivo. *Exp Hematol.*, 2002 Jan, 30(1), 42-48.

[113] Le Blanc, K; Rasmusson, I; Götherström, C; Seidel, C; Sundberg, B; Sundin, M; Rosendahl, K; Tammik, C; Ringdén, O. Mesenchymal stem cells inhibit the expression of CD25 (interleukin-2 receptor) and CD38 on phytohaemagglutinin-activated lymphocytes. *Scand J Immunol*, 2004 Sep, 60(3), 307-15.

[114] Arinzeh, TL; Peter, SJ; Archambault, MP; van den Bos, C; Gordon, S; Kraus, K; Smith, A; Kadiyala, S. Allogeneic mesenchymal stem cells regenerate bone in a critical-sized canine segmental defect. *J Bone Joint Surg Am.*, 2003 Oct, 85-A(10), 1927-1935.

[115] Horwitz, EM; Prockop, DJ; Fitzpatrick, LA; Koo, WW; Gordon, PL; Neel, M; Sussman, M; Orchard, P; Marx, JC; Pyeritz, RE; Brenner, MK. Transplantability and therapeutic effects of bone marrow-derived mesenchymal cells in children with osteogenesis imperfecta. *Nat Med.*, 1999 Mar, 5(3), 309-13.

[116] Horwitz, EM; Gordon, PL; Koo, WK; Marx, JC; Neel, MD; McNall, RY; Muul, L; Hofmann, T. Isolated allogeneic bone marrow-derived mesenchymal cells engraft and stimulate growth in children with osteogenesis imperfecta: Implications for cell therapy of bone. *Proc Natl Acad Sci*, U S A., 2002 Jun, 99(13), 8932-8937.

[117] Goshima, J; Goldberg, VM; Caplan, AI. The osteogenic potential of culture-expanded rat marrow mesenchymal cells assayed in vivo in calcium phosphate ceramic blocks. *Clin Orthop Relat Res.*, 1991 Jan, (262), 298-311.

[118] Kern, S; Eichler, H; Stoeve, J; Klüter, H; Bieback, K. Comparative analysis of mesenchymal stem cells from bone marrow, umbilical cord blood, or adipose tissue. *Stem Cells.*, 2006 May, 24(5), 1294-301.

[119] Teplyashin, AS; Chupikova, NI; Sharifullina, SZ; Korjikova, SV; Rostovskaya, MS; Vasyunina, NY; Teplyashin, ZA; Savtchenkova, IP. A comparison of bone marrow-derived and adipose tissue-derived multipotent mesenchymal stromal cells as perspective material for engineering osteo- and chondrogenesis in vitro. *Proceedings of the European Chapter Meeting of Tissue Engineering and Regenerative Medicine International Society*. Rotterdam, Netherlands, 2006 Oct, 80 (95).

[120] Teplyashin, AS; Tchupikova, NI; Sharifullina, SZ; Rostovskaya, MS; Korjikova, SV; Abdrakhmanov, I; Dunaev, A; Petrin, SA; Eremenko, VS; Savchenkova, IP. Characterization of human MSC-like cells isolated from bone marrow, adipose tissue, skin and placenta. *Proceedings of the Annual*

fall Meeting German Society for Biochemistry and Molecular Biology, Symposium: ECTS". Münster, Germany, 2004 Sept, 19-22, 24.

[121] Kruyt, MC; de Bruijn, JD; Yuan, H; van Blitterswijk, CA; Verbout, AJ; Oner, FC; Dhert, WJA. Optimization of bone tissue engineering in goats: a peroperative seeding method using cryopreserved cells and localized bone formation in calcium phosphate scaffolds. *Transplantation*, 2004 Feb, 77(3), 359-365.

[122] Arinzeh, TL; Tran, T; Mcalary, J; Daculsi, G. A comparative study of biphasic calcium phosphate ceramics for human mesenchymal stem-cell-induced bone formation. *Biomaterials*, 2005 Jun, 26(17), 3631-8.

[123] Papadimitropoulos, A; Mastrogiacomo, M; Peyrin, F; Molinari, E; Komlev, VS; Rustichelli, F; Cancedda, R. Kinetics of in vivo bone deposition by bone marrow stromal cells within a resorbable porous calcium phosphate scaffold: an X-ray computed microtomography study. *Biotechnol Bioeng.*, 2007 Sep 1, 98(1), 271-81.

[124] Cedola, A; Masrogiacomo, M; Burghammer, M; Komlev, V; Giannoni, P; Favia, A; Cancedda, R; Rustichelli, F; Lagomarsino, S. Engineered bone from bone marrow stromal cells: a structural study by an advanced x-ray microdiffraction technique. *Phys Med Biol.*, 2006 Mar 21, 51(6), 109-16.

[125] Teplyashin, AS; Chupikova, NI; Korjikova, SV; Sharifullina, SZ; Rostovskaya, MS; Vasyunina, NY; Savtchenkova, IP. Influence of Seeding Concentration of Cells on Efficiency of Occupation of Ca-P 3D Scaffolds with Bone Marrow Stromal Stem Cells and Differentiation in Bone-like Tissue. Eur. Cell Mater. *Proceedings of the Biosurf VI* «Surface - Tissue Interaction».Zurich, Switzerland. 2005 Sept, 10(5), STE13.

[126] Концевая, СЮ; Кузнецов, АВ; Тепляшин, АС; Чупикова, НИ; Шарифуллина, СЗ; Дерхо, МА. Влияние стволовых клеток на течение остеогенеза. *Ветеринарный доктор*, 2007 август, 16-19.

[127] Hosseinkhani, H; Hosseinkhani, M; Gabrielson, NP; Pack, DW; Khademhosseini, A; Kobayashi, H. DNA nanoparticles encapsulated in 3D tissue-engineered scaffolds enhance osteogenic differentiation of mesenchymal stem cells. *J Biomed Mater Res A.*, 2008 Apr, 85(1), 47-60.

[128] Komlev, VS; Peyrin, F; Mastrogiacomo, M; Cedola A; Papadimitropoulos, A; Rustichelli, F; Cancedda, R. Kinetics of In Vivo Bone Deposition by Bone Marrow Stromal Cells into Porous Calcium Phosphate Scaffolds: An X-Ray Computed Microtomography Study. *Tissue Engineering*, 2006 Dec, 12(12), 3449-3458.

[129] Leboy, PS; Beresford, NJ; Devlin, C; Owen ME. Dexamethasone induction of osteoblast mRNAs in rat marrow stromal cell cultures. *Journal of Cellular Physiology*, 2005 Feb, 146(3), 370-378.

[130] Gao, J; Dennis, JE; Solchaga, LA; Awadallah, SA; Goldberg, VM; Caplan AI. Tissue-Engineered Fabrication of an Osteochondral Composite Graft Using Rat Bone Marrow-Derived Mesenchymal Stem Cells. *Tissue Engineering.*, 2001 Aug, 7(4), 363-371.

[131] Stevenson, S; Cunningham, N; Toth, J; Davy, D; Reddi, AH. The effect of osteogenin (a bone morphogenetic protein) on the formation of bone in orthotopic segmental defects in rats. *The Journal of Bone and Joint Surgery*, 1994, 76(11), 1676-1687.

[132] Wozney, JM; Rosen, V; Celeste, AJ; Mitsock, LM; Whitters, MJ; Kriz, RW; Hewick, RM; Wang, EA. Novel regulators of bone formation: molecular clones and activities. *Science*, 1988 Dec 16, 242(4885), 1528-34.

[133] Shang, Q; Wang, Z; Liu, W; Shi, Y; Cui, L; Cao, Y. Tissue-engineered bone repair of sheep cranial defects with autologous bone marrow stromal cells. *J Craniofac Surg.*, 2001 Nov, 12(6), 586-93, discussion 594-5.

[134] Ren, LL; Ma, DY; Feng, X; Mao, TQ; Liu, YP; Ding, Y. A novel strategy for prefabrication of large and axially vascularized tissue engineered bone by using an arteriovenous loop. *Med Hypotheses.*, 2008 Nov, 71(5), 737-40.

[135] Leach, JK; Kaigler, D; Wang, Z; Krebsbach, PH; Mooney, DJ. Coating of VEGF-releasing scaffolds with bioactive glass for angiogenesis and bone regeneration. *Biomaterials.*, 2006 Jun, 27(17), 3249-55.

[136] Chen, MY; Lie, PC; Li, ZL; Wei, X. Endothelial differentiation of Wharton's jelly-derived mesenchymal stem cells in comparison with bone marrow-derived mesenchymal stem cells. *Exp Hematol.*, 2009 May, 37(5), 629-40.

[137] Gang, EJ; Jeong, JA; Han, S; Yan, Q; Jeon, CJ; Kim H. In vitro endothelial potential of human UC blood-derived mesenchymal stem cells. *Cytotherapy*, 2006, 8(3), 215-27.

In: Bone Regeneration: Growth Factors... ISBN: 978-1-60876-621-5
Editors: V. Legard et al., pp. 147-176

Chapter V

Platelet Rich Plasma: Highlights and Shadows from the Bench to the Bedside

***E. Cenni*[1], *L. Savarino*[1], *F. Perut*[1], *C. Fotia*[1,2], *S. Avnet*[1] *and N. Baldini*[1,2]**

[1]Laboratory for Orthopaedic Pathophysiology and Regenerative Medicine, Istituto Ortopedico Rizzoli, Bologna, Italy.

[2]Department of Human Anatomy and Musculoskeletal Pathophysiology, University of Bologna, Bologna, Italy.

Abstract

Autologous activated platelet rich plasma (PRP) is used to accelerate bone repair as a possible alternative to recombinant growth factors (GF), through the osteogenic GF released from α-granules. The advantages of PRP, which is usually prepared by adding thrombin and calcium to a platelet concentrate, lie in its mimicking the GF effects of the physiological bone healing and regenerative processes, in addition to a relatively simple and low cost technique. Moreover, if autologous PRP is used, immunological reactions are avoided. However, the efficacy of PRP in orthopaedic applications is still controversial because of the unpredictability of GF concentration in different platelet preparations, due both to individual factors and to different techniques. Moreover, the effects of PRP are not dependant on a single GF, but on the combination of the effects of some molecules released by platelets, each of them acting on the different cells of bone microenvironment. The cells, in consequence of the stimulation with GF, in their turn influence other cells with various effects. In *in vitro* systems, PRP stimulated osteogenic differentiation of bone marrow stromal cells, while it inhibited osteoclastogenesis from the peripheral blood precursors, particularly complete osteoclast differentiation and activation. Moreover, PRP favored endothelial cell proliferation, tubulogenesis and expression of pro-osteogenic functions. PRP induced also the expression of ICAM-1, which is an adhesion molecule for monocytes. In experimental animals and in clinical application, the PRP efficiency to improve bone regeneration was increased by the combination with bone allografts, acting as scaffolds, and by the addition of bone marrow stromal cells.

Keywords: Platelet rich plasma, growth factor, bone regeneration.

Introduction

Platelet activation in response to vessel damage participates in the formation of a blood clot to provide haemostasis. The composition of this naturally occurring haematoma is 95% erythrocytes, 4% platelets and 1% leukocytes. If platelets have been concentrated before inducing coagulation, the clot is composed of 95% platelets, 4% erythrocytes and a similar amount of leukocytes [1]. Platelet concentrates are used for transfusion to patients with thrombocytopenia, in an effort to prevent or treat bleeding. A more recent application of platelet concentrates is the local use in wound and bone healing, which involves physiologic cascades in which many cellular and hormonal molecules play pivotal roles. Growth factors (GF) are peptides that promote migration, proliferation and differentiation of various cells involved in wound healing and bone growth. GF are synthesized by different cell types; they are also produced by platelets and released during blood clotting. To improve the physiological effect of the GF released by platelets, in the clinical application for tissue repair, platelet concentrates in a small volume of plasma (platelet rich plasma or PRP) are prepared. The release of GF is induced by thrombin and calcium or other substances, which determine also fibrin generation, resulting in a viscous clot (platelet gel), which can be exogenously applied to soft wound tissues or bone.

The combination of GF in PRP is in an optimal level and ratio, and seems to be more efficient than the single rhGF, owing to the synergistic effect of transforming growth factor β (TGF-β), platelet-derived growth factor (PDGF) and insulin-like growth factor I (IGF-I) [2]. The technique for obtaining GF from PRP is relatively simple and cheaper than rhGF. Because of its autologous origin, PRP does not hold any risk of immunological reactions and transmissible diseases, such as HIV and hepatitis. When the drawing of autologous PRP is contra-indicated for anemia, thrombocypenia, defects in platelet functionality, homologous PRP, collected by donors that are ABO- and RhD matched, and HIV, HBV, HCV and Lues-negative, has been proposed. Homologous PRP should also be leukocyte-depleted and irradiated to prevent immunological reactions [3]. As immunological rejection and pathogen contamination cannot be completely excluded, some authors proposed to separate the GF fractions in homologous PRP by gel chromatography and vacuum freeze-drying, to reduce immunogenicity [4].

The drawbacks of PRP are the high variability of GF concentrations in different preparations, due both to individual factors and to different preparation methods. In order to better understand the mechanisms underlying the advantages and the drawbacks of PRP, for a more addressed clinical use, it is necessary to know the mechanism of platelet release reaction, to analyze the preparation methods, to evaluate the in vitro and in vivo effects and critically to discuss the clinical studies performed to date.

Platelet Adhesion and Aggregation

The physiological role of platelets is the participation in blood clotting, in response to a vessel injury with exposure of subendothelial collagen. Platelets adhere to collagen through their receptors glycoproteins VI and Ia. They also adhere to von Willebrand factor through glycoprotein Ib. In the meanwhile, thrombin is rapidly generated from circulating prothrombin. Thrombin does not only generate fibrin, but also represents a potent platelet activator. Platelet response to thrombin is largely mediated through the protease-activated receptor 1 (PAR1), which couples to members of G protein family. The G_q signaling leads to Ca^{2+} release from the endoplasmic reticulum; G_i inhibits adenylate cyclase, decreasing cyclic AMP (cAMP) (Figure 1).

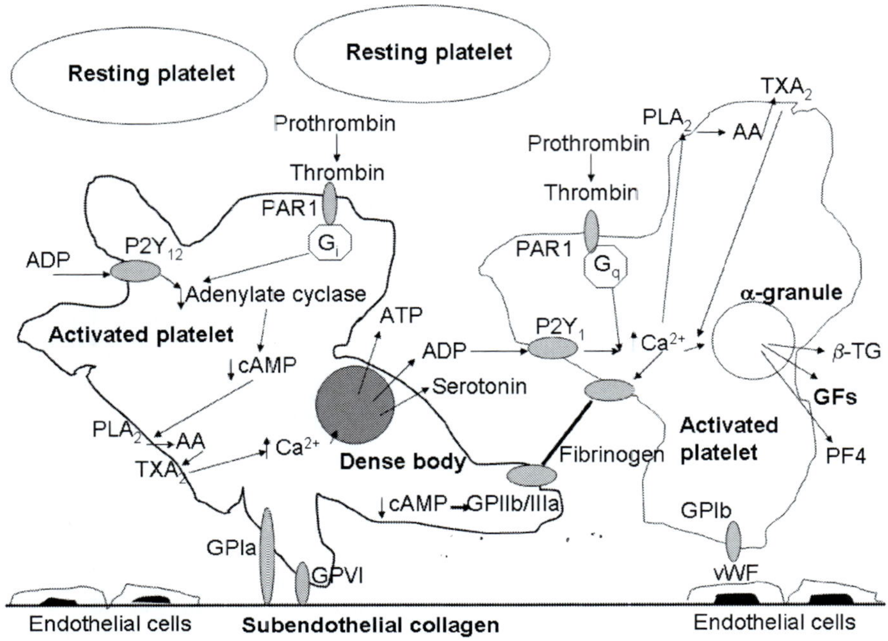

Figure 1. Pathways of platelet activation. TXA_2: thromboxane A_2; PLA_2: phospholipase A_2; AA: arachidonic acid; PAR1: protease-activated receptor 1; β-TG: β-thromboglobulin; GFs: growth factors; PF4: platelet factor 4; GP: glycoprotein; vWF: von Willebrand factor; cAMP: cyclic AMP.

The increase of Ca^{2+} and the decrease of cAMP stimulate phospholipase A_2 to cleave arachidonic acid from platelet membrane phospholipids. The final product is thromboxane A_2 (TXA_2), which diffuses across the platelet membrane. The binding of TXA_2 to its receptors couples to G proteins, which activate phospholipase C. This enzyme degrades the membrane phosphoinositides, releasing diacylglycerol, which activates protein kinase C, inducing protein phosphorylation. Moreover, the degradation of phosphoinositides releases inositol triphosphate, which further increases intracellular Ca^{2+}. The increased intracellular Ca^{2+} triggers also the actin-myosin interaction and the contraction of microtubule and microfilaments, determining the shape change from a discoid form to a spherical one, the pseudopods generation and the release of the content of dense bodies and α-granules into the platelet canalicular systems and into the surrounding blood.

ADP released by dense bodies, together with ADP produced by the surrounding erythrocytes, binds to the receptor $P2Y_{12}$, inhibiting adenylate cyclase and causing a decrease in the cAMP level. Moreover ADP binds also to the receptor $P2Y_1$, increasing the intracellular Ca^{2+} level.

The decrease of cAMP and the increase of intracellular Ca^{2+} determine also the binding of the glycoprotein IIb/IIIa integrin to its ligand, i.e. the small-peptide sequence arginine–glycine–aspartic acid (RGD). Fibrinogen, which contains two RGD sequences, bridges glycoprotein IIb/IIIa between platelets. Therefore, a permanent linkage forms among adjacent platelets to ensure stable and irreversible aggregate formation [5].

Platelet Release Reaction

Platelets contain dense bodies and α-granules. Dense bodies contain nucleotides (ADP, ATP), amines (serotonin) and cations (calcium ions). Platelet α-granules contain β-thromboglobulin, thrombospondin, fibrinogen, fibronectin [6], vitronectin [7] and GF, such as TGF-β1, TGF-β2, PDGF-AA, PDGF-BB, PDGF-AB, vascular endothelial growth factor (VEGF), IGF-I, connective tissue growth factor (CTGF) [8] and epidermal growth factor (EGF) [9]. Platelet α-granules release also inflammatory mediators, such as interleukin-1β [10], matrix metalloproteinase-2 [11] and its inhibitors [12], and chemokines, such as platelet factor 4, also known as CXCL4 [13] (Table 1).

Table 1. Main molecules released by platelets and their role in tissue repair.

Molecule	Location in platelets	Functions
Transforming growth factor β (TGF-β)	α-granules	Immunosuppression; angiogenesis; stimulation of cell growth, differentiation and extracellular matrix synthesis; induction of collagen formation by fibroblasts; stimulation of undifferentiated mesenchymal cell proliferation; osteoblast precursor recruiting; osteoblast and chondrocyte differentiation (but inhibition of terminal differentiation); bone matrix production; recruitment of osteoclast precursors but inhibition of terminal differentiation and induction of apoptosis

Table 1. (Continued)

Molecule	Location in platelets	Functions
Platelet-derived growth factor (PDGF)	α-granules	Stimulation of connective tissue cell proliferation; stimulation of monocyte, macrophage and smooth muscle cell migration; angiogenesis. Macrophage activation. Stimulation of osteoprogenitor migration, proliferation and differentiation. Osteoclastogenesis.
Vascular endothelial growth factor (VEGF)	α-granules	Angiogenesis; endothelial cell migration; conversion of cartilage into bone; osteoblast proliferation and differentiation; induction of RANK expression in osteoclast precursors
Insulin-like growth factor (IGF)	α-granules	Regulation of growth hormone biological activity; stimulation of osteoblast proliferation and bone matrix synthesis; bone resorption
Connective tissue growth factor (CTGF)	α-granules	Stimulation of fibroblast proliferation, migration, adhesion, and extracellular matrix formation; angiogenesis; fibrosis; cartilage regeneration
Epidermal growth factor (EGF)	α-granules	Stimulation of epidermal cell growth and wound healing; stimulation of osteoblast recruitment and proliferation; increase of osteoprotegerin and decrease of RANKL
Angiopoietin	α-granules	Angiogenesis
β-thromboglobulin	α-granules	Unknown
Platelet factor 4	α-granules	Monocyte recruitment, E-selectin expression; protein C inhibition
Fibronectin	α-granules	Cell adhesion; migration; osteoblast migration and adhesion
Vitronectin	α-granules	Cell adhesion, migration. osteoblast migration and adhesion
Glycoprotein IIb/IIIa	α-granules	Platelet adhesion to fibrinogen
Thrombospondin	α-granules	Apoptosis induction; inhibition of angiogenesis
Von Willebrand factor	α-granules	Platelet adhesion to the vessel wall

Molecule	Location in platelets	Functions
Factor XI	α-granules	Clotting
Factor VIII	α-granules	Clotting
Factor V	α-granules	Clotting
Fibrinogen	α-granules	Clotting
Factor XIII	α-granules	Clotting
PAI-1	α-granules	Fibrinolysis inhibition
C1 inhibitor	α-granules	Inhibition of complement, fibrinolysis, clotting and kinins
Matrix metalloproteinase-2	α-granules	Degradation of extracellular matrix
Inhibitors of matrix metalloproteinases	α-granules	Inhibition of matrix metalloproteinases
Interleukin-1β	α-granules	Inflammation, fever, osteolysis
ADP	Dense bodies	Induction of platelet aggregation
ATP	Dense bodies	Energy source for the majority of cellular functions
Serotonin	Dense bodies	Neurotransmission; vasoconstriction, fibroblast and macrophage migration; osteoblast and osteoclast activation
Calcium	Dense bodies	Clotting, keratinocyte proliferation and differentiation, epidermal cell migration
P-selectin	α-granule membrane	Monocyte recruitment, dendritic cell generation
PECAM-1	Cell membrane	Intercellular adhesion of endothelial cells
Platelet factor 3	Cell membrane	Clotting
CD40 ligand	Cell membrane	Binding to CD40 on antigen-presenting cells
Thromboxane A2	-	Platelet aggregation
Prostaglandin E2	-	Vasodilation, smooth muscle relaxation; bone resorption

During release reaction, P-selectin (CD62P or GMP140), a protein of the α-granules membrane, moves from the internal cell location to the platelet surface. P-selectin, an adhesion molecule for leukocytes, promotes also generation of dendritic-like cells from peripheral blood monocytes and inhibits macrophage differentiation [14].

During activation, CD40 ligand, which is stored in the cytoplasm of resting platelets, appears on the surface. CD40 ligand is cleaved to generate a soluble, functional fragment, that is released into the extracellular environment, where it induces an inflammatory phenotype in endothelial cells [15].

Platelet Concentrates for in vitro and in vivo Application

PRP is a concentrate of platelets from freshly drawn whole blood and activated with a mixture of thrombin and calcium [16]. PRP contains GF, fibrin mesh, adhesion molecules, such as fibronectin and vitronectin, in a small volume of plasma. PRP production usually follows a two-step procedure. Firstly, a platelet concentrate is prepared through centrifugation of anticoagulated whole blood; then, platelet activation and growth factor release, as well as fibrin formation, are induced by adding calcium and thrombin.

The effectiveness of PRP lies in the higher concentration of GF in comparison with normal clot, owing to the higher platelet number. The optimal number of platelets was supposed to be about 1×10^9 platelets/ml [17]. Below this number, the effect is suboptimal; beyond this concentration, there may be a paradoxically inhibitory effect [18]. Usually, blood is collected and platelet concentrate is prepared the day before activation at maximum, because GF level decreases in the long run. After activation, PRP should be used immediately.

Some protocols have changed the original preparation method of PRP, either using another anticoagulant or platelet activator, or depleting leukocytes.

Anticoagulants

As anticoagulants, citrate-phosphate-dextrose (CPD) or acid-citrate-dextrose (ACD) or sodium citrate are generally used, as they affect release reaction only to a minimal extent. EDTA should not be used for PRP preparation, because it inhibits platelet activation. Platelet concentrates without any anticoagulant have also been proposed (platelet-rich fibrin or PRF) [19], but they require the blood centrifugation immediately after collection. In few minutes, for the absence of anticoagulant, platelets and coagulation factors are activated by the contact with the tube glass wall. The fibrin network generated by this slow polymerization seems to be more similar to the natural one.

Platelet Activators

Platelets are generally activated by a solution of thrombin and calcium chloride or gluconate. Bovine thrombin, which is largely used for in vitro and in vivo research, may expose the patient to the risk of prion transmission and to the potential development of antibodies against the bovine thrombin. Autologous thrombin, generated upon the addition of calcium to the plasma of the patient himself, was suggested for clinical application [20]. In other protocols, clotting is triggered by calcium chloride without thrombin (preparation rich in growth factors or PRGF) [21]. Through this method a more sustained release of GF and the formation of a three-dimensional fibrin scaffold can be obtained serving as a matrix for progenitor cells. Batroxobin, a serine protease derived from the venom of Bothrops atrox, has also been proposed. Batroxobin does not induce platelet aggregation but passively entraps platelets within the fibrin network and ensures a slower release reaction, a longer clot retraction and a longer in loco availability of GF [22]. In other formulations PRP was not activated at all [23].

These protocols have different applications in orthopaedics. Platelet gel, obtained with thrombin or calcium or through the contact with the tube glass wall, is ideal when added to a scaffold, but for its high viscosity cannot be delivered via a small gauge needle for intra-articular or percutaneous injection. In this case, a non activated platelet concentrate is injected, which will be activated by the collagen that is abundant in musculoskeletal tissue.

Leukocytes

Another variable is the presence of leukocytes. The removal of leukocytes was suggested to avoid the release of their proteases, acid hydrolases and proinflammatory cytokines [24, 25]. However, other authors guess that leukocytes represent an additional source of GF and have an important role in the host's immune defense [26]. Myeloperoxidase contained in neutrophils and monocytes generates hypochlorous acid and reactive oxygen derivatives, that act as potent bactericidal oxidants toxic to microorganisms and fungi and may be helpful to prevent post-surgical infections [27]. Platelet-leukocyte gel has been proposed as an adjuvant for the prophylaxis and treatment of infection-related delayed healing and non-union.

Devices for Preparing PRP

The use of PRP by clinicians has encouraged the introduction of devices for controlled manufacturing procedures. These closed systems reduce manipulation, decrease the risk of infection, allow a higher standardization of the manufacturing procedure and can be directly used in the operating room. These devices have different operating principles, which may result in different platelet yield, leukocyte number and GF content. Each preparation method should be analysed for the resulting platelet number, GF concentration and platelet activation during the gel manufacturing. In fact, if platelets become activated by handling before the final centrifugation, GF may be lost with the discarded supernatant [28, 29, 30, 31, 32].

Classification of PRP Protocols

From the considerations previously reported, a classification of platelet concentrates may be performed on the basis of three main sets of parameters: the preparation protocol, the presence of leukocyte and the quality of the fibrin network. Using these sets of parameters, actual available methods can then be classified in four main categories: pure PRP (P-PRP), leukocyte-rich PRP (L-PRP), pure platelet-rich fibrin (P-PRF) and leukocyte-rich PRF (L-PRF) [33].

The denomination "concentrated PRP" (cPRP) concerns the high platelet number in a small quantity of plasma [34]. The names platelet-rich gel or platelet gel or platelet-rich plasma-gel or platelet-rich plasma-clot, plasma-rich in growth factors or preparation rich in growth factors [35] or platelet-leukocyte gel (PLG) [36] concern the thrombus-like gelatinous substance, in which activated platelets are trapped on the fibrin network whereas the bioactive substances diffuse into the surroundings.

Owing to the variety of preparation protocols, each PRP preparation should be ideally provided with a certificate of analysis, which reported the preparation method, the platelet number of whole blood and of PRP, the storage conditions and the levels of GF [37]. The record of these data will allow to evaluate the therapeutic effect of PRP. Guidelines should be established for the production of

PRP, considering that GF assay is long, expensive and unfeasible for all PRP preparations, even though the levels of GF are the most useful parameters for evaluating the clinical results.

Osteogenic Differentiation and Platelet Growth Factors

The use of PRP was proposed to provide a microenvironment for the orchestration of the sequential process of bone regeneration involving migration, proliferation and differentiation of osteogenic cells. The mitogenic effect of PRP was demonstrated by in vitro studies on different cell types, particularly on human trabecular osteoblasts [38] and human MSC [39, 40]. A mitogenic activity on human trabecular osteoblasts was demonstrated not only by platelet gel, but also by native platelets, microparticles, platelet membranes and the supernatant obtained by centrifugation after platelet clot retraction, i.e. platelet released supernatant (PRS) or releasate [41]. Particularly, the use of PRS may be advantageous in testing cell cultures, owing to the difficulty in handling a gel. Among the GF released by platelets, the highest contribute to the proliferation of osteoblasts was given by PDGF and TGF-β1 [42].

In vivo platelet effects are likely restricted to the early stages of bone regeneration until the blood clot is replaced by granulation tissue, which is rich in osteogenic progenitor cells and blood vessels. Similarly, in vitro PRS increased MSC proliferation but inhibited differentiation [43]. PRS seems to have opposite effects than bone morphogenetic proteins (BMP). PRS stimulated migration and proliferation of the osteogenic cell line MC_3T_3-E1, while BMP did not have any effect on migration and induced only a moderate increase of proliferation. However, PRS suppressed alkaline phosphatase activity in response to BMP. Together, these findings indicate that activated platelets can influence the microenvironment of the blood clot in such a way that osteogenic differentiation is suppressed. Once platelet GF have lost their activity, osteogenic cells can regain their responsiveness to BMP [44].

In bone engineering, PRP represents a source of GF to be added to MSC seeded on a scaffold. Therefore, it is interesting to evaluate how different scaffolds could modulate the effects of PRP. Leukocyte-depleted PRP favored the adhesion and proliferation of MSC on three-dimensional ceramic scaffolds with high specific surface area, such as calcium-deficient hydroxyapatite (HA), in comparison with scaffolds with lower specific surface area, such as β-tricalcium phosphate (β-TCP) [45].

The combination of PRP and freeze-dried bone allografts accelerated MSC differentiation, as demonstrated by the significantly higher osterix expression at 15 days, and sustained an autonomous production of GF by the cells themselves, as shown by the higher level of FGF-2 in the conditioned culture medium after 11 days (Figure 2) [46].

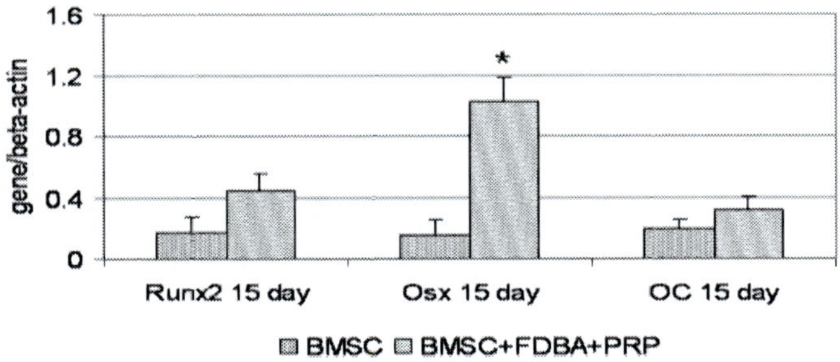

(a)

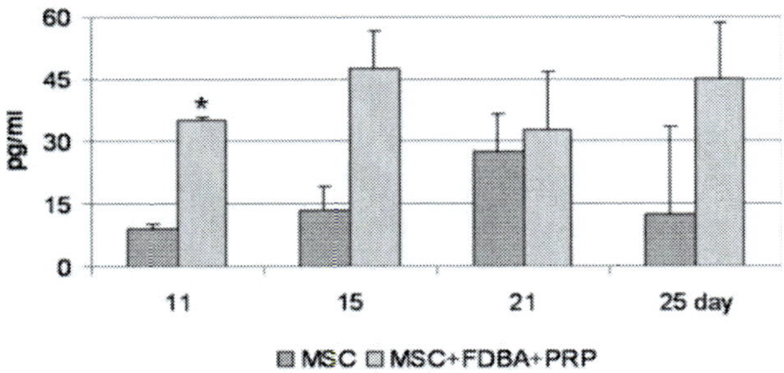

(b)

a. Runx2, osterix and osteocalcin gene expression by MSC incubated with PRP combined with FDBA. The results are expressed as mean ± standard error of the gene of interest/β-actin ratio. Osx: osterix; OC: osteocalcin.

b. FGF-2 concentration in the supernatants of MSC incubated with PRP combined with FDBA.

Figure 2. Gene expression and FGF-2 production in human MSC incubated with PRP combined with freeze-dried bone allografts (FDBA). * $p < 0.05$.

An interesting application is the replacement of fetal calf serum (FCS) in MSC cultures on artificial scaffolds before implantation, due to the necessity of avoiding animal-derived materials [47]. For this application, platelet GF can be obtained through platelet aggregation or platelet lysis. Platelet lysates, prepared by freezing/thawing, had higher levels of GF than platelet aggregates and were more efficient than FCS in increasing MSC proliferation [48], thus reducing the duration of ex vivo manipulation. After seeding onto a HA scaffold, MSC incubated with platelet lysates were able to differentiate into osteogenic lineage, even though more slowly than on plastic dishes [49].

Osteoclast Generation and Platelet Growth Factors

Bone remodelling is controlled by the joined action of osteoblasts and osteoclasts. Osteoclasts, that are responsible of bone resorption, are multinucleated cells formed by the fusion of cells of the monocyte/macrophage lineage. Osteoclast formation requires the presence of the receptor activator of nuclear factor-κB ligand (RANKL) and macrophage colony-stimulating factor (M-CSF), which are produced by neighbouring stromal cells and osteoblasts. RANKL binds RANK, which is expressed on the surface of osteoclasts. This binding triggers the activation of osteoclasts. Therefore, a cell-to-cell interaction between osteoclast precursors and osteoblasts is required for osteoclastogenesis. On the other hand, osteoblasts secrete osteoprotegerin, that is a decoy receptor for RANKL, preventing its interaction with RANK. Osteoclast generation is further regulated by hormones, cytokines and GF. Activated platelets release different modulators of osteoclast differentiation, such as TGF-β1, IGF-I, PDGF [50] and prostaglandin E_2. Also VEGF favored osteoclast differentiation in the presence of M-CSF [51].

However, conflicting results in the effect of PRP on osteoclasts have been reported. In vitro, PRS was shown to favor osteoclast differentiation of mouse bone marrow cells, particularly the generation of tartrate-resistant acid phosphatase (TRACP) positive multinucleated cells, the formation of bone pits and the expression of mRNA for cathepsin K and matrix metalloproteinase-9 [52, 53]. Instead, other authors found that PRP decreased the generation of TRACP positive multinucleated cells from rat bone marrow and enhanced the secretion of osteoprotegerin from osteoblasts [54]. From a research on the osteoclast precursors of human peripheral blood, PRP at 10% interfered with the complete differentiation process and osteoclast activation, as it was demonstrated by the decrease of bone collagen type I degradation. At high dosage PRP affected osteoclast formation also at an early stage of differentiation, particularly it impaired the fusion of precursors and the formation of TRACP positive multinucleated cells [55]. These results suggest a possible inhibitory effect of the substances released by PRP on osteoclast activation. In vivo, dogs with mandibular defects treated with autologous bone grafts and with PRP showed initially an increase of osteoclast number, but after 2 months there were no significant difference between PRP- and non-PRP-added grafts [56]. On the basis of these contrasting results, we could suppose that PRP had opposite effects on osteoclast generation and differentiation through the simultaneous interaction of several released molecules with osteoclast precursors (Figure 3). TGF-β could be a possible inhibitor of terminal differentiation of osteoclasts. In fact, TGF-β1 favored the recruitment of hematopoietic precursors [57] but inhibited bone resorption [58].

Endothelial Cells and Platelet Growth Factors

Angiogenesis is crucial for a correct bone healing: newly formed vessels mediate delivery of cell precursors, secrete GF and transport nutrients and oxygen [59]. A role for endothelium in the regulation of bone remodelling has been postulated: endothelial cells synthesize many molecules known to influence bone cells, including cytokines, chemokines, prostanoids, growth factors, adhesion molecules and matrix constituents [60]. Endothelial cells produce osteogenic molecules such as BMP-2 [61] and endothelin-1 [62], which favor differentiation of osteoprogenitor cells.

Activated platelets express chemotactic agents for endothelial cell migration, such as phosphatidylserine, phosphatidic acid and lysophosphatidyl ethanolamine [63]. The GF released by platelets, such as VEGF, TGF-β1 and PDGF, have proangiogenic properties and enhance migration and proliferation of endothelial cells [64], tube formation and the synthesis of matrix-metalloproteinase-9 [65]. Moreover, PRP induced endothelial cells to express ICAM-1, which mediates the adhesion of leukocytes to endothelium during inflammation [66]. Therefore, PRP may also contribute to the macrophage response in the inflammatory phase, which is essential for a correct bone repair. PRP induced PDGF-B mRNA expression from endothelial cells, whereas PDGF-A was not affected (Figure 4). Therefore, PRP may turn the balance between PDGF-A and PDGF-B mRNA and induce a higher synthesis of PDGF-BB, which is considered to have a higher chemotactic effect for human MSC than the other isoforms of PDGF [67]. In fact, MSC migration was higher in HUVEC incubated with PRP in comparison with HUVEC incubated with low serum medium or with PPP [66]. Thus, PRP may induce HUVEC to recruit MSC to the site of bone repair.

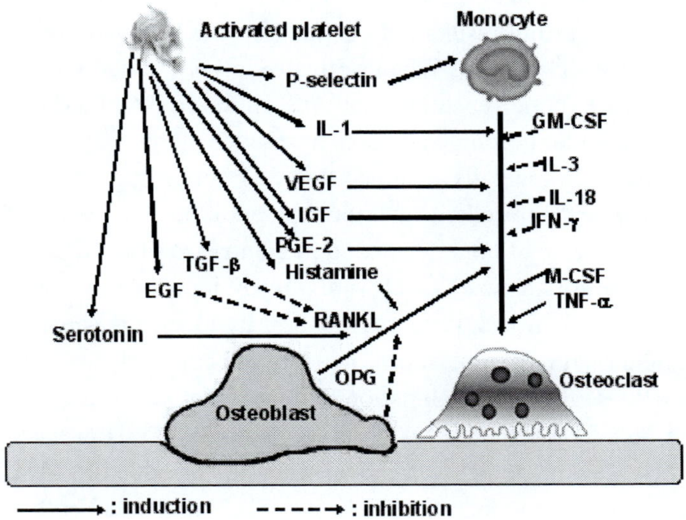

Figure 3. Effects on osteoclast differentiation of the different substances released by activated platelets.

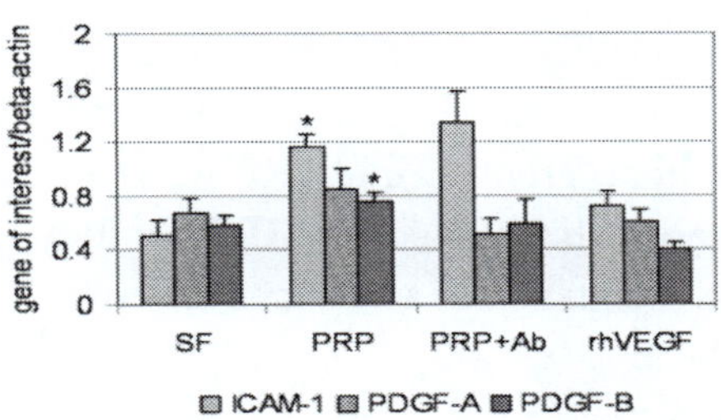

Figure 4. Gene expression by HUVEC incubated with PRP. SF: low-serum medium; PRP: platelet-rich plasma; PRP+Ab: PRP added with a neutralizing antibody anti-VEGF; rhVEGF: recombinant VEGF. The results are expressed as mean and standard error of the gene of interest/β-actin ratio. * $p < 0.05$.

PRP Effects on Experimental Animals

In vivo studies differ with regard to animals, treated bone, the presence of MSC and of scaffolds. Newly formed bone and biomechanical stability were the main evaluated outcome parameters. In experimental animals PRP can be used as alone or in combination with bone substitutes. In combination with bovine cancellous bone, PRP increased bone formation in a non-critical-size bone defect in rabbit cranium [68]. However, other authors noticed that the use of bone graft material in combination with PRP was not superior to bone allograft alone [69, 70, 71]. These contrasting results may depend on the fact that critical-size bone defects were not ever assayed. When critical-size bone defects were tested, the combination of PRP, bone substitutes and MSC induced the highest improvement in healing compared to the single component assayed alone [72].

Even though it is generally assumed that PRP might support osteogenesis in the presence of precursor cells, sometimes the addition of MSC to PRP did not further increase the rate of bone formation in comparison to PRP alone. Probably MSC added to the critical size defect were already induced towards osteogenesis and PRP could not significantly increase this. Moreover, the same injury due to osteotomy might recruit osteoblast precursors of the bone marrow [73].

When PRP was used in combination with artificial scaffolds, such as β-TCP, more intense bone regeneration was observed in dogs [74]. However, in goats no effect of platelet-leukocyte gel was shown when it was added to MSC seeded on scaffolds made of the combination of HA and β-TCP implanted both at orthotopic and ectopic location [75].

The effect of PRP combined with BMSC was also tested on experimental models for dental implants. In a dog mandible model of bone defects, the combination of canine MSC and PRP favored a well-formed mature bone formation and neo-vascularization comparable with that induced by the autologous cancellous bone group, while PRP alone performed poorly [76]. In other experiments, the combination of MSC-PRP-fibrin gel, used as grafting material for alveolar augmentation in adult dogs, determined the highest degree of bone-implant contact [77].

From the analysis of the results of in vivo research, we can assert that PRP promotes bone healing [78], particularly in the early phase and in combination with bone grafts [79, 80] and MSC. In the later course of bone healing, the PRP effects were less evident [81, 82]. However, a beneficial effect of PRP on bone healing was not demonstrated in every animal model and with every bone substitute. Factors that could influence the efficacy are the animal species, the preparation method of PRP, the physical and chemical properties of the scaffold.

The concentration of GF in platelets seems to be species dependent, being highest in humans, followed by goat and rat [83]. These differences may be reflected on the experimental results. In goats, the addition of PRP to autogenous cancellous bone did not improve bone healing of a critical-size defect [84]. In pigs, PRP failed to improve bone healing in anterior spinal fusion [85]. In dogs, PRP decreased bone formation around titanium implants in tibia defects [86]. Moreover, the concentration of platelets and GF, and consequently their osteogenic capacity, was significantly influenced by the preparation method [87].

Also the physical and chemical properties of the scaffold may affect PRP performance. After subcutaneous implantation in mice, ectopic bone formation was higher when PRP was combined with calcium-deficient hydroxyapatite (CDHA) than with β-TCP [88]. As a subcutaneous implant cannot be compared to a bone defect, CDHA in combination with PRP was also investigated in a critical-size diaphyseal bone defect. PRP yielded better bone formation than the empty CDHA scaffold, but the combination of MSC and PRP did not further improve bone healing [89]. Also the rate of resorption of the scaffold may affect the performance of PRP.

In summary, the in vivo results on the role of PRP are controversial. Several researches did show a positive effect, mainly on early bone healing, while others did not. However, many studies could hardly be compared because the protocol design and the method for measuring bone formation were very different.

The Clinical Use of Platelet Rich Plasma for Bone Repair

In orthopaedics, PRP is generally considered a tool for promoting bone repair, even though with discordant results. PRP efficacy generally improved when it was combined with bone allografts [90], eventually mixed to autologous cryoprecipitate, obtained from fresh frozen plasma by freeze-thaw precipitation of proteins and subsequent resuspension in plasma [20]. Also in spine fusion surgery, platelet concentrates improved the results when added to autologous bone graft [91], even though some authors did not observe any clear benefit [92]. PRP improved the fusion rates also in total ankle replacement [93]. PRP seems to be useful particularly in clinical situations where the bone repair is compromised, such as in diabetic fractures [94, 95].

The discrepancies on the clinical effects of PRP could derive from the low number of studies with control group comparison, as well as the different preparation methods and the poor characterization of the platelet concentrate. Randomized controlled clinical trials are needed to definitively show the effect of PRP on the promotion of bone growth and functional recovery. A recent systematic literature review on the use of PRP in orthopaedic surgery, after the exclusion of the applications in periodontal and maxillofacial surgery, identified 39 articles concerning clinical trials [96]. Only one was a randomized controlled study [97], which demonstrated that PRP, particularly if combined with MSC, increased the osteogenic potential of lyophilized bone chips (Figure 5). The strength of this clinical research was the randomization and the comparison among different treatments (freeze-dried bone allograft alone or in combination with PRP with or without MSC); its weakness was the lack of the evaluation of PRP effect in isolation and of a clear primary outcome.

Another clinical study reported a controlled case series of children, but was limited by the selection criteria of the patients, which differed in pathology [98].

In a prospective randomised clinical study on 120 patients, PRP was compared with rhBMP-7 for the treatment of persistent long bone non-unions. The preliminary results supported the view that the application of rhBMP-7 as a bone-stimulating agent was superior than PRP [99].

An interesting application of PRP is the percutaneous treatment of delayed unions or nonunions. PRP application without bone allografts improved healing only if the treatment was performed before 11 months from the initial surgery. After 11 months, simultaneous application of PRP and autologous bone grafts could be required [100].

PRP was largely applied for promoting bone regeneration also in odontostomatology and maxillofacial surgery. MSC and PRP administered contemporary to titanium implants provided stable and predictable results in terms of implant success [101]. Implants of β-HA seeded with MSC and PRP for maxillary sinus floor augmentation were clinically stable 12 months post-loading [102]. A systematic review on human controlled clinical trials designed to treat maxillofacial bony defects with application of PRP found evidence for beneficial effects in the treatment of periodontal defects, but only scarce effects in sinus elevation. No conclusions could be drawn about other applications of PRP in dentistry [103].

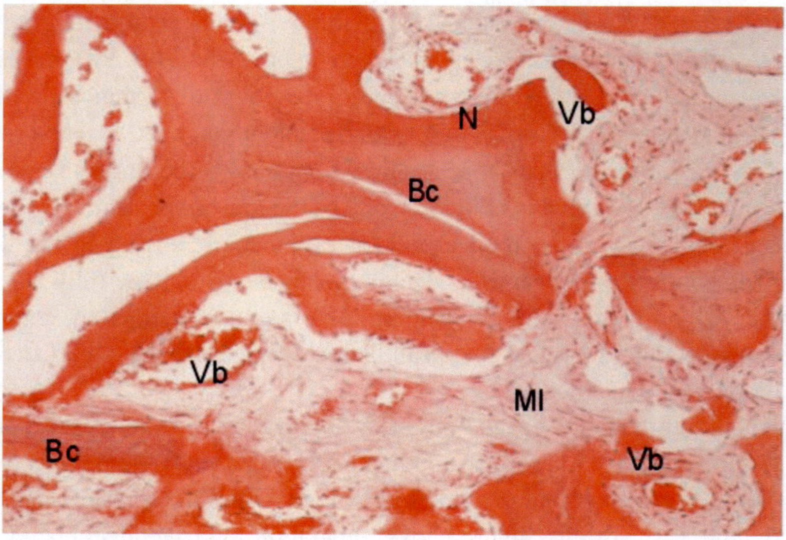

Figure 5. Light micrograph of a needle-biopsy from a patient who underwent a opening-wedge high tibial osteotomy for genu varum, six weeks after surgery. The defect was filled with lyophilized bone chips added with platelet gel and bone marrow stromal cells. Active osteogenetic process with vascular buds (Vb), new bone (N), and marrow-like (Ml) regions around bone chips (Bc) was shown at the level of the bone graft/host bone junction (hematoxylin and eosin; original magnification, ×10).

From this critical discussion on the clinical application of PRP effects, we infer that PRP could have a significant potential in accelerating bone repair, but further basic science and clinical research is needed to define the clinical indications and the methods of administration, before widespread use of this technology can take place. Unfortunately, also for practical and ethical reasons, clinical situations often lack proper controls, frustrating the efforts of data interpretation.

Conclusion

PRP has been evaluated in various experimental and clinical models of orthopaedic surgery, with conflicting results. PRP appears to be safe and feasible and may have a beneficial therapeutic effect, particularly in the early phases of bone formation. The best results are obtained when PRP is added with MSC and bone allografts or an osteogenic scaffold. PRP increases endothelial cell proliferation and the expression of functions favoring the migration of monocytes/macrophages and of osteoblast precursors. Moreover, PRP may induce target cells to synthesize GF and other substances, able to sustain bone healing. Many clinical studies report a significant acceleration of the repair process induced by PRP, but the final bone healing is similar to the untreated group. In delayed fracture healing, which could be one of the most important application of PRP, at present there is no clinical evidence of benefit, owing to the lack of randomized controlled clinical trials. Other aspects that should be fully elucidated by further experimental and clinical research are the interaction among the various platelet GF during tissue healing and the antimicrobial effect of the preparations rich in leukocytes.

The conflicting results depend on the variability in GF concentration of PRP, linked to the variability in proliferation and differentiation of MSC when they are added. Other parameters that can contribute to the conflicting results depend on the differences in the techniques of preparation and application of PRP. Consensus should be obtained regarding the production and characterization of PRP, in order to select the best standards that can help in the planning of experimental and clinical research producing comparable results.

Acknowledgments

This study was supported by the Istituto Ortopedico Rizzoli, ''Ricerca corrente'', the ''Fondazione Monte dei Paschi'' of Siena (Italy) and the Emilia-Romagna district (Italy) ''Progetto di Ricerca Regione-Università: Regenerative Medicine in Osteo-articular Diseases.''

References

[1] Mehta, S; Watson, JT. Platelet rich concentrate: basic science and current clinical applications. *J Orthop Trauma*, 2008, 22, 433-438.

[2] Bianco, P; Riminucci, M, Gronthos, S; Robey, PG. Bone marrow stromal stem cells: nature, biology, and potential applications. *Stem Cells*, 2001, 19, 180-192.

[3] Smrke, D; Gubina, B; Domanoviç, D; Rozman, P. Allogeneic platelet gel with autologous cancellous bone graft for the treatment of a large bone defect. *Eur Surg Res*, 2007, 39, 170-174.

[4] Huang, Q; Wang, YD; Wu, T; Jiang, S; Hu, YL; Pei, GX. Preliminary separation of the growth factors in platelet-rich plasma: effects on the proliferation of human marrow-derived mesenchymal stem cells. *Chin Med J (Engl)*, 2009,122, 83-87.

[5] Davì, G; Patrono, C. Platelet activation and atherothrombosis. *N Engl J Med*, 2007, 357, 2482-2494.

[6] Suzuki, H; Katagiri, Y, Tsukita, S; Tanoue, K; Yamazaki, H. Localization of adhesive proteins in two newly subdivided zones in electron-lucent matrix of human platelet alpha-granules. *Histochemistry*, 1990, 94, 337-344.

[7] Røger, M; Halstensen, TS; Høgåsen, K; Mollnes, TE; Solum, NO; Hovig, T. Platelets and vitronectin: immunocytochemical localization and platelet interaction with exogenously added vitronectin. *Nouv Rev Fr Hematol*, 1992, 34, 47-54.

[8] Moussad, EE; Brigstock, DR. Connective tissue growth factor: what's in a name? *Mol Genet Metab*, 2000, 71, 276-292.

[9] Zheng, MH; Wood, DJ; Papadimitriou, JM. What's new in the role of cytokines on osteoblast proliferation and differentiation? *Pathol Res Pract*, 1992, 188, 1104-1121.

[10] Nurden, AT; Nurden, P; Sanchez, M; Andia, I; Anitua E. Platelets and wound healing. *Front Biosci*, 2008, 13, 3532-3548.

[11] Martinez, A; Salas, E; Radomski, A; Radomski, MW. Matrix metalloproteinase-2 in platelet adhesion to fibrinogen: interactions with nitric oxide. *Med Sci Monit*, 2001, 7, 646-651.

[12] Villeneuve, J; Block, A; Le Bousse-Kerdilès, MC; Lepreux, S; Nurden, P; Ripoche, J; Nurden, AT. Tissue inhibitors of matrix metalloproteinases in platelets and megakaryocytes: a novel organization for these secreted proteins. *Exp Hematol*, 2009, 37, 849-856.

[13] Gear, AR; Camerini, D. Platelet chemokines and chemokine receptors: linking hemostasis, inflammation, and host defense. *Microcirculation*, 2003, 10, 335-350.

[14] Li, G; Kim, YJ; Mantel, C; Broxmeyer, HE. P-selectin enhances generation of CD14+CD16+ dendritic-like cells and inhibits macrophage maturation from human peripheral blood monocytes.*J Immunol*, 2003, 171, 669-677.

[15] Inwald, DP; McDowall, A; Peters, MJ; Callard, RE; Klein, NJ. CD40 is constitutively expressed on platelets and provides a novel mechanism for platelet activation. *Circ Res*, 2003, 92, 1041-1048.

[16] Marx, RE; Carlson, ER; Eichstaedt, RM; Schimmele, SR; Strauss, JE; Georgeff, KR Platelet-rich plasma: Growth factor enhancement for bone grafts. *Oral Surg Oral Med Oral Pathol Oral Radiol Endod*, 1998, 85, 638-646.

[17] Marx RE. Platelet-rich plasma: evidence to support its use. *J Oral Maxillofac Surg*, 2004, 62, 489-496.

[18] Weibrich, G; Hansen, T; Kleis, W; Buch, R; Hitzler, WE. Effect of platelet concentration in platelet-rich plasma on peri-implant bone regeneration. *Bone*, 2004, 34, 665-671.

[19] Dohan, DM; Choukrou, J; Diss, A; Dohan, SL; Dohan, AJ; Mouhyi, J; Gogly, B. Platelet-rich fibrin (PRF): a second-generation platelet concentrate. Part I: technological concepts and evolution. *Oral Surg Oral Med Oral Pathol Oral Radiol Endod*, 2006, 101, e37-44.

[20] Savarino, L; Cenni, E; Tarabusi, C; Dallari, D; Stagni, C; Cenacchi, A; Fornasari, PM; Giunti, A; Baldini, N. Evaluation of bone healing enhancement by lyophilized bone grafts supplemented with platelet gel: a standardized methodology in patients with tibial osteotomy for genu varus. *J Biomed Mater Res B Appl Biomater*, 2006, 76, 364-372.

[21] Anitua, E; Sánchez, M; Orive, G; Andía, I. The potential impact of the preparation rich in growth factors (PRGF) in different medical fields. *Biomaterials*, 2007, 28, 4551-4560.

[22] Mazzucco, L; Balbo, V; Cattana, E; Borzini, P. Platelet-rich plasma and platelet gel preparation using Plateltex. *Vox Sang*, 2008, 94, 202-208.

[23] Mishra, A; Tummala, P; King, A; Lee, B; Kraus, M; Tse, V; Jacobs, CR. Buffered platelet-rich plasma enhances mesenchymal stem cell proliferation and chondrogenic differentiation. *Tissue Engineering Part C Methods,* 2009, 15, 431-435.

[24] Anitua, E; Aguirre, JJ; Algorta, J; Ayerdi, E; Cabezas, AI; Orive, G; Andia, I. Effectiveness of autologous preparation rich in growth factors for the treatment of chronic cutaneous ulcers. *J Biomed Mater Res B Appl Biomater*, 2008, 84, 415-421.

[25] Trindade-Suedam, IK; Leite, FR; de Morais, JA; Leite, ER; Marcantonio, EJ; Leite, AA. Avoiding leukocyte contamination and early platelet activation in platelet-rich plasma. *J Oral Implantol*, 2007, 33, 334-339.

[26] Moojen, DJ; Everts, PA; Schure, RM; Overdevest, EP; van Zundert, A; Knape, JT; Castelein, RM; Creemers, LB; Dhert, WJ. Antimicrobial activity of platelet-leukocyte gel against Staphylococcus aureus. *J Orthop Res*, 2008, 26, 404-410.

[27] Dohan, DM; Choukroun, J; Diss, A; Dohan, SL; Dohan, AJ; Mouhyi, J; Gogly, B. Platelet-rich fibrin (PRF): a second-generation platelet concentrate. Part III: leucocyte activation: a new feature for platelet concentrates? *Oral Surg Oral Med Oral Pathol Oral Radiol Endod*, 2006, 101, e51-55.

[28] Everts, PA; Mahoney, CB; Hoffmann, JJ; Schonberger, JP; Box, HA; van Zundert, A; Knape, JT. Platelet-rich plasma preparation using three devices: implications for platelet activation and platelet growth factor release. *Growth factors*, 2006, 24, 165-171.

[29] Everts, PA; Hoffmann, J; Weibrich, G; Mahoney, CB; Schönberger, JP; van Zundert, A; Knape, JT. Differences in platelet growth factor release and leucocyte kinetics during autologous platelet gel formation. *Transfus Med*, 2006, 16, 363-368.

[30] Leitner, GC; Gruber, R; Neumuller, J; Wagner, A; Kloimstein, P; Hocker, P; Kormoczi, GF; Buchta, C. Platelet content and growth factor release in platelet-rich plasma: a comparison of four different systems. *Vox Sang*, 2006, 91, 135-139.

[31] Weibrich, G; Kleis, WK; Hitzler, WE; Hafner, G. Comparison of the platelet concentrate collection system with the plasma-rich-in-growth-factors kit to produce platelet-rich plasma: a technical report. *Int J Oral Maxillofac Implants*, 2005, 20, 118-123.

[32] Weibrich, G; Kleis, WK; Hafner, G; Hitzler, WE; Wagner, W. Comparison of platelet, leukocyte, and growth factor levels in point-of-care platelet-enriched plasma, prepared using a modified Curasan kit, with preparations received from a local blood bank. *Clin Oral Implants Res*, 2003, 14, 357-362.

[33] Dohan Ehrenfest, DM; Rasmusson, L; Albrektsson, T. Classification of platelet concentrates: from pure platelet-rich plasma (P-PRP) to leucocyte- and platelet-rich fibrin (L-PRF). *Trends Biotechnol*, 2009, 27, 158-167.

[34] Dugrillon, A; Eichler, H; Kern, S; Kluter, H. Autologous concentrated platelet-rich plasma (cPRP) for local application in bone regeneration. *Int J Oral Maxillofac Surg*, 2002, 31, 615-619.

[35] Anitua, E; Sánchez, M; Nurden, AT; Nurden, P; Orive, G; Andía, I. New insights into and novel applications for platelet-rich fibrin therapies. *Trends Biotechnol*, 2006, 24, 227-234.

[36] Everts, PA; Overdevest, EP; Jakimowicz, JJ; Oosterbos, CJ; Schönberger, JP; Knape, JT; van Zundert, A. The use of autologous platelet-leukocyte gels to enhance the healing process in surgery, a review. *Surg Endosc*, 2007, 21, 2063-2068.

[37] Borzini, P; Mazzucco, L; Giampaolo, A; Hassan, HJ. Platelet gel-the Italian way: a call for procedure standardization and quality control. *Transfus Med*, 2006, 16, 303-304.

[38] Gruber, R; Varga, F; Fischer, MB; Watzek, G. Platelets stimulate proliferation of bone cells: involvement of platelet-derived growth factor, microparticles and membranes. *Clin Oral Implants Res*, 2002, 13, 529-535.

[39] Vogel, JP; Szalay, K; Geiger, F; Kramer, M; Richter, W; Kasten, P. Platelet-rich plasma improves expansion of human mesenchymal stem cells and retains differentiation capacity and in vivo bone formation in calcium phosphate. *Platelets*, 2006, 17, 462-469.

[40] Lucarelli, E; Beccheroni, A; Donati, D; Sangiorgi, L; Cenacchi, A; Del Vento, AM; Meotti, C; Bertoja, AZ; Giardino, R; Fornasari, PM; Mercuri, M; Picci, P. Platelet-derived growth factors enhance proliferation of human stromal stem cells. *Biomaterials*, 2003, 24, 3095-3100.

[41] Gruber, R; Varga, F; Fischer, MB; Watzek, G. Platelets stimulate proliferation of bone cells: involvement of platelet-derived growth factor, microparticles and membranes. *Clin Oral Implants Res*, 2002, 13, 529-535.

[42] Ogino, Y; Ayukawa, Y; Kukita, T; Koyano, K. The contribution of platelet-derived growth factor, transforming growth factor-beta1, and insulin-like growth factor-I in platelet-rich plasma to the proliferation of osteoblast-like cells. *Oral Surg Oral Med Oral Pathol Oral Radiol Endod*, 2006, 101, 724-729.

[43] Gruber, R; Karreth, F; Kandler, B; Fuerst, G; Rot, A; Fischer, MB; Watzek, G. Platelet-released supernatants increase migration and proliferation, and decrease osteogenic differentiation of bone marrow-derived mesenchymal progenitor cells under in vitro conditions. *Platelets*, 2004, 15, 29-35.

[44] Gruber, R; Kandler, B; Fischer, MB; Watzek, G. Osteogenic differentiation induced by bone morphogenetic proteins can be suppressed by platelet-released supernatant in vitro. *Clin Oral Implants Res*, 2006, 17, 188-193.

[45] Kasten, P; Vogel, J; Beyen, I; Weiss, S; Niemeyer, P; Leo, A; Lüginbuhl, R. Effect of platelet-rich plasma on the in vitro proliferation and osteogenic

differentiation of human mesenchymal stem cells on distinct calcium phosphate scaffolds: the specific surface area makes a difference. *J Biomater Appl*, 2008, 23, 169-188.

[46] Cenni, E; Perut, F; Ciapetti, G; Savarino, L; Dallari, D; Cenacchi, A; Stagni, C; Giunti, A; Fornasari, PM; Baldini, N. In vitro evaluation of freeze-dried bone allografts combined with platelet rich plasma and human bone marrow stromal cells for tissue engineering. *J Mater Sci Mater Med*, 2009, 20, 45-50.

[47] Kocaoemer, A; Kern, S; Klueter, H; Bieback, K. Human AB-serum as well as thrombin-activated platelet-rich-plasma are suitable alternatives to fetal calf serum for the expansion of mesenchymal stem cells. *Stem Cells*, 2007, 25, 1270-1278.

[48] Doucet, C; Ernou, I; Zhang, Y; Llense, JR; Begot, L; Holy, X; Lataillade, JJ. Platelet lysates promote mesenchymal stem cell expansion: a safety substitute for animal serum in cell-based therapy applications. *J Cell Physiol*, 2005, 205, 228-236.

[49] Salvadè, A; Della Mina, P; Gaddi, D; Gatto, F; Villa, A; Bigoni, M; Perseghin, P; Serafini, M; Zatti, G; Biondi, A; Biagi, E. Characterization of platelet lysate (PL)-cultured mesenchymal stromal cells (MSC) and their potential use in tissue-engineered osteogenic devices for the treatment of bone defects. *Tissue Engineering Part C Methods*, 2009 May 26 [Epub ahead of print].

[50] Zhang, Z; Chen, J; Jin, D. Platelet-derived growth factor (PDGF)-BB stimulates osteoclastic bone resorption directly: the role of receptor beta. *Biochem Biophys Res Commun*, 1998, 251, 190-194.

[51] Niida, S; Kaku, M; Amano, H; Yoshida, H; Kataoka, H; Nishikawa, S; Tanne, K; Maeda, N; Nishikawa, S; Kodama, H. Vascular endothelial growth factor can substitute for macrophage colony-stimulating factor in the support of osteoclastic bone resorption. *J Exp Med*, 1999, 190, 293-298.

[52] Weicht, B; Maitz, P; Kandler, B; Fischer, MB; Watzek, G; Gruber, R. Activated platelets positively regulate RANKL-mediated osteoclast differentiation. *J Cell Biochem*, 2007, 102, 1300-1307.

[53] Gruber, R; Karreth, F; Fischer, MB; Watzek, G. Platelet-released supernatants stimulate formation of osteoclast-like cells through a prostaglandin/RANKL-dependent mechanism. *Bone*, 2002, 30, 726-732.

[54] Ogino, Y; Ayukawa, Y; Kukita, T; Atsuta, I; Koyano, K. Platelet-rich plasma suppresses osteoclastogenesis by promoting the secretion of osteoprotegerin. *J Periodontal Res*, 2009, 44, 217-224.

[55] Cenni, E; Avnet, S; Fotia, C; Salerno, M; Baldini, N. Platelet-rich plasma impairs osteoclast generation from human precursors of peripheral blood. *J Orthop Res*, in press.

[56] Gerard, D; Carlson, ER; Gotcher, JE; Jacobs, M. Effects of platelet-rich plasma at the cellular level on healing of autologous bone-grafted mandibular defects in dogs. *J Oral Maxillofac Surg*, 2007, 65, 721-727.

[57] Pilkington, MF; Sims SM; Dixon, SJ. Transforming growth factor-β induces osteoclast ruffling and chemotaxis: potential role in osteoclast recruitment. *J Bone Miner Res*, 2001, 16, 1237-1247.

[58] Takai, H; Kanematsu, M; Yano, K; Tsuda, E; Higashio, K; Ikeda, K; Watanabe, K; Yamada Y. Transforming growth factor-β stimulates the production of osteoprotegerin/osteoclastogenesis inhibitory factor by bone marrow stromal cells. *J Biol Chem*, 1998, 273, 27091-27096.

[59] Towler, DA. The osteogenic-angiogenic interface: novel insights into the biology of bone formation and fracture repair. *Curr Osteoporos Rep*, 2008, 6, 67-71.

[60] Brandi, ML; Collin-Osdoby, P. Vascular biology and the skeleton. *J Bone Miner Res*, 2006, 21, 183-192.

[61] Bouletreau, PJ; Warren, SM; Spector, JA; Peled, ZM; Gerrets, RP; Greenwald, JA; Longaker, MT. Hypoxia and VEGF up-regulate BMP-2 mRNA and protein expression in microvascular endothelial cells: implications for fracture healing. *Plast Reconstr Surg*, 2002, 109, 2384-2397.

[62] von Schroeder, HP; Veillette, CJ; Payandeh, J; Qureshi, A; Heersche, JN. Endothelin-1 promotes osteoprogenitor proliferation and differentiation in fetal rat calvarial cell cultures. *Bone*, 2003, 33, 673-684.

[63] Xiao, H; Siddiqui, RA; Al-Hassani, MH; Sliva, D; Kovacs, RJ. Phospholipids released from activated platelets improve platelet aggregation and endothelial cell migration. *Platelets*, 2001, 12, 163-170.

[64] Kilian, O; Flesch, I; Wenisch, S; Taborski, B; Jork, A; Schnettler, R; Jonuleit, T. Effects of platelet growth factors on human mesenchymal stem cells and human endothelial cells in vitro. *Eur J Med Res*, 2004, 9, 337-344.

[65] Kandler, B; Fischer, MB; Watzek, G; Gruber, R. Platelet-released supernatant increases matrix metalloproteinase-2 production, migration, proliferation and tube formation of human umbilical vascular endothelial cells. *J Periodontol*, 2004, 75, 1255-1261.

[66] Cenni, E; Ciapetti, G; Granchi, D; Fotia, C; Perut, F; Giunti, A; Baldini, N. Endothelial cells incubated with platelet-rich plasma express PDGF-B and

ICAM-1 and induce bone marrow stromal cell migration. *J Orthop Res,* 2009, 27, 1493-1498.

[67] Fiedler, J; Etzel, N; Brenner, RE. To go or not to go: migration of human mesenchymal progenitor cells stimulated by isoforms of PDGF. *J Cell Biochem*, 2004, 93, 990-998.

[68] Aghaloo, TL; Moy, PK; Freymiller, EG. Evaluation of platelet-rich plasma in combination with anorganic bovine bone in the rabbit cranium: a pilot study. *Int J Oral Maxillofac Implants*, 2004, 19, 59-65.

[69] Jensen, TB; Rahbek, O; Overgaard, S; Søballe, K. Platelet rich plasma and fresh frozen bone allograft as enhancement of implant fixation. An experimental study in dogs. *J Orthop Res*, 2004, 22, 653-658.

[70] Roldán, JC; Jepsen, S; Miller, J; Freitag, S; Rueger, DC; Açil, Y; Terheyden, H. Bone formation in the presence of platelet-rich plasma vs. bone morphogenetic protein-7. *Bone*, 2004, 34, 80-90.

[71] Yamada, Y; Ueda, M; Naiki, T; Takahashi, M; Hata, K; Nagasaka, T. Autogenous injectable bone for regeneration with mesenchymal stem cells and platelet-rich plasma: tissue-engineered bone regeneration. *Tissue Eng*, 2004, 10, 955-964.

[72] Dallari, D; Fini, M; Stagni, C; Torricelli, P; Nicoli Aldini, N; Giavaresi, G; Cenni, E; Baldini, N; Cenacchi, A; Bassi, A, Giardino, R; Fornasari, PM; Giunti, A. In vivo study on the healing of bone defects treated with bone marrow stromal cells, platelet-rich plasma, and freeze-dried bone allografts, alone and in combination. *J Orthop Res*, 2006, 24, 877-888.

[73] Lieberman, JR; Daluiski, A; Einhorn, TA. The role of growth factors in the repair of bone. Biology and clinical applications. *J Bone Joint Surg Am*, 2002, 84-A, 1032-1044.

[74] Suba, Z; Takacs, D; Gyulai-Gaal, S; Kovacs, K. Facilitation of β-tricalcium phosphate-induced alveolar bone regeneration by platelet-rich plasma in beagle dogs: a histologic and histomorphometric study. *Int J Oral Maxillofac Implants*, 2004, 19, 832-838.

[75] Geuze, RE; Everts, PA; Kruyt, MC. Orthotopic location has limited benefit from allogeneic or autologous multipotent stromal cells seeded on ceramic scaffolds. *Tissue Engineering: Part A,* 2009 Apr 17 Epub ahead of print].

[76] Yamada, Y; Ueda, M; Naiki, T; Nagasaka, T. Tissue-engineered injectable bone regeneration for osseointegrated dental implants. *Clin Oral Implants Res*, 2004, 5, 589-597.

[77] Ito, K; Yamada, Y; Naiki, T; Ueda, M. Simultaneous implant placement and bone regeneration around dental implants using tissue-engineered bone

with fibrin glue, mesenchymal stem cells and platelet-rich plasma. *Clin Oral Implants Res*, 2006, 5, 579-586.

[78] Fuerst, G; Gruber, R; Tangl, S; Sanroman, F; Watzek, G. Enhanced bone-to-implant contact by platelet-released growth factors in mandibular cortical bone: a histomorphometric study in minipigs. *Int J Oral Maxillofac Implants*, 2003, 18, 685-690.

[79] Wiltfang, J; Kloss, FR; Kessler, P; Nkenke, E; Schultze-Mosgau, S; Zimmermann, R; Schlegel, KA. Effects of platelet-rich plasma on bone healing in combination with autogenous bone and bone substitutes in critical-size defects. *Clin Oral Implants Res*, 2004, 15, 187-193.

[80] Fennis, JP; Stoelinga, PJ; Jansen, JA. Mandibular reconstruction: a histological and histomorphometric study on the use of autogenous scaffolds, particulate cortico-cancellous bone grafts and platelet rich plasma in goats. *Int J Oral Maxillofac Surg*, 2004, 33, 48-55.

[81] Choi, BH; Im, CJ; Huh, JY; Suh, JJ; Lee, SH. Effect of platelet-rich plasma on bone regeneration in autogenous bone graft. *Int J Oral Maxillofac Surg*, 2004, 33, 56-59.

[82] Thor, A; Franke-Stenport, V; Johansson, CB; Rasmusson, L. Early bone formation in human bone grafts treated with platelet-rich plasma: preliminary histomorphometric results. *Int J Oral Maxillofac Surg*, 2007, 36, 1164-1171.

[83] van den Dolder, J; Mooren, R; Vloon, AP; Stoelinga, PJ; Jansen, JA. Platelet-rich plasma: quantification of growth factor levels and the effect on growth and differentiation of rat bone marrow cells. *Tissue Eng*, 2006, 12, 3067-3073.

[84] Mooren, RE; Merkx, MA; Bronkhorst, EM; Jansen, JA; Stoelinga, PJ. The effect of platelet-rich plasma on early and late bone healing: an experimental study in goats. *Int J Oral Maxillofac Surg*, 2007, 36, 626-631.

[85] Li, H; Zou, X; Xue, Q; Egund, N; Lind, M; Bunger C. Anterior lumbar interbody fusion with carbon fiber cage loaded with bioceramics and platelet-rich plasma. An experimental study on pigs. *Eur Spine J*, 2004, 13, 354-358.

[86] You, TM; Choi, BH; Li, J; Jung, JH; Lee, HJ; Lee, SH; Jeong, SM. The effect of platelet-rich plasma on bone healing around implants placed in bone defects treated with Bio-Oss: a pilot study in the dog tibia. *Oral Surg Oral Med Oral Pathol Oral Radiol Endod*, 2007, 103, e8-12.

[87] Weibrich, G; Kleis, WK; Hafner, G; Hitzler, WE; Wagner, W. Comparison of platelet, leukocyte, and growth factor levels in point-of-care platelet-

enriched plasma, prepared using a modified Curasan kit, with preparations received from a local blood bank. *Clin Oral Implants Res*, 2003, 14, 357-362.

[88] Kasten, P; Vogel, J; Luginbuhl, R; Niemeyer, P; Weiss, S; Schneider, S; Kramer, M; Leo, A; Richter, W. Influence of platelet-rich plasma on osteogenic differentiation of mesenchymal stem cells and ectopic bone formation in calcium phosphate ceramics. *Cells Tissues Organs*, 2006, 183, 68-79.

[89] Kasten, P; Vogel, J; Geiger, F; Niemeyer, P; Luginbuhl, R; Szalay, K. The effect of platelet-rich plasma on healing in critical-size long-bone defects. *Biomaterials*, 2008, 29, 3983-3992.

[90] Ilgenli, T; Dundar, N; Kal, BI. Demineralized freeze-dried bone allograft and platelet-rich plasma vs. platelet-rich plasma alone in infrabony defects: a clinical and radiographic evaluation. *Clin Oral Investig*, 2007, 11, 51-59.

[91] Lowery, GL; Kulkarni, S; Pennisi, AE. Use of autologous growth factors in lumbar spinal fusion. *Bone*, 1999, 25, 47S-50S.

[92] Weiner, BK; Walker, M. Efficacy of autologous growth factors in lumbar intertransverse fusions. *Spine*, 2003, 28, 1968-1970.

[93] Coetzee, JC; Pomeroy, GC; Watts, JD; Barrow, C. The use of autologous concentrated growth factors to promote syndesmosis fusion in the Agility total ankle replacement. A preliminary study. *Foot Ankle Int*, 2005, 26, 840-846.

[94] Gandhi, A; Doumas, C; O'Connor, JP; Parsons, JR, Lin, SS. The effects of local platelet rich plasma delivery on diabetic fracture healing. *Bone*, 2006, 38, 540-546.

[95] Grant, WP; Jerlin, EA; Pietrzak, WS; Tam, HS. The utilization of autologous growth factors for the facilitation of fusion in complex neuropathic fractures in the diabetic population. *Clin Podiatr Med Surg*, 2005, 22, 561-584, vi.

[96] Griffin, XL; Smith, CM; Costa, ML. The clinical use of platelet-rich plasma in the promotion of bone healing: a systematic review. *Injury*, 2009, 40, 158-162.

[97] Dallari, D; Savarino, L; Stagni, C; Cenni, E; Cenacchi, A; Fornasari, PM; Albisinni, U; Rimondi, E; Baldini, N; Giunti, A. Enhanced tibial osteotomy healing with use of bone grafts supplemented with platelet gel or platelet gel and bone marrow stromal cells. *J Bone Joint Surg Am*, 2007, 89, 2413-2420.

[98] Kitoh, H; Kitakoji, T; Tsuchiya, H; Katoh, M; Ishiguro, N. Transplantation of culture expanded bone marrow cells and platelet rich plasma in distraction osteogenesis of the long bones. *Bone*, 2007, 40, 522-528.
[99] Calori, GM; Tagliabue, L; Gala, L, d'Imporzano, M; Peretti, G; Albisetti, W. Application of rhBMP-7 and platelet-rich plasma in the treatment of long bone non-unions: a prospective randomised clinical study on 120 patients. *Injury*, 2008, 39, 1391-1402.
[100] Bielecki, T; Gazdzik, TS; Szczepanski, T. Benefit of percutaneous injection of autologous platelet-leukocyte-rich gel in patients with delayed union and nounion. *Eur Surg Res*, 2008, 40, 289-296.
[101] Yamada, Y; Ueda, M; Hibi, H; Nagasaka, T. Translational research for injectable tissue-engineered bone regeneration using mesenchymal stem cells and platelet-rich plasma: from basic research to clinical case study. *Cell Transplant*, 2004, 4, 343-355.
[102] Ueda, M; Yamada, Y; Ozawa, R; Okazaki, Y. Clinical case reports of injectable tissue-engineered bone for alveolar augmentation with simultaneous implant placement. *Int J Periodont Restorative Dent*, 2005, 2, 129-137.
[103] Plachokova, AS; Nikolidakis, D; Mulder, J; Jansen, JA; Creugers, NH. Effect of platelet-rich plasma on bone regeneration in dentistry: a systematic review. *Clin Oral Implants Res*, 2008, 19, 539-545.

In: Bone Regeneration: Growth Factors... ISBN: 978-1-60876-621-5
Editors: V. Legard et al., pp. 177-202

Chapter VI

Bone Regeneration by Octacalcium Phosphate and Collagen Composites

Shinji Kamakura[1*], Takahisa Anada[2] and Osamu Suzuki[2]

[1]Bone Regenerative Engineering Laboratory, Graduate School of Biomedical Engineering, Tohoku University, Aoba-ku, Sendai, Japan.

[2]Division of Craniofacial Function Engineering (CFE), Tohoku University Graduate School of Dentistry, Aoba-ku, Sendai, Japan.

Abstract

Octacalcium phosphate ($Ca_8H_2(PO_4)_6 \cdot 5H_2O$; OCP) is a calcium phosphate and has been suggested to be a precursor of biological apatite crystals in bones and teeth. The importance of synthetic OCP as a bone substitute material was recently recognized in the field of tissue engineering for bone regeneration. Our previous studies elucidated the below-mentioned phenomena by using a rat calvarial defect model. OCP irreversibly converts

* Corresponding author: Bone Regenerative Engineering Laboratory, Graduate School of Biomedical Engineering, Tohoku University, 2-1 Seiryo-Machi, Aoba-Ku, Sendai 980-8574, Japan, E-mail: kamakura@m.tains.tohoku.ac.jp., Phone: +81-22-717-8235; Fax: +81-22-717-8235.

to the apatitic phase, and the process of OCP-apatite conversion is involved in bone regeneration. Implanted OCP enhances bone regeneration more than hydroxyapatite (HA) and β-tricalcium phosphate (β-TCP), which have already been used clinically as a scaffold for bone regeneration. OCP can be used as an effective scaffold for growth factors, such as bone morphogenetic protein-2 (BMP-2) and transforming growth factor-β_1 (TGF-β_1), to enhance bone regeneration. Although OCP possesses many desirable properties, OCP cannot be molded by a sintering process, unlike bioceramics, such as HA or β-TCP. To improve the performance of bone regeneration and handling performance, a composite synthetic bone substitute constructed of synthetic OCP and porcine atelocollagen sponge (OCP/Col) has been developed. The synergistic effects of OCP/Col on bone regeneration are the result of bone nucleation by OCP and cell infiltration by collagen, and OCP/Col shows stable bone regeneration. OCP/Col significantly enhances bone regeneration more than OCP *per se*, β-TCP collagen composite, and HA collagen composite if implanted into a critical-sized defect model in a rat calvarium. During optimization of OCP/Col, it was confirmed that bone regeneration by OCP/Col was influenced by dehydrothermal treatment in the fabrication of OCP/Col or the composition of OCP and collagen. As OCP/Col is applicable without cell transplantation and exogenous osteogenic cytokines, cost-effective bone regenerative therapy is expected in the future.

Keywords: Bone regeneration, Octacalcium phosphate, Collagen, Calcium phosphate, Bone.

1. Biomaterials for Regeneration of Bone Tissues

The reconstruction of critical-sized bone defects resulting from trauma, neoplasms, or infection is a major problem in reconstructive surgery [1-3]. Autogeneous bone grafting has been a well-accepted treatment modality to repair the function and structure of bone defects, although it has disadvantages, such as restricted availability and morbidity associated with harvesting bone from a second operative site [4, 5]. To overcome the disadvantages of autogeneous bone grafting, various materials have been used to regenerate bone tissues [6-8]. Biodegradable polymer matrices have been widely investigated for bone regeneration and are classified into natural polymers, such as proteins (collagen, fibrin) and polysaccharides (starch, alginate, chitosan, hyaluronic acid), and synthetic polymers, such as poly(lactic acid) (PLA), poly(glycolic acid) (PGA), and their copolymers (PLGA) [6].

Calcium phosphates are known to be bioactive materials, analogous with biological apatites [7]. Sintered hydroxyapatite (HA) ceramic, having a stoichiometric Ca/P molar ratio of 1.67, usually obtained by synthesis and then sintering above 1000°C, is a widely recognized biocompatible and non-resorbable bioceramic for bone substitute [5, 9]. Synthetic tricalcium phosphate (TCP) is known to a biodegradable bioceramic, and is clinically used in a β form for bone defects [5, 10]. Tetracalcium phosphate (TTCP) is another artificial material used as a component in calcium phosphate cement with α-TCP to form HA [11]. Bioglass® [12] and HA/TCP bi-phase ceramic [13] are already applied clinically; however, no bone regenerative material comparable to autogeneous bone has been developed so far.

2. Octacalcium Phosphate

Interstitial bone material consists of two major components, organic matrices, such as collagen, and minerals. The prototype for the mineral in mature bone and tooth is usually considered to be the basic calcium phosphate hydroxyapatite ($Ca_{10}(PO_4)_6(OH)_2$; HA) [14, 15]. The mechanism of the initiation of biomineralization is supposed to be via a direct precipitation process [16-18] or via precursor phases, such as amorphous calcium phosphate (ACP; $Ca_3(PO_4)_2 \cdot nH_2O$), dicalcium phosphate dihydrate ($CaHPO_4 \cdot 2H_2O$; DCPD), and octacalcium phosphate ($Ca_8H_2(PO_4)_6 \cdot 5H_2O$; OCP) [19-22]. OCP has been suggested to be the direct precursor of HA because in vitro studies showed that ACP and DCPD could be hydrolyzed to HA via OCP [23, 24]. OCP consists of apatite layers stacked with alternating hydrated layers that have structural similarity with HA [20], and favorably converts to HA due to its thermodynamic instability in the physiological environment [25, 26]. This is based on the general consensus that the transition of OCP to HA is promoted thermodynamically under physiological conditions and that, once initiated, it is likely to be spontaneous and irreversible [20, 27, 28], and can proceed via in situ hydrolysis utilizing apatitic layers [20, 29, 30] and de novo precipitation of apatite crystals on OCP crystal templates [25]. Because several studies have reported that OCP is present as an intermediate to HA in human dentine [31], porcine enamel [32] and in sutures undergoing premature suture closure during intramembranous mineralization of mouse calvaria [33], the physiological importance of OCP is a concern.

3. Osteogenic Properties of OCP

It has been well documented that mature teeth and bone minerals are composed of non-stoichiometric apatite crystals containing various impurities, such as carbonate and acid phosphate [14, 15]. It is also becoming clear that the major incorporation of impurities (e.g., carbonate ions) into biominerals takes place during the hydrolysis process of OCP [27, 34, 35]. We have elucidated the non-stoichiometric composition of OCP in a hot solution synthesis condition [36, 37], and the structural changes and osteogenic effects of OCP were analyzed in an in vitro study.

Apatitic conversion of OCP in culture medium was confirmed until 11 days by X-ray diffraction (XRD) and Fourier transform infrared spectroscopy (FTIR) [38]. During the conversion of OCP into the apatitic phase, chemical analyses of the supernatant of the OCP coating showed that the concentration of Ca^{2+} decreased and inorganic phosphate (Pi) increased in a dose-dependent manner, whereas that of HA coating was lower than the control, and was not dose-dependent [39]. Anada et al. reported a dose-dependent osteogenic effect on synthetic OCP or synthetic sintered ceramic HA when an osteoblastic cell line (ST-2 cells) was cultured on a different amount of coating. OCP coating enhanced the expression of osteogenic markers, including type Ia collagen, alkaline phosphatase, and osterix in a dose-dependent manner, whereas HA coating did not affect the expressions of these markers. Therefore, OCP is capable of inducing osteoblastic cell differentiation of stromal cells to pre-osteoblasts, accompanied by the progressive conversion of OCP to an apatite structure [39].

4. Bone Regeneration by Synthetic OCP

Although several studies have reported that synthetic OCP had potential as a bone regenerative scaffold in various forms, such a coatings on metallic implants [40-46], the microscaffold self-assembled [47, 48], osteoconductive property of synthetic OCP was first found in the subperiosteal implantation of mouse calvaria [49]. Several subsequent *in vivo* studies succeeded in implanting the granules of

OCP (300 - 500 μm in diameter) to assess whether synthetic OCP could be an effective bone substitute to regenerate bone. During OCP implantation, critical-sized rat calvarial defects were mostly developed. This model has advantages for quantitative analysis and is able to assess the efficiency of various bone regenerative materials, because the potential for bone regeneration without treatment in the created cranial bone defect is very low [50]. The process of bone regeneration has been investigated by radiographic, histological, and histomorphometrical examinations.

It was reported that OCP acts on osteoblasts, bone lining cells and/or their closely committed progenitors on the bone surface to express the phenotype and deposit new bone on the OCP implant [51]. If OCP granules were implanted into the calvarial defect model, isolated granulous radiopacity was observed 4 weeks after implantation, whereas OCP granules alone had no radiopacity under standardized conditions. The isolated radiopacity was amalgamated at 12 weeks, and became plate-like at 24 weeks, whereas the radiopacity was observed along defect margins in untreated defects [52]. This radiographic transition seemed to correspond to bone regeneration and/or apatitic conversion from OCP. It was consistent that newly formed bone was histologically observed in two locations 4 weeks after OCP implantation: on implanted OCP away from the margin and new bone formation from the defect margin. Examination using serial sections showed that the implanted OCP provided the core for nucleating multiple osteogenic sites and orthotopically induced bone formation by both histological identification and the bone-specific protein, osteocalcin [53]. Concentric bone regeneration at 12 weeks and fusion of concentric bone mass at 24 weeks were confirmed by histological sections [52].

The implanted OCP was surrounded by multinucleated giant cells (MNGCs) before bone regeneration. MNGCs on implanted OCP were tartrate-resistant acid phosphatase (TRAP)-positive cells and had the ultrastructural characteristics of a ruffled border-like structure and a clear zone-like structure that was characteristic of osteoclasts [54-56]. The surface of implanted OCP was frayed where it was in contact with the ruffled border-like structure of MNGC [55]. This suggests that implanted OCP could be resorbed by MNGCs that shared ultrastructural characteristics with osteoclasts.

As OCP is known to be more physicochemically resorbable than HA or β-tricalcium phosphate (β-TCP) under physiological conditions [57], we investigated whether implanted OCP is more resorbable than other commonly used bone graft substitutes of synthetic calcium phosphates, such as non-synthetic sintered ceramic HA and synthetic sintered ceramic β-TCP 6 months after implantation [58]. The percentages of newly formed bone (n-Bone%) or remaining implants (r-Imp%) in the defect in the selected area are statistically analyzed using light micrographs of sections stained with hematoxylin and eosin. The r-Imp% of OCP (6.58% ± 1.46%) were significantly lower than those of β-TCP (23.4% ± 6.82%) and HA (39.0% ± 6.86%), whereas the n-Bone% of OCP (63.3% ± 7.52%) was significantly higher than those of β-TCP (42.8% ± 13.6%) and HA (24.2% ± 7.82%). Also, histomorphometric analysis of bone regeneration by OCP was significantly higher than that of synthetic sintered ceramic HA, if implanted into the bone marrow of the rabbit femur [54]. These results suggest that implanted OCP is resorbed and replaced by newly formed bone more than β-TCP and HA.

It has been accepted that the three main components for bone regeneration are stem cells or osteoprogenitors, an appropriate biological scaffold, and signal molecules [59]. The application of signal molecules or osteoprogenitors with an appropriate scaffold has been proposed to regenerate bone efficiently because bone regeneration was not assured if signal molecules or osteoprogenitors were implanted independently [60]. Recombinant human bone morphogenetic protein-2 (rhBMP-2) cloned by Wozney and coworkers [61] induces both the differentiation of mesenchymal cells in subcutaneous or muscle tissue into osteoblasts and bone formation [62, 63]; however BMP-2 in the bloodstream is rapidly diffused, and an appropriate delivery system is needed to elicit the powerful bone regenerative property of BMP-2 [64].

To confirm that OCP would be an effective scaffold for BMP-2, OCP combined with BMP-2 (OCP/BMP) was prepared by a freeze-drying process, and the synergistic effect of OCP/BMP on bone regeneration was investigated [65]. Fifteen milligrams of OCP combined with 10 μg rhBMP-2 (OCP/BMP 10 μg), 15 mg OCP combined with 1 μg rhBMP-2 (OCP/BMP 1 μg), or OCP was implanted into a critical-sized defect of the rat calvarium and fixed 4 or 8 weeks after implantation. The defects treated with OCP/BMP were mostly filled with plate-like radiopacity and newly formed bone occupied the defect 8 weeks after implantation (Figure 1). Defects treated with OCP showed the amalgamation of isolated radiopacity and regenerated bone was concentrically formed around the implanted OCP (Figure 1). N-Bone% with OCP/BMP at 8 weeks was significantly higher than that with OCP (30.5% ± 10.8%), while the difference in n-Bone% between OCP/BMP-10 μg (68.4% ± 10.9%) and OCP/BMP-1 μg (59.2% ± 17.3%) was not significant. These results suggest that OCP would be applicable as an effective scaffold for BMP-2 and this OCP delivery system may reduce the standard effective dose of BMP-2, which would be beneficial because even low doses (less than 100 pg/μg OCP) of BMP-2 enhanced bone regeneration.

5. Synthetic OCP and Collagen Composites

(1) Preparation of Synthetic OCP and Collagen Composites

Although OCP possesses many desirable properties as a scaffold, OCP cannot be molded using sintering processes, unlike bioceramics such as HA or β-TCP, because of the intrinsic crystal structure of OCP, including a large number of water molecules within the layered HA structure [20]. Also, the inherent brittleness of OCP makes hard to maintain its shape under pressure [66] and precludes its use in load-bearing or large bone defects such as ridge augmentation or a defect after segmental osteotomy. To resolve the disadvantages of OCP, we attempted to make a new material composed of OCP and other material.

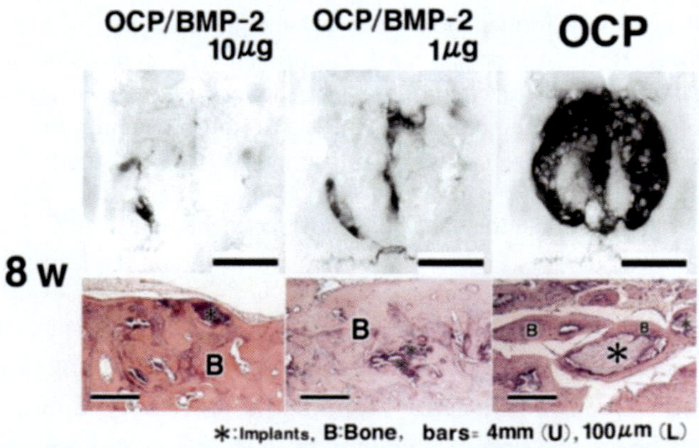

In OCP/BMP 10 μg and OCP/BMP 1 μg groups, the defect is mostly filled with plate-like radiopacity. Regenerated bone (B) shows a mosaic pattern and the bone matrix is compact. In the OCP group, larger radiopacity is formed by the amalgamation of isolated radiopacity. Regenerated bone (B) directly surrounds implanted OCP (*) and shows a lamellar structure. *: implants, B: newly formed bone. (Reprinted with permission and reorganized from Kamakura S. et al.: New scaffold for recombinant human bone morphogenetic protein-2. J Biomed Mater Res 71A: 299-307, 2004, John Wiley & Sons, Inc.)

Figure 1. Radiographs and histological findings of treatment with OCP/BMP 10 μg, OCP/BMP 1 μg, and OCP implantation at 8 weeks.

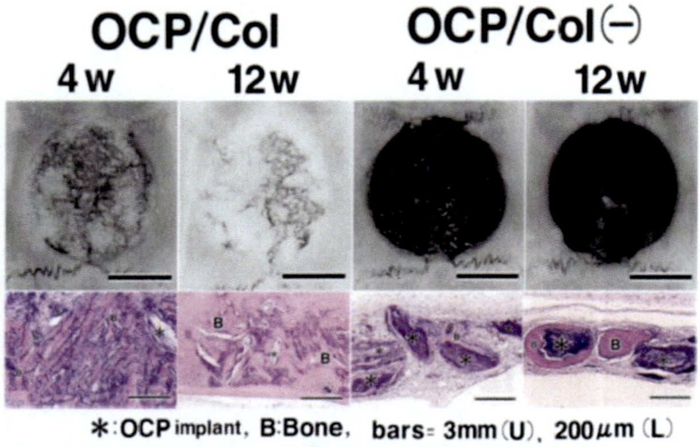

In OCP/Col, amalgamated thorn-like radiopacity forms in large defect areas in week 4 and condenses in weeks 12. Bone matrix (B) occupies in the reticulum of OCP/Col in week 4, and the newly formed bone (B) matures and increases, and bone remodeling with the resorption of OCP and implanted collagen is observed in week 12. In OCP/Col (-), isolated granulous radiopacity is scattered in the defect in week 4, and new bone formation (B) is limited around implanted OCP. Some implanted OCP (*) is surrounded by multinucleated giant cells. In week 12, larger radiopacity is formed by the amalgamation of isolated radiopacity, and implanted OCP (*) is concentrically surrounded by newly formed bone with a lamellar structure. *: implanted OCP of OCP/Collagen, B: newly formed bone. (Reprinted with permission and reorganized from Kamakura S. et al.: Dehydrothermal treatment of collagen influences on bone regeneration by octacalcium phosphate (OCP) collagen composites. J Tissue Eng Regen Med1:450-6, 2007, Blackwell Publishing Ltd.)

Figure 2. Radiographs and histological findings of treatment with OCP/Col and OCP/Col (-) at 4 and 12 weeks.

Bone matrix is mainly composed of inorganic components, such as bone minerals and organic components, such as type I collagen. Collagen is a clinically well-established and Food and Drug Administration (FDA) -approved matrix in the treatment of burns and chronic wounds [8, 67]. It is well known that collagen is a cell infiltration matrix with excellent biocompatibility [68], which degrades into physiological end products. A natural extracellular matrix such as collagen has a structural function to resist tensile and compressive stress [67] and can induce the nucleation of orientated apatite crystals [69, 70]. Althogh the bone-tissue reaction of OCP and collagen-conjugated sponge, which contains a relatively low content of differently manufactured OCP, was briefly reported previously [71], we have developed a synthetic biomimetic composite constructed of synthetic OCP and porcine atelocollagen sponge [72].

Firstly, OCP and collagen composites (OCP/Collagen) with dehydrothermal treatment (DHT) (OCP/Col) or OCP/Collagen without DHT (OCP/Col (-)) were prepared. OCP was directly synthesized by mixing calcium solution and phosphate solution [49], and the dried cake was ground and sieved into granules (particle sizes of 300-500 μm). The granules were sterilized by heating at 120˚C for 2 h. Such heating does not affect physical properties such as the crystalline structure or specific surface area of OCP granules [73]. Collagen was prepared from NMP collagen PS (Nippon Meat Packers, Inc., Tsukuba, Ibaraki, Japan), a lyophilized powder of pepsin-digested atelocollagen isolated from the porcine dermis that contains type I and III collagen. NMP collagen PS was dissolved in distilled water and adjusted to 3 % of the final concentration at pH 7.4 from the acidic solution. OCP/Collagen was prepared from NMP collagen PS and OCP granules. OCP was added to the concentrated collagen and mixed. The OCP/Collagen mixture was then lyophilized and a disk molded (9 mm diameter 1 mm thick: 14 mg, OCP 10.5 mg, Collagen 3.5 mg) (OCP/Col (-)). OCP/Col was prepared under DHT (150 ˚C, 24 h) for molded OCP/Col (-) in a vacuum drying oven. Finally, molded disks of OCP/Col or OCP/Col (-) were sterilized using electron beam irradiation (5 kGy). The prepared OCP/Collagen had better handling performance than granules of OCP.

The X-ray diffraction (XRD) patterns of both OCP/Col and OCP/Col (-) retained their characteristic (100) reflection OCP at 2θ = 4.7 [74]. OCP/Col (-) showed a reduced primary (100) peak of OCP. OCP/Col revealed not only collapse of the (100) peak with a shift from 4.7 to 5.3 degrees but also significant reduction of the (700) peak around 33.6 degrees, and had a HA-like pattern. Reduction of the (100) peak may correspond to the enhancement of hydrolysis, and simultaneous reduction of the (700) peak of OCP has been reported when OCP was heated from 150 ˚C to 200 ˚C, which was attributed to the partial loss of water molecules in OCP [75]. Because the reduction of the (700) peak and fusion with the (420) peak around 31.5 degrees of OCP was involved in the change through the hydrated layer of the OCP structure [36], the structural change of the hydrated layer of OCP would occur during DHT. In Fourier transform infrared (FTIR) spectroscopy patterns of OCP/Collagen, two distinct bands derived from amide at 1554 and 1650 cm^{-1} were recognized in both OCP/Col and OCP/Col (-). Spectral peaks originating from orthophosphate around 1030-1130 cm^{-1} in the original OCP tended to be indistinct in OCP/Col, whereas they remained unchanged in OCP/Col (-). This suggests that the conditions adopted in this DHT (150 ˚C, 24 h) affected the partial dehydration of OCP [75], which was also detected by XRD examination. While it was reported that dry heat of at least

160°C irreversibly damaged the helices of collagen [68], the collagen structure could be maintained unchanged even after DHT at 150°C.

The production of OCP/Collagen resulted in stable calcified collagen. When a disk of OCP/Col was stained with modified alizarin red S staining [76] to detect calcified tissue [77], the entire disk of OCP/Col was positively stained, including collagen and OCP granules [78]. This suggests that collagen in OCP/Col had been coated with the *de novo* precipitation of calcium phosphate, possibly re-precipitation of OCP, on collagen as a result of the interaction of soluble OCP with collagen.

(2) Influences of Dehydrothermal Treatment of Collagen in OCP/Col

It was reported that cross-linking of collagen, such as dehydrothermal treatment (DHT), is applied to optimize the mechanical properties and slow the biodegradation rate of collagen scaffolds [68, 79]. On the other hand, it has been reported that a slowly degrading collagen sponge at the defect site may physically obstruct the regeneration of bone tissue [80]. To examine whether DHT during the fabrication of OCP/Collagen might influence bone regeneration, OCP/Collagen with DHT (OCP/Col) or OCP/Collagen without DHT (OCP/Col (-)) was implanted into critical-sized defects in the rat crania. Bone regeneration by OCP/Col or OCP/Col (-) was investigated by radiographic, histological, and histomorphometrical analysis [74].

In OCP/Col, thorn-like radiopaque masses in week 4, and radiopaque condensation in weeks 12 were observed in a large area of the defect, whereas radiographic appearance of OCP/Col (-) was similar to that of OCP alone [52] (Figure 2). Bone regeneration by the implantation of OCP/Collagen composites was markedly influenced by the DHT, whereas calcium deposition on OCP/Collagen composites before implantation was similar with or without dehydrothermal treatment. In week 4, n-Bone% ± SD in OCP/Col and OCP/Col (-) was 29.2 ± 12.8 and 4.56 ± 3.19, respectively. In week 12, n-Bone% ± SD in OCP/Col and OCP/Col (-) was 52.5 ± 13.7 and 15.4 ± 13.7, respectively. N-Bone% treated with OCP/Col was significantly higher than with OCP/Col (-) 4 and 12 weeks after implantation, because rapid degradation of the implanted collagen of OCP/Col (-) elicited the disappearance of the scaffold for bone regeneration. This suggests that the collapse of implanted collagen would make it difficult to sustain the framework of newly formed bone, and the stiffness of calcified collagen in OCP/Collagen would be more important than the existence of calcified collagen to enhance bone regeneration by OCP/Collagen composites.

Because a recent study showed that synthetic OCP facilitates in vitro osteoblastic cell differentiation in an OCP dose-dependent manner [38, 39, 47, 48], the competence of cell attachment and proliferation was verified by mouse bone marrow-derived stromal ST-2 cells which were cultured in dishes pre-coated with OCP/Col or OCP alone with different OCP contents. OCP/Col promoted the proliferation and attachment of ST-2 cells more than OCP alone regardless of OCP content, and Fourier transform infrared spectroscopy analysis of the coating after incubation indicated that OCP tended to convert to apatite regardless of the presence of collagen [81].

(3) Bone Regeneration by Synthetic OCP and Collagen Composites

Bone regeneration by OCP/Col was compared with OCP granules and collagen disks (Col). OCP/Col disks, 15 mg OCP granules, or collagen disks (9 mm diameter 1 mm thick: 3.5 mg) prepared from NMP collagen PS with DHT (150 ˚C, 24 h) were implanted into rat calvarial critical-sized defects. Radiographic examination showed rapid amalgamation and condensation of radiopacity in OCP/Col, whereas OCP/Col itself had no radiopacity. In week 2 of OCP/Col, numerous cells invaded and adhered to the reticulum of OCP/Col, and bone matrix was secreted on the implanted collagen of OCP/Col. In week 4, bone matrix filled up the reticulum of OCP/Col. In week 8, the newly formed bone matured and increased (Figure 3). Bone remodeling with the resorption of OCP and implanted collagen was observed. The resorption of implanted collagen appeared to be delayed compared to the implantation of collagen. The synergistic effects of OCP/Col were the initiation of bone regeneration on stable calcified collagen in OCP/Col, which resulted from the interaction of OCP with collagen, and bone nucleation by OCP. This correlated with the characteristic bone regenerative pattern of OCP/Col.

Histomorphometrical analysis showed that n-Bone% treated with OCP/Col (48.4 ± 5.14%) was significantly higher than with OCP (27.6 ± 4.04%) or collagen (27.4 ± 5.69%) in week 8. In the OCP-treated group, isolated radiopacity amalgamated and increased, and implanted OCP was concentrically surrounded by newly formed bone with a lamellar structure in week 8. The amount of OCP remaining in the defect seemed to have decreased. In the collagen-treated group, a small amount of isolated radiopacity and newly formed bone were scattered throughout the defect in week 8 (Figure 4). The implanted collagen in OCP/Col was secure and difficult to resorb, whereas the implanted collagen in collagen was detached and tended to degrade. The collapse of implanted collagen would make it difficult to sustain the framework for newly formed bone.

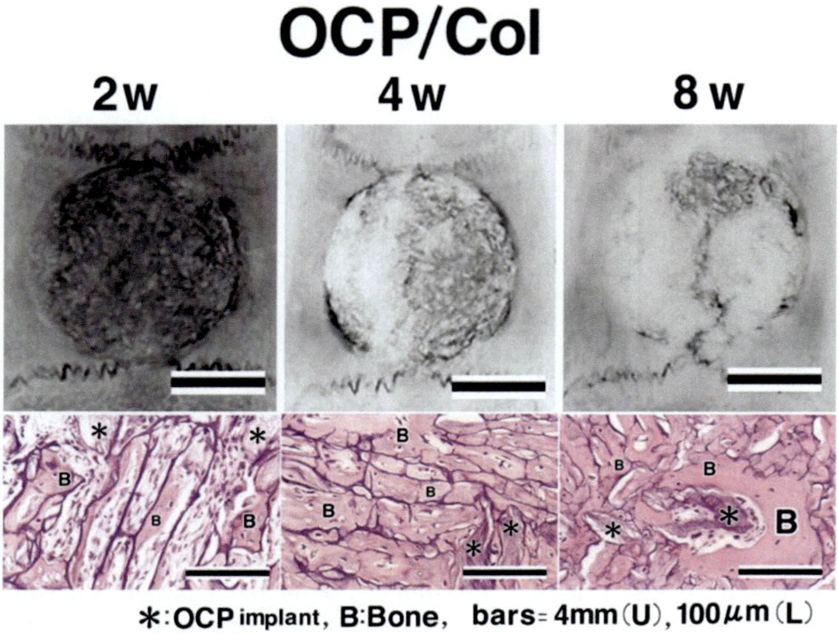

Thorn-like radiopacity forms in large defect areas in week 2, and the amalgamated radiopacity condenses and increases in weeks 4 and 8. Numerous cells invade the reticulum of OCP/Col and bone matrix is mostly secreted on implanted collagen of OCP/Col in week 2. Bone matrix, which is secreted by osteoblasts, fills up the space in the reticulum of OCP/Col in week 4. The increased newly formed bone (B) matures and bone remodeling with the resorption of OCP/Col is observed in week 8. * implanted OCP of OCP/Col, B: newly formed bone. (Reprinted with permission and reorganized from Kamakura S. et al.: Octacalcium phosphate combined with collagen orthotopically enhances bone regeneration. J Biomed Mater Res B Appl Biomater. 79:210-217, 2006, John Wiley & Sons, Inc.)

Figure 3. Radiographs and histological findings of treatment with OCP/Col at 2, 4 and 8 weeks.

We have also investigated whether regenerated bone by OCP/Col would be stable for a long period [82]. After 24 weeks of OCP/Col implantation, radiopaque figure and newly formed bone were observed throughout the defect in

OCP/Col (Figure 5), suggesting that the regenerated bone was stable after 24 weeks of OCP/Col implantation. The implanted OCP/Col tended to be resorbed and was replaced by newly formed bone where OCP/Col was presented. The reticulum of the implanted collagen degraded and became indistinct, whereas the implanted granules of OCP became small but persisted. Degradation of the implanted collagen reticulum suggests that implanted OCP/Col would be gradually resorbed and replaced by newly formed bone; however, the remaining granules of implanted OCP might resist resorption, because they converted to the apatitic phase, which is less resorbable than OCP itself.

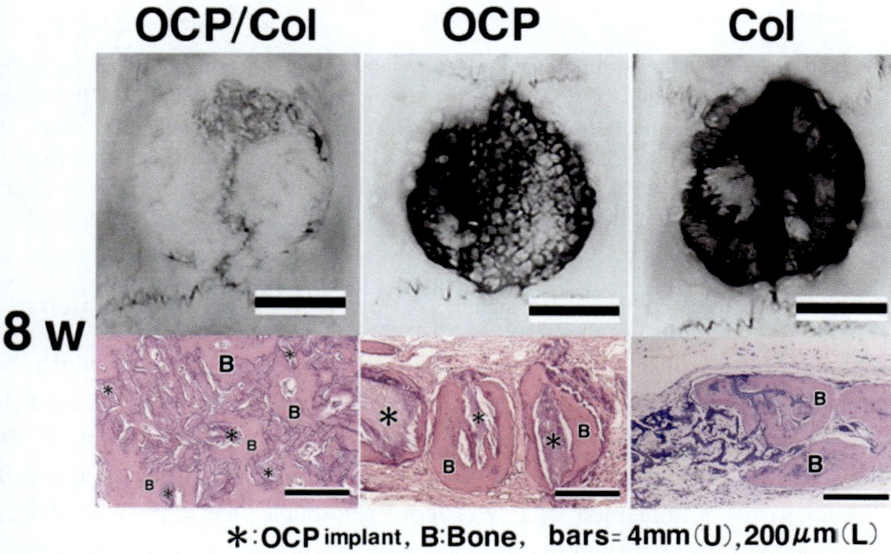

In OCP/Col, a larger area of condensed radiopacity is apparent and newly formed bone (B) with the resorption of OCP/Col is observed. In OCP, radiopacity is formed by the amalgamation of isolated radiopacity and the implanted OCP is concentrically surrounded by newly formed bone with a lamellar structure. In collagen-treated group (Col), isolated radiopacity is scattered throughout the defect, and newly formed bone is sporadically observed with collapse of the reticulum of collagen. * implanted granules of OCP, B: newly formed bone. (Reprinted with permission and reorganized from Kamakura S. et al.: Octacalcium phosphate combined with collagen orthotopically enhances bone regeneration. J Biomed Mater Res B Appl Biomater. 79:210-217, 2006, John Wiley & Sons, Inc.)

Figure 4. Radiographs and histological findings of treatment with OCP/Col, OCP, and Col at 8 weeks.

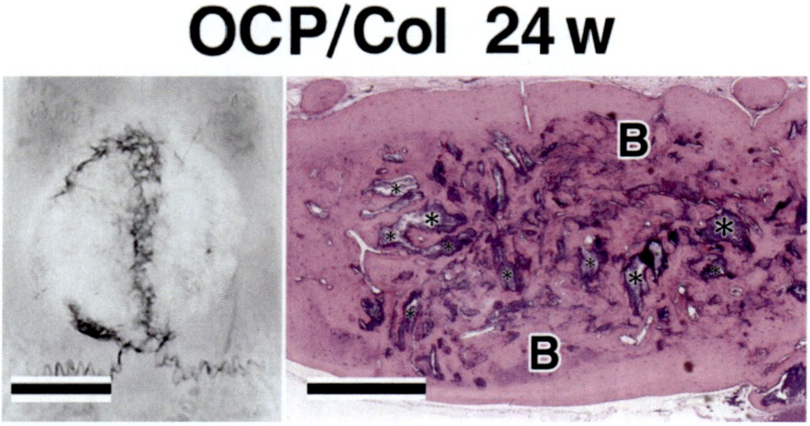

Radiopaque figure occupy throughout the defect, and the regenerated bone (B) is stable and matured. Implanted OCP/Col tends to be resorbed and the reticulum of implanted collagen degrades and becomes indistinct. Bars = 4 mm (left), 500 μm. (right), * implanted OCP of OCP/Col, B: newly formed bone. (Reprinted with permission and reorganized from Kamakura S. et al.: Stability of regenerated bone by octacalcium phosphate (OCP) combined with collagen. Key Eng Mater 330-332:1315-8, 2007. Trans Tech Publications)

Figure 5. Radiographs and histological findings of treatment with OCP/Col at 24 weeks.

Sintered synthetic HA ceramics and β-TCP are bone substitutes which are already applied for clinical use [7], and a sintered synthetic HA ceramic collagen composite or β-TCP collagen composite has exhibited excellent properties [83-85]. Because synthetic OCP enhanced bone regeneration more than HA and β-TCP, and likewise implanted OCP was more resorbable than implanted β-TCP and HA [58], bone regeneration by OCP/Col, β-TCP collagen composite (β-TCP/Col) or synthetic sintered HA collagen composite (HA/Col) was investigated by implanting in critical-sized rat calvarial defects [78]. Osteogenecity among calcium phosphate composites should be assessed by filling the same volume of composite, especially the same volume of calcium phosphate, to the same volume of the bone defect, whereas the evaluation of biomaterial itself may affect the nature of Ca-P via its microstructure, its dissolution rate, and its specific interactions with body fluids.

Sintered synthetic HA ceramics and β-TCP were derived from commercially available bone substitutes. OCP/Col, β-TCP/Col, or synthetic sintered HA/Col, which have the same size (particle sizes of 300-500 μm) and volume (9 mm diameter 1 mm thick) of each calcium phosphate granule, was prepared because the amount of these biomaterials in the clinical setting would be defined by the volume of bone defects. The prepared materials then underwent dehydrothermal treatment (150 ˚C, 24 h) and electron beam irradiation (5 kGy).

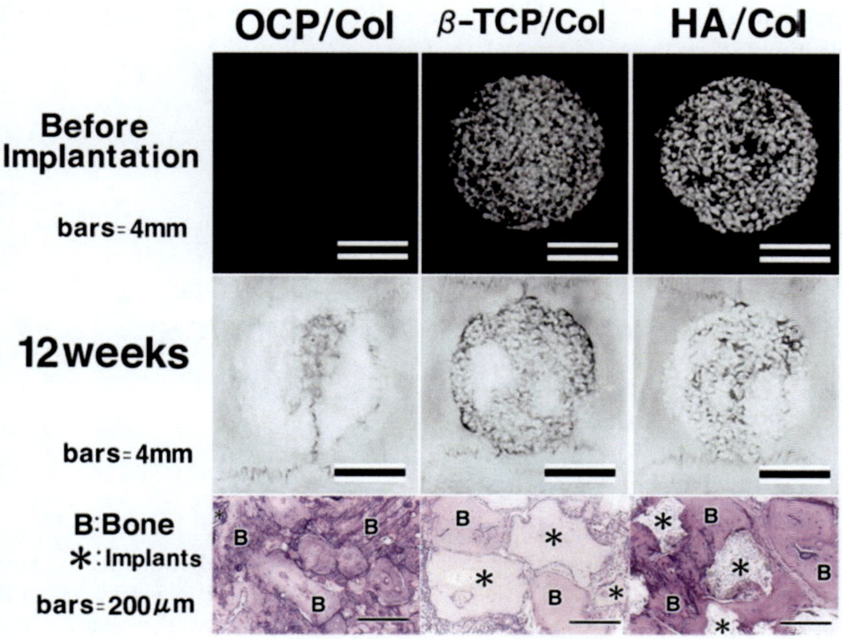

Under standardized conditions, OCP/Col disks before implantation show no radiopacity, whereas β-TCP/Col or HA/Col disks before implantation show granulous radiopacity throughout the disk. In OCP/Col, thorn-like radiopaque masses are condensed, and newly formed bone (B) is abundantly observed with active bone remodeling and resorption of implanted OCP/Col. In β-TCP/Col and HA/Col, larger radiopacity is formed by the amalgamation of isolated radiopacity. Newly formed bone (B) is observed between the implanted granules of β-TCP (*) or HA (*), and resorption of implanted β-TCP (*) or HA (*) is not clearly observed. (Reprinted with permission and reorganized from Kamakura S. et al.: The primacy of octacalcium phosphate collagen composites in bone regeneration. J Biomed Mater Res A83:725-33, 2007, John Wiley & Sons, Inc.)

Figure 6. Radiographs and histological findings of treatment with OCP/Col, β-TCP/Col or HA/Col at 12 weeks.

During the fabrication of OCP/Col, β-TCP/Col or synthetic sintered HA/Col, it was ascertained that the XRD pattern derived from OCP/Col showed a collapsed and reduced primary (100) peak of OCP due to partial loss of water molecules in OCP [75], although that derived from β-TCP/Col or HA/Col was almost the same as the original granules of β-TCP or HA, respectively. This suggests that OCP would interact with collagen during the fabrication of OCP/Col more than β-TCP or HA. OCP would be deposited on mixed collagen during the fabrication of OCP/Col [86], because collagen is a good substrate for OCP nucleation [87, 88].

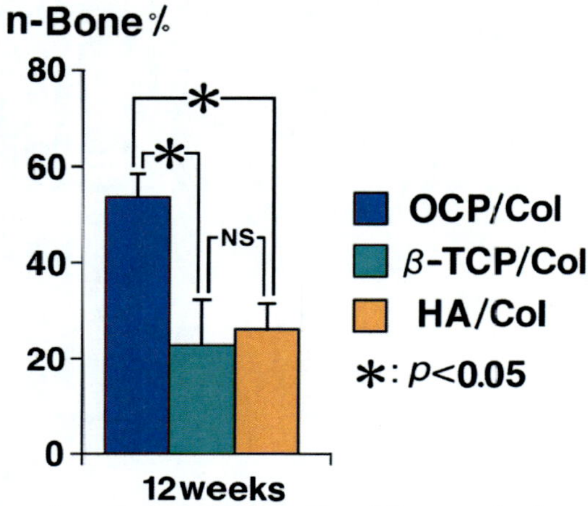

The percentage of new bone in the defect (n-Bone%) ± standard error (SE) is shown in OCP/Col, β-TCP/Col, and HA/Col. A significant difference is seen between OCP/Col and β-TCP/Col, and between OCP/Col and HA/Col. Data are the means ± SE of 5 specimens. * $p < 0.05$ (Reprinted with permission and reorganized from Kamakura S. et al.: The primacy of octacalcium phosphate collagen composites in bone regeneration. J Biomed Mater Res A83:725-33, 2007, John Wiley & Sons, Inc.)

Figure 7. Quantitative analysis of newly formed bone in defects at 12 weeks.

Under standardized conditions, OCP/Col disks before implantation showed no radiopacity, whereas β-TCP/Col or HA/Col disks before implantation showed granulous radiopacity throughout the disk (Figure 6). In OCP/Col, condensed radiopaque masses were observed throughout the defect in week 12, whereas isolated granulous radiopacity was intermingled with larger amalgamated radiopacity in week 12 of β-TCP/Col and HA/Col. In OCP/Col-treated defects, newly formed bone was abundantly observed by nucleating secured calcified collagen, accompanied with active bone remodeling and resorption of implanted OCP/Col (Figure 6). Newly formed bone in β-TCP/Col-treated defects was mainly based on poorly calcified collagen, relatively less than in OCP/Col. Although the implanted β-TCP was surrounded by many cells, including multinucleated giant cells, resorption of implanted β-TCP did not clearly progress. Similarly, new bone was observed in HA/Col-treated defects relatively less than in OCP/Col. Implanted HA retained an irregular shape without resorptive activity. This suggests that bone regeneration by OCP/Col was based on secured calcified collagen and bone nucleation by OCP, whereas bone regeneration by β-TCP/Col or HA/Col was initiated by poorly calcified collagen and osteoconductivity by β-TCP or HA.

Histomorphometrical analysis showed that the percentage of newly formed bone in defects (n-Bone%) treated with OCP/Col (53.6 ± 4.65 %) was significantly higher than with β-TCP/Col (23.0 ± 9.05%) and HA/Col (26.2 ± 5.20) in week 12 (Figure 7), although there were no significant differences among them (OCP/Col: 28.0 ± 10.1%, β-TCP/Col: 19.4 ± 3.12%, HA/Col : 20.9 ± 2.99%) in week 4. The percentage of remaining granules of implants in the defect (r-GI%) treated with OCP/Col was significantly lower than with β-TCP/Col and HA/Col in week 4 (OCP/Col: 9.43 ± 1.36%, β-TCP/Col: 30.9 ± 2.28%, HA/Col : 29.2 ± 0.91%) and 12 (OCP/Col: 6.22 ± 0.39%, β-TCP/Col: 24.7 ± 1.09%, HA/Col : 24.8 ± 2.02%). These results suggest that OCP/Col would better enhance bone regeneration than β-TCP/Col or HA/Col in week 12, and that implanted OCP granules in OCP/Col are more resorbable than β-TCP in β-TCP/Col or HA in HA/Col. These results are linked with the point that the difference in sintering temperature influenced the extent of bone regeneration [89] and a composite consisting of precipitated bone-like HA and collagen could be biodegradable [90], whereas OCP biodegradation was faster than non-sintered HA [58] or sintered HA ceramic [54] used in this study.

5. Conclusion

The presented outcome suggests that OCP/Col would be a better bone regenerative material than clinically applied bone substitutes, such as HA or β-TCP, and their composites with collagen. OCP/Col is easy to handle and could promptly enhance bone regeneration without cell transplantation and exogenous osteogenic cytokines. Also, OCP/Col is involved in osteoblastic cell differentiation and has biodegradable property. Because it was recently reported that bone regeneration of a critical-sized bone defect in dog calvaria by OCP/Col was enhanced for 3 to 6 months [91], the clinical application of cost-effective bone regenerative therapy by OCP/Col is expected in the future.

Since subtle compositional and structural changes of OCP or the microstructure of OCP would affect their bone regenerative properties [92, 93] and the osteoconductive characteristics of OCP/Col are exhibited in an OCP dose-dependent manner [81], development of a new type of OCP/Col with more advanced bone regenerative properties is anticipated.

Acknowledgments

We wish to thank Mr. Kazuo Sasaki, R & D Center, Nippon Meat Packers, Inc., Tsukuba, Ibaraki, Japan, for his excellent technical advice and assistance in this study.

References

[1] Boyne, PJ. Application of bone morphogenetic proteins in the treatment of clinical oral and maxillofacial osseous defects. *J Bone Joint Surg Am.*, 2001, 83A, S146-S50.

[2] Alsberg, E; Hill, EE; Mooney, DJ. Craniofacial tissue engineering. *Crit Rev Oral Biol Med.*, 2001, 12, 64-75.

[3] Petite, H; Viateau, V; Bensaid, W; Meunier, A; de Pollak, C; Bourguignon, M; et al. Tissue-engineered bone regeneration. *Nat Biotechnol.*, 2000, 18, 959-63.

[4] Block, JE; Thorn, MR. Clinical indications of calcium-phosphate biomaterials and related composites for orthopedic procedures. *Calcif Tissue Int.*, 2000, 66, 234-8.

[5] LeGeros, RZ. Properties of osteoconductive biomaterials: calcium phosphates. *Clin Orthop Relat Res.*, 2002, 395, 81-98.

[6] Burg, KJL; Porter, S; Kellam, JF. Biomaterial developments for bone tissue engineering. *Biomaterials*, 2000, 21, 2347-59.

[7] Kokubo, T; Kim, HM; Kawashita, M. Novel bioactive materials with different mechanical properties. *Biomaterials*, 2003, 24, 2161-75.

[8] Lee, CH; Singla, A; Lee, Y. Biomedical applications of collagen. *Int J Pharm.*, 2001, 221, 1-22.

[9] Bucholz, RW. Nonallograft osteoconductive bone graft substitutes. *Clin Orthop Relat Res.*, 2002, 44-52.

[10] Ogose, A; Hotta, T; Kawashima, H; Kondo, N; Gu, WG; Kamura, T; et al. Comparison of hydroxyapatite and beta tricalcium phosphate as bone substitutes after excision of bone tumors. *J Biomed Mater Res B Appl Biomater.*, 2005, 72B, 94-101.

[11] Chow, L. Calcium phosphate cements. In: L. Chow, & E. Eanes (Eds.), *Octacalcium phosphate*. Basel: Karger, 2001, 148-64.

[12] Oonishi, H; Hench, LL; Wilson, J; Sugihara, F; Tsuji, E; Matsuura, M; et al. Quantitative comparison of bone growth behavior in granules of Bioglass (R), A-W glass-ceramic, and hydroxyapatite. *J Biomed Mater Res.*, 2000, 51, 37-46.

[13] Daculsi, G; Legeros, RZ; Heughebaert, M; Barbieux, I. Formation of carbonate-apatite crystals after implantation of calcium-phosphate ceramics *Calcif Tissue Int.*, 1990, 46, 20-7.

[14] Aoba, T. Recent observations on enamel crystal formation during mammalian amelogenesis. *Anat Rec.*, 1996, 245, 208-18.

[15] Kim, HM; Rey, C; Glimcher, MJ. Isolation of calcium-phosphate crystals of bone by non-aqueous methods at low temperature. *J Bone Miner Res.*, 1995, 10, 1589-601.

[16] Kuhn, LT; Wu, Y; Rey, C; Gerstenfeld, LC; Grynpas, MD; Ackerman, JL; et al. Structure, composition, and maturation of newly deposited calcium-phosphate crystals in chicken osteoblast cell cultures. *J Bone Miner Res.*, 2000, 15, 1301-9.

[17] Wu, Y; Ackerman, JL; Strawich, ES; Rey, C; Kim, HM; Glimcher, MJ. Phosphate ions in bone: identification of a calcium-organic phosphate complex by 31P solid-state NMR spectroscopy at early stages of mineralization. *Calcif Tissue Int.*, 2003, 72, 610-26.

[18] Wu, Y; Glimcher, MJ; Rey, C; Ackerman, JL. A unique protonated phosphate group in bone mineral not present in synthetic calcium phosphates. Identification by phosphorus-31 solid state NMR spectroscopy. *J Mol Biol.*, 1994, 244, 423-35.

[19] Betts, F; Blumenthal, NC; Posner, AS. Bone mineralization. *J Cryst Growth*, 1981, 53, 63-73.

[20] Brown, W; Smith, J; Lehr, J; Frazier, A. Crystallographic and chemical relations between octacalcium phosphate and hydroxyapatite. *Nature*, 1962, 196, 1050-5.

[21] Brown, WE. Crystal growth of bone mineral. *Clin Orthop Relat Res.*, 1966, 44, 205-20.

[22] Johnsson, MS; Nancollas, GH. The role of brushite and octacalcium phosphate in apatite formation. *Crit Rev Oral Biol Med.*, 1992, 3, 61-82.

[23] Brown, WE; Mathew, M; Tung, MS. Crystal-chemistry of octacalcium phosphate. *Progress in Crystal Growth and Characterization of Materials*, 1981, 4, 59-87.

[24] Meyer, JL; Eanes, ED. A thermodynamic analysis of the secondary transition in the spontaneous precipitation of calcium phosphate. *Calcif Tissue Res.*, 1978, 25, 209-16.

[25] LeGeros, RZ; Daculsi, G; Orly, I; Abergas, T; Torres, W. Solution-mediated transformation of octacalcium phosphate (OCP) to apatite. *Scanning Microsc.*, 1989, 3, 129-37, discussion 37-8.

[26] Tomazic, B; Tung, M; Gregory, T; Brown, W. Mechanism of hydrolysis of octacalcium phosphate. *Scanning Microsc.*, 1989, 3, 119-27.

[27] Brown, WE; Eidelman, N; Tomazic, B. Octacalcium phosphate as a precursor in biomineral formation. *Adv Dent Res.*, 1987, 1, 306-13.

[28] Suzuki, O; Imaizumi, H; Kamakura, S; Katagiri, T. Bone regeneration by synthetic octacalcium phosphate and its role in biological mineralization. *Curr Med Chem.*, 2008, 15, 305-13.

[29] Iijima, M; Tohda, H; Suzuki, H; Yanagisawa, T; Moriwaki, Y. Effects of F- on apatite-octacalcium phosphate intergrowth and crystal morphology in a model system of tooth enamel formation. *Calcif Tissue Int.*, 1992, 50, 357-61.

[30] Tseng, YH; Mou, CY; Chan, JC. Solid-state NMR study of the transformation of octacalcium phosphate to hydroxyapatite: a mechanistic model for central dark line formation. *J Am Chem Soc.*, 2006, 128, 6909-18.

[31] Bodier-Houlle, P; Steuer, P; Voegel, JC; Cuisinier, FJG. First experimental evidence for human dentine crystal formation involving conversion of octacalcium phosphate to hydroxyapatite. *Acta Crystallogr Sect D-Biol Crystallogr.*, 1998, 54, 1377-81.

[32] Tohda, H; Yamada, M; Yamaguchi, Y; Yanagisawa, T. High-resolution electron microscopical observations of initial enamel crystals. *J Electron Microsc (Tokyo)*, 1997, 46, 97-101.

[33] Crane, N; Popescu, V; Morris, M; Steenhuis, P; Ignelzi, MJ. Raman spectroscopic evidence for octacalcium phosphate and other transient mineral species deposited during intramembranous mineralization. *Bone*, 2006, 39, 434-42.

[34] Chickerur, NS; Tung, MS; Brown, WE. A mechanism for incorporation of carbonate into apatite. *Calcif Tissue Int.*, 1980, 32, 55-62.

[35] Siew, C; Gruninger, SE; Chow, LC; Brown, WE. Procedure for the study of acidic calcium-phosphate precursor phases in enamel mineral formation. *Calcif Tissue Int.*, 1992, 50, 144-8.

[36] Suzuki, O; Yagishita, H; Amano, T; Aoba, T. Reversible structural changes of octacalcium phosphate and labile acid phosphate. *J Dent Res.*, 1995, 74, 1764-9.

[37] Suzuki, O; Yagishita, H; Yamazaki, M; Aoba, T. Adsorption of bovine serum-albumin onto octacalcium phosphate and its hydrolyzates. *Cells Mater.*, 1995, 5, 45-54.

[38] Suzuki, O; Kamakura, S; Katagiri, T; Nakamura, M; Zhao, BH; Honda, Y; et al. Bone formation enhanced by implanted octacalcium phosphate involving conversion into Ca-deficient hydroxyapatite. *Biomaterials*, 2006, 27, 2671-81.

[39] Anada, T; Kumagai, T; Honda, Y; Masuda, T; Kamijo, R; Kamakura, S; et al. Dose-dependent osteogenic effect of octacalcium phosphate on mouse bone marrow stromal cells. *Tissue Eng Part A.*, 2008, 14, 965-78.

[40] Barrere, F; Layrolle, P; van Blitterswijk, CA; de Groot, K. Biomimetic calcium phosphate coatings on Ti6AI4V: a crystal growth study of octacalcium phosphate and inhibition by Mg^{2+} and $HCO3^{-}$. *Bone*, 1999, 25, 107S-11S.

[41] Barrere, F; van der Valk, CM; Dalmeijer, RA; Meijer, G; van Blitterswijk, CA; de Groot, K; et al. Osteogenecity of octacalcium phosphate coatings applied on porous metal implants. *J Biomed Mater Res A.*, 2003, 66, 779-88.

[42] Barrere, F; van der Valk, CM; Dalmeijer, RA; van Blitterswijk, CA; de Groot, K; Layrolle, P. In vitro and in vivo degradation of biomimetic octacalcium phosphate and carbonate apatite coatings on titanium implants. *J Biomed Mater Res A.*, 2003, 64, 378-87.

[43] Bigi, A; Bracci, B; Cuisinier, F; Elkaim, R; Fini, M; Mayer, I; et al. Human osteoblast response to pulsed laser deposited calcium phosphate coatings. *Biomaterials*, 2005, 26, 2381-9.

[44] Dekker, RJ; de Bruijn, JD; Stigter, M; Barrere, F; Layrolle P, van Blitterswijk, CA. Bone tissue engineering on amorphous carbonated apatite and crystalline octacalcium phosphate-coated titanium discs. *Biomaterials*, 2005, 26, 5231-9.

[45] Habibovic, P; Van der Valk, CM; Van Blitterswijk, CA; De Groot, K; Meijer, G. Influence of octacalcium phosphate coating on osteoinductive properties of biomaterials. *J Mater Sci Mater Med.*, 2004, 15, 373-80.

[46] Lin, S; LeGeros, RZ; LeGeros, JP. Adherent octacalciumphosphate coating on titanium alloy using modulated electrochemical deposition method. *J Biomed Mater Res A.*, 2003, 66, 819-28.

[47] Liu, Y; Cooper, PR; Barralet, JE; Shelton, RM. Influence of calcium phosphate crystal assemblies on the proliferation and osteogenic gene expression of rat bone marrow stromal cells. *Biomaterials*, 2007, 28, 1393-403.

[48] Shelton, RM; Liu, Y; Cooper, PR; Gbureck, U; German, MJ; Barralet, JE. Bone marrow cell gene expression and tissue construct assembly using octacalcium phosphate microscaffolds. *Biomaterials*, 2006, 27, 2874-81.

[49] Suzuki, O; Nakamura, M; Miyasaka, Y; Kagayama, M; Sakurai, M. Bone formation on synthetic precursors of hydroxyapatite. *Tohoku J Exp Med.*, 1991, 164, 37-50.

[50] Bosch, C; Melsen, B; Vargervik, K. Guided bone regeneration in calvarial bone defects using polytetrafluoroethylene membranes. *Cleft Palate Craniofac J.*, 1995, 32, 311-7.

[51] Sasano, Y; Kamakura, S; Homma, H; Suzuki, O; Mizoguchi, I; Kagayama, M. Implanted octacalcium phosphate (OCP) stimulates osteogenesis by osteoblastic cells and/or committed osteoprogenitors in rat calvarial periosteum. *Anat Rec.*, 1999, 256, 1-6.

[52] Kamakura, S; Sasano, Y; Homma, H; Suzuki, O; Kagayama, M; Motegi, K. Implantation of octacalcium phosphate (OCP) in rat skull defects enhances bone repair. *J Dent Res.*, 1999, 78, 1682-7.

[53] Kamakura, S; Sasano, Y; Homma, H; Suzuki, O; Kagayama, M; Motegi, K. Implantation of octacalcium phosphate nucleates isolated bone formation in rat skull defects. *Oral Dis.*, 2001, 7, 259-65.

[54] Imaizumi, H; Sakurai, M; Kashimoto, O; Kikawa, T; Suzuki, O. Comparative study on osteoconductivity by synthetic octacalcium phosphate and sintered hydroxyapatite in rabbit bone marrow. *Calcif Tissue Int.*, 2006, 78, 45-54.

[55] Kamakura, S; Sasano, Y; HommaOhki, H; Nakamura, M; Suzuki, O; Kagayama, M; et al. Multinucleated giant cells recruited by implantation of octacalcium phosphate (OCP) in rat bone marrow share ultrastructural characteristics with osteoclasts. *J Electron Microsc (Tokyo)*, 1997, 46, 397-403.

[56] Kikawa, T; Kashimoto, O; Imaizumi, H; Kokubun, S; Suzuki, O. Intramembranous bone tissue response to biodegradable octacalcium phosphate implant. *Acta Biomater.*, 2008, 5, 1756-66.

[57] Gazit, D; Turgeman, G; Kelley, P; Wang, E; Jalenak, M; Zilberman, Y; et al. Engineered pluripotent mesenchymal cells integrate and differentiate in regenerating bone: a novel cell-mediated gene therapy. *J Gene Med.*, 1999, 1, 121-33.

[58] Kamakura, S; Sasano, Y; Shimizu, T; Hatori, K; Suzuki, O; Kagayama, M; et al. Implanted octacalcium phosphate is more resorbable than beta-tricalcium phosphate and hydroxyapatite. *J Biomed Mater Res.*, 2002, 59, 29-34.

[59] Rose, FR; Oreffo, RO. Bone tissue engineering: hope vs hype. *Biochem Biophys Res Commun.*, 2002, 292, 1-7.

[60] Cowan, CM; Shi, YY; Aalami, OO; Chou, YF; Mari, C; Thomas, R; et al. Adipose-derived adult stromal cells heal critical-size mouse calvarial defects. *Nat Biotechnol.*, 2004, 22, 560-7.

[61] Wozney, JM; Rosen, V; Celeste, AJ; Mitsock, LM; Whitters, MJ; Kriz, RW; et al. Novel regulators of bone formation: molecular clones and activities. *Science*, 1988, 242, 1528-34.

[62] Wang, EA; Israel, DI; Kelly, S; Luxenberg, DP. Bone morphogenetic protein-2 causes commitment and differentiation in C3H10T1/2 and 3T3 cells. *Growth Factors*, 1993, 9, 57-71.

[63] Wang, EA; Rosen, V; D'Alessandro, JS; Bauduy, M; Cordes, P; Harada, T; et al. Recombinant human bone morphogenetic protein induces bone formation. *Proc Natl Acad Sci*, USA, 1990, 87, 2220-4.

[64] Uludag, H; Gao, T; Porter, TJ; Friess, W; Wozney, JM. Delivery systems for BMPs: factors contributing to protein retention at an application site. *J Bone Joint Surg Am.*, 2001, 83-A Suppl 1, S128-35.

[65] Kamakura, S; Nakajo, S; Suzuki, O; Sasano, Y. New scaffold for recombinant human bone morphogenetic protein-2. *J Biomed Mater Res A.*, 2004, 71A, 299-307.

[66] Kamakura, S; Sasano, Y; Nakamura, M; Suzuki, O; Ohki, H; Kagayama, M; et al. Initiation of alveolar ridge augmentation in the rat mandible by subperiosteal implantation of octacalcium phosphate. *Arch Oral Biol.*, 1996, 41, 1029-38.

[67] Lutolf, MP; Hubbell, JA. Synthetic biomaterials as instructive extracellular microenvironments for morphogenesis in tissue engineering. *Nat Biotechnol.*, 2005, 23, 47-55.

[68] Geiger, M; Li, RH; Friess, W. Collagen sponges for bone regeneration with rhBMP-2. *Adv Drug Deliv Rev.*, 2003, 55, 1613-29.

[69] Chaikof, EL; Matthew, H; Kohn, J; Mikos, AG; Prestwich, GD; Yip, CM. Biomaterials and scaffolds in reparative medicine. *Ann NY Acad Sci.*, 2002, 961, 96-105.

[70] Landis, W. An overview of vertebrate mineralization with emphasis on collagen-mineral interaction. *Gravit Space Biol Bull.*, 1999, 12, 15-26.

[71] Sugihara, F; Oonishi, H; Minamigawa, K; Maidai, Y; Tsuji, E; Yoshikawa, M; et al. Bone tissue reaction of octacalcium phosphate-collagen conjugated sponge. *Bioceramics*, 1996, 9, 399-402.

[72] Kamakura, S; Sasaki, K; Honda, Y; Anada, T; Suzuki, O. Octacalcium phosphate combined with collagen orthotopically enhances bone regeneration. *J Biomed Mater Res B Appl Biomater.*, 2006, 79, 210-7.

[73] Suzuki, O; Nakamura, M; Miyasaka, Y; Kagayama, M; Sakurai, M. *Maclura pomifera* agglutinin-binding glycoconjugates on converted apatite from synthetic octacalcium phosphate implanted into subperiosteal region of mouse calvaria. *Bone Miner.*, 1993, 20, 151-66.

[74] Kamakura, S; Sasaki, K; Honda, Y; Anada, T; Matsui, K; Echigo, S; et al. Dehydrothermal treatment of collagen influences on bone regeneration by octacalcium phosphate (OCP) collagen composites. *J Tissue Eng Regen Med.*, 2007, 1, 450-6.

[75] Nelson, DG; McLean, JD. High-resolution electron microscopy of octacalcium phosphate and its hydrolysis products. *Calcif Tissue Int.*, 1984, 36, 219-32.

[76] Kagayama, M; Sasano, Y; Sato, H; Kamakura, S; Motegi, K; Mizoguchi, I. Confocal microscopy of dentinal tubules in human tooth stained with alizarin red. *Anat Embryol (Berl)*, 1999, 199, 233-8.

[77] Bareggi, R; Grill, V; Sandrucci, MA; Baldini, G; Depol, A; Forabosco, A; et al. Developmental pathways of vertebral centra and neural arches in human embryos and fetuses. *Anat Embryol (Berl)*, 1993, 187, 139-44.

[78] Kamakura, S; Sasaki, K; Homma, T; Honda, Y; Anada, T; Echigo, S; et al. The primacy of octacalcium phosphate collagen composites in bone regeneration. *J Biomed Mater Res A.*, 2007, 83, 725-33.

[79] Ma, L; Gao, CY; Mao, ZW; Zhou, J; Shen, JC. Biodegradability and cell-mediated contraction of porous collagen scaffolds: The effect of lysine as a novel crosslinking bridge. *J Biomed Mater Res A.*, 2004, 71, 334-42.

[80] Ueda, H; Nakamura, T; Yamamoto, M; Nagata, N; Fukuda, S; Tabata, Y; et al. Repairing of rabbit skull defect by dehydrothermally crosslinked collagen sponges incorporating transforming growth factor beta 1. *J Control Release*, 2003, 88, 55-64.

[81] Kawai, T; Anada, T; Honda, Y; Kamakura, S; Matsui, K; Matsui, A; et al. Synthetic octacalcium phosphate augments bone regeneration correlated with its content in collagen scaffold. *Tissue Eng Part A.*, 2009, 15, 23-32.

[82] Kamakura, S; Sasaki, K; Honda, Y; Anada, T; Kawai, T; Matsui, K; et al. Stability of regenerated bone by octacalcium phosphate (OCP) combined with collagen. *Key Eng Mater.*, 2007, 330-332, 1315-8.

[83] Itoh, S; Kikuchi, M; Takakuda, K; Koyama, Y; Matsumoto, HN; Ichinose, S; et al. The biocompatibility and osteoconductive activity of a novel hydroxyapatite/collagen composite biomaterial, and its function as a carrier of rhBMP-2. *J Biomed Mater Res A.*, 2001, 54, 445-53.

[84] Matsuno, T; Nakamura, T; Kuremoto, K; Notazawa, S; Nakahara, T; Hashimoto, Y; et al. Development of beta-tricalcium phosphate/collagen sponge composite for bone regeneration. *Dent Mater J.*, 2006, 25, 138-44.

[85] Tenhuisen, KS; Martin, RI; Klimkiewicz, M; Brown, PW. Formation and properties of a synthetic bone composite - hydroxyapatite-collagen. *J Biomed Mater Res.*, 1995, 29, 803-10.

[86] Honda, Y; Anada, T; Kamakura, S; Nakamura, M; Sugawara, S; Suzuki, O. Elevated extracellular calcium stimulates secretion of bone morphogenetic protein 2 by a macrophage cell line. *Biochem Biophys Res Commun.*, 2006, 345, 1155-60.

[87] Combes, C; Rey, C; Freche, M. In vitro crystallization of octacalcium phosphate on type I collagen: influence of serum albumin. *J Mater Sci Mater Med.*, 1999, 10, 153-60.

[88] Iijima, M; Iijima, K; Moriwaki, Y; Kuboki, Y. Oriented growth of octacalcium phosphate crystals on type-I collagen fibrils under physiological conditions. *J Cryst Growth.*, 1994, 140, 91-9.

[89] Habibovic, P; Yuan, HP; van der Valk, CM; Meijer, G; van Blitterswijk, CA; de Groot, K. *3D microenvironment as essential element for osteoinduction by biomaterials*, 2005, 26, 3565-75.

[90] Kikuchi, M; Itoh, S; Ichinose, S; Shinomiya, K; Tanaka, J. Self-organization mechanism in a bone-like hydroxyapatite/collagen nanocomposite synthesized in vitro and its biological reaction in vivo. *Biomaterials*, 2001, 22, 1705-11.

[91] Kawai, T; Matsui, K; Iibuchi, S; Anada, T; Honda, Y; Sasaki, K; et al. Reconstruction of Critical-Sized Bone Defect in Dog Skull by Octacalcium Phosphate Combined with Collagen. *Clin Implant Dent Relat Res.*, 2009, in press (PMID: 19438952).

[92] Honda, Y; Anada, T; Kamakura, S; Morimoto, S; Kuriyagawa, T; Suzuki, O. The Effect of Microstructure of Octacalcium Phosphate on the Bone Regenerative Property. *Tissue Eng Part A.*, 2009, 15, 1965-73.

[93] Miyatake, N; Kishimoto, KN; Anada, T; Imaizumi, H; Itoi, E; Suzuki, O. Effect of partial hydrolysis of octacalcium phosphate on its osteoconductive characteristics. *Biomaterials*, 2009, 30, 1005-14.

In: Bone Regeneration: Growth Factors... ISBN: 978-1-60876-621-5
Editors: V. Legard et al., pp. 203-231

Chapter VII

Gelatin Based Microcarriers for Bone Tissue Engineering

***H. Declercq*[1], *E. Schacht*[2], *G. Vertenten*[3], *L. Vlaminck*[3], *F. Gasthuys*[3] *and M. Cornelissen*[1]**

[1]Department of Basic Medical Sciences, Faculty of Medicine and Health Sciences, Ghent University, De Pintelaan 185 (6B3), B-9000 Ghent, Belgium.

[2]Polymer Chemistry and Biomaterial Research Group, Department of Organic chemistry, Faculty of Sciences, Ghent University, Krijgslaan 281 (S4), B-9000 Ghent, Belgium.

[3]Department of Surgery and Anaesthiology of Domestic Animals, Faculty of Veterinary Medicine, Ghent University, Salisburylaan 133, B-9820 Merelbeke, Belgium.

Abstract

Bone tissue engineering requires a suitable osteoconductive and – inductive matrix and, dependent on the severity of the defect, additional sources of osteogenic cells. Traditionally, the osteoconductive matrix is a prefabricated three-dimensional scaffold (3D) composed of degradable polymers and/or ceramics. However, the increasing popularity of arthroscopic procedures and the requirement to bridge irregular bone defects resulted in great interest in fixation materials that are injectable, *in situ* forming and biodegradable.

To restore damaged bone defects additional bone-forming cells are needed. *In situ* forming materials (e.g. moldable calcium phosphate cements, injectable polymers hardening by redox- or photocrosslinking) capable of filling irregular bone defects are mostly inconvenient to mix directly with cells. The use of cell delivery systems which can be colonized by cells in which they are protected during the polymerization process of the biomaterial could be advantageous. The use of macroporous gelatin-based microcarriers in tissue engineering offers several advantages: (1) cells can be expanded in vitro, (2) biodegradable, (3) cells are in a "natural" 3D-environment and (4) cells within the interior of the microcarrier are in a protected environment.

In our study, bone marrow cells were cultured on *in situ* photopolymerizable scaffolds based on D, L-lactide, ε-caprolactone and trimethylene carbonate. Bone formation was demonstrated at the polymer interface. Cellular infiltration was possible but was limited to the edge of the polymer scaffold.

In the second part of this study, bone marrow cells were seeded on macroporous gelatin microcarriers (CultiSpher S) (diameter 130-380 μm). The cells proliferated and differentiated well on the microcarriers and were completely colonized after 14 days in culture. The polymer paste allowed to mix the cell seeded microcarriers and the porogen (e.g. gelatin) properly. After the photopolymerization process, microcarriers and gelatin were evenly distributed throughout the scaffold. After leaching out the gelatin, a porous scaffold was obtained. The microcarriers remained in the scaffold even after 7 days which demonstrates that they were well entrapped in the polymer. Dependent on the composite type, the encapsulated cells survived the handling of the viscous paste and the photopolymerization proces.

These cell seeded microcarriers combined with different combinations of polymer and alfa-tricalcium phosphate were tested in a unicortical tibial defect model in goats. The results demonstrated cell survival and proliferation in the polymer-substituted bone defects. Hence cell seeded microcarriers combined with *in situ* photopolymerizable materials are promising for bone regeneration.

1. Introduction

The increasing popularity of arthroscopic procedures in orthopaedics and the requirement to bridge irregular bone defects resulted in great interest in fixation materials that are injectable, *in situ* forming and biodegradable. The injectable material should be administered by a physician in the desired shape within the wound before the material hardens. Such material should not only be biocompatible but also bioactive. This enables the integration with the residual bone tissue and therefore facilitates and promotes the replacement or the recovery of the bone within a short period of time.

The materials of choice in this research project are injectable, *in situ* forming composites based on photopolymerizable methacrylate-endcapped polyesters and *in situ* setting calcium phosphate cements (Figure 1 and 2).

In the Polymer Materials Research Group, a variety of *in situ*, photopolymerizable 3D scaffolds were synthesized with variables of polymer type (polyethylene glycol, lactide, glycolide, ε-caprolactone, trimethylene carbonate,...), copolymer ratio and initiator systems. The scaffolds were compared regarding degradation time, mechanical strength and biocompatibility. Some materials were selected based on methycrylate-endcapped polyesters (D, L-lactide, ε-caprolactone and trimethylene carbonate) [Jackers G *et al.*, 2002; Schacht E *et al.*, 2004; Gorski T *et al.*, 2005].

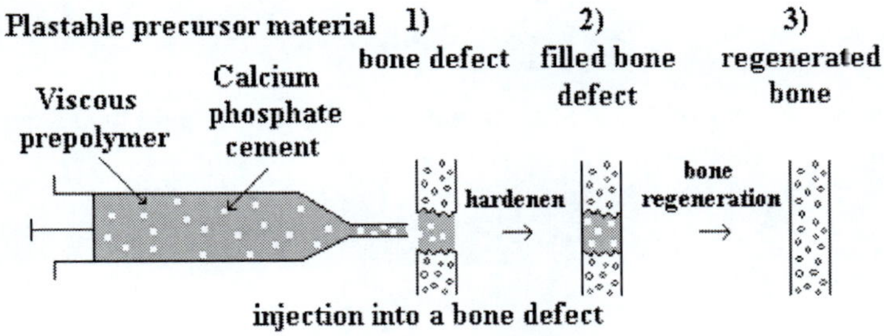

Figure 1. Principle of injectable, photopolymerizable composite materials for bone tissue engineering. The composite is applied in the bone cavity, is hardened in situ and subsequently promote bone regeneration.

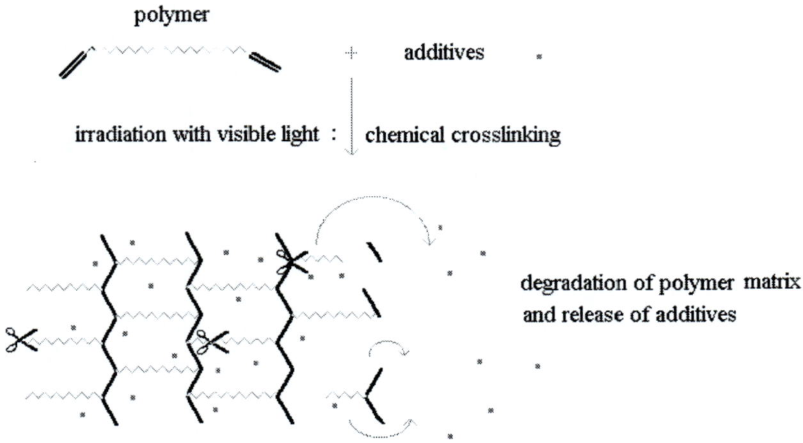

Figure 2. Principle of the photopolymerization process. Methacrylate-endcapped polyesters are crosslinked leading to a network structure. Additives can include: calcium phosphate cements, porogens e.g. gelatin, bioactive compontents, cells.

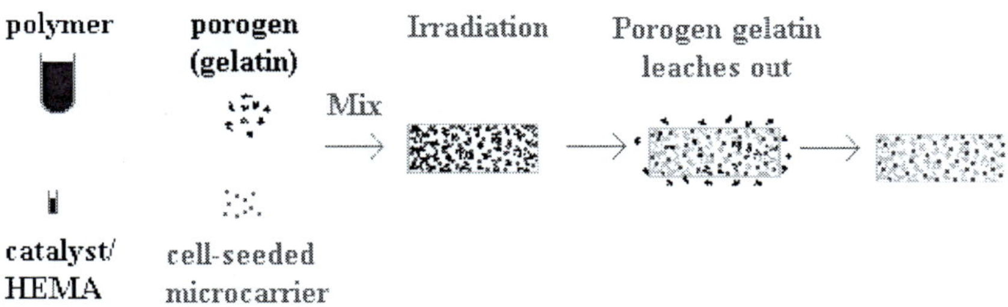

Figure 3. Encapsulation of cell seeded microcarriers into injectable, in situ, photopolymerizable 3D scaffolds.

Although the aim of the research is to obtain an injectable, *in situ* forming composite material, the first step in this study is to study how osteogenic cells behave on preformed photopolymerizable 3D scaffolds based on methacrylate-endcapped polyesters (D, L-lactide, ε-caprolactone and trimethylene carbonate).

To restore damaged critical-size bone defects, additional bone-forming cells are needed. *In situ* forming materials capable of filling irregular bone defects are mostly inconvenient to mix directly with cells. The use of cell delivery systems which can be colonized by cells in vitro and in which they are protected during the polymerization process of the biomaterial could be advantageous (Figure 3). The use of macroporous gelatin-based microcarriers in tissue engineering offers several advantages: (1) cells can be expanded in vitro, (2) biodegradable, (3) cells are in a "natural" 3D environment and (4) cells within the interior of the microcarrier are in a protected environment. The expansion and differentiation of osteogenic cells on macroporous gelatin microcarriers was studied *in vitro*.

The cells seeded microcarriers were subsequently encapsulated in in situ photopolymerizable materials. The cell survival was studied in vitro and in vivo.

2. Materials and Methods

2.1. Osteogenic Cell Cultures

2.1.1. Culture of UMR-106 osteosarcoma cells and MC3T3-E1 cells

UMR-106 rat osteosarcoma cells (ETCC) were routinely grown in monolayer culture in DMEM GlutaMAX-I containing 10 vol % fetal bovine serum (FBS), 0,5 vol % penicillin-streptomycin (10000 U/ml-10000 μg/ml) and 1 vol % Fungizone® at 5 % CO_2/95 % air, 37°C.

MC3T3-E1 clone 14 (ATCC) were routinely grown in monolayer culture in MEM Alpha medium containing 10 vol% FBS, 0,5 vol % penicillin-streptomycin and 1 vol% Fungizone® at 5 % CO_2/95 % air, 37°C. Culture media and additives were purchased from Invitrogen, Merelbeke, Belgium.

2.1.2. Isolation and culture of rat bone marrow derived cells

Bone marrow cells were obtained from the tibiae and femora of young (6 weeks old) adult male Wistar rats. Skin, soft connective tissue and periosteum were removed. Tibiae and femora were washed and the diaphyses were cut free of epiphyseal cartilage. The bone marrow was flushed out repeatedly with MEM Alpha medium (containing 10 vol % FBS, 0,5 vol % penicillin-streptomycin, 1 vol % Fungizone®) and centrifuged (10 minutes, 1000 rpm). The cell pellet was resuspended in MEM Alpha medium supplemented with 100 μM L-ascorbic acid 2-phosphate and 10 nM dexamethasone (both from Sigma-Aldrich NV/SA, Bornem, Belgium) and seeded in T75 tissue culture dishes (Greiner bio-one, Wemmel, Belgium). After 24 hours, the medium was changed to remove nonadherent cells. The cells were cultured until confluence (5% CO_2/95% air, 37°C).

2.1.3. Isolation and culture of goat bone marrow derived cells

Bone marrow was aspirated by a Jamshidi needle (11 gauge, 10 cm; DMS, Hampshire, United Kingdom) under sterile conditions from the iliac crest of female anesthetized goats. The bone marrow was collected in a 10 ml lithium heparin tube (Venoject, Leuven, Belgium) in which 1 ml of ACT-A anti coagulant (Anticoagulant Citrate Dextrose Solution Formula A; Baxter, Brussels, Belgium) was added. The needle was abundantly flushed with ACD-A anticoagulant and lithium heparin before the aspiration of the bone marrow (Vertenten G *et al.*, 2009). After removal of blood clots, the bone marrow was thoroughly washed with MEM Alpha medium (containing 10 vol % FBS, 0,5 vol % penicillin-streptomycin, 1 vol % Fungizone® and centrifuged (10 minutes, 1000 rpm). The cell pellet was resuspended in osteogenic differentiation medium, that is, MEM alpha medium supplemented with 10 vol% FBS, 0.5 vol% penicillin-streptomycin, 1 vol% Fungizone®, 100 μM L-ascorbic acid 2-phosphate, and 10 nM dexamethasone, and seeded in five T75 tissue culture dishes. The culture dishes were placed in a humidified incubator (37°C, 5% CO2/95% air), and the medium was renewed after 2 days, leading to the removal of the nonadherent cells (mainly hematopoietic cells). Stromal cells were kept in culture for 2 weeks (renewal of the medium twice a week) until they reached confluence.

2.2. Preparation of Scaffolds

2.2.1. Synthesis of methacrylate-endcapped oligomer

Methacrylate-endcapped oligomers were prepared by esterification of the corresponding hydroxyl-terminated oligomers with methacryloyl chloride in the presence of triethylamine at room temperature.

Poly(D,L-lactide-co-ε-caprolactone)-HXD (LA/CL-HXD) (P1) respectively poly(D,L-lactide-co-ε-caprolactone)-DPENT (LA/CL-DPENT) (P2) and poly(D,L-lactide-co-trimethylencarbonate-HXD (LA/TMC-HXD) (P3) were synthesized as desribed below [Jackers G et al., 2002; Schacht E et al., 2004, Gorski T et al., 2005].

Linear and star-shaped, telechelic poly(D,L-lactide-co-ε-caprolactone) respectively poly(D,L-lactide-co-trimethylene carbonate) with hydroxyl groups at the termini was synthesized by the ring-opening polymerization of D,L-lactide (LA) and ε-caprolactone (CL) respectively D,L-lactide and trimethylene carbonate (TMC) in the presence of 1,6-hexanediol (HXD) as an initiator and zinc acetate as a catalyst. A variable was synthesized by replacing HXD into dipentaerythritol (DPENT). Polymerization was performed in bulk at 140°C for 48h (on a 10-20 g scale. A glass tube was silanized with dichlorodimethylsilane, dried in an oven, and cooled in a dessicator. A typical experimental procedure was as follows: the glass tube was charged with predetermined amounts of comonomers (LA/CL(TMC) molar ratio 50/50), the initiator (monomer/initiator molar ratio 20/1), and the catalyst (1 mol % per total amount of hydroxyl groups derived from the alcohol initiator). The target number average molecular weight was 2700 $gmol^{-1}$. The tube was degassed three times, sealed, and placed in a constant-temperature oven. Polymerization was terminated by cooling the tube in a refrigerator.

In the second step, the corresponding oligomers with hydroxyl groups at the ends of the polymer chains were functionalised using excess of methacryloyl chloride (2 folds excess to the total number of hydroxyl groups derived from the oligomer) in the presence of triethylamine (1.5 molar excess to the amount of methacryloyl chloride) in methylene chloride (2:1 volume ratio) to give the corresponding methacrylate-endcapped oligomer. The esterification proceeded at room temperature for 24 h. The triethylamine hydrochloride produced from the reaction was removed by filtration and the remaining solution was concentrated and stored in a freezer for one hour, and again a white solid was filtrated. Finally, the remaining solution was dialysed (Spectra/Por® Membrane, MWCO: 1000) in dry acetone for 2 days. The yield was in the range 50-80 %.

Oligomers based on TMC were dissolved in dichloromethane (10 wt %) and mixed with 15 % of triacetin (plasticizer). The solvent was evaporated and the polymer was vacuum-dried at room temperature for 24 hours.

2.2.2. Scaffold preparation

Scaffolds based on D, L-lactide, ε-caprolactone and trimethylene carbonate with apparent porosity of 70 were prepared with gelatin as porogen (particle size 250-355 μm), respectively LA/CL-HXD (P1), LA/CL-DPENT (P2) and

LA/TMC-HXD (P3) as described below [Jackers G *et al.*, 2002; Schacht E *et al.*, 2004; Gorski T *et al.*, 2005].

The methacrylate-endcapped copolymers can be converted into a solid, 3-D polymer network by visible-light irradiation in the presence of D,L-camphorquinone (CQ)/ethyl 4-dimethylaminobenzoate (EDMAB) initiator system. By adding leachable particulates (gelatin), a porous 3-D network can be obtained. The porogens (gelatin) was sieved with ASTM E11-70 standard testing sieves with openings of 250 μm and 355 μm, placed on the Retsch AS 200 sieved shaker until particulates with a defined particle size (250-355 μm) were obtained.

Hydroxy ethyl methacrylate (HEMA) (15 wt %) was added to the katalysator (0,6 wt % CQ and 0,7 wt % EDMAB) until dissolved. The HEMA/katalysator solution followed by the porogen is added to the viscous prepolymer and mixed thoroughly. The viscous paste is then brought into cylindrical Teflon molds (diameter 5 mm, height 1 mm) and the prepolymer is photopolymerized for 80 s at each side with the dental lamp curing unit (500 mW/cm^2) (3M Unitek^{TM}, Ortholux^{TM} XT) (3M Dental Products). After polymerization, the scaffolds are removed from the mold and immersed in distilled water (37°C, bench shaker, 55 rpm) for a period of 3 days (to leach out the porogen). The scaffolds were dried under reduced pressure in a vacuum oven (3 days, room temperature) and sterilized by ethylene oxide (12 hours, 60°C, 48 hours aerated).

After the comparison of these 3 scaffolds (P1, P2 and P3), further experiments were performed with P1. As an alternative for HEMA, triacetin was added to the catalyst until dissolved. In a parallel experiment, the porogen gelatin was omitted.

2.2.3. Cell seeding and differentiation on porous scaffolds

Before cell seeding, the scaffolds were immersed in serum-free MEM Alpha medium in Eppendorf tubes. Air was removed from their pores by generating vacuum with a 20 ml syringe equipped with an 18-gauge needle. The scaffolds were left in medium on a bench shaker (37°C, 55 rpm). After 24 hours, the scaffolds were placed into 96-well tissue culture dishes (for suspension culture).

The scaffolds were seeded with 435000 rat bone marrow cells/90 μl/scaffold (2,2. 10^6 cells/cm^2) and incubated for 4 hours. 160 μl medium was added to each well and the seeded scaffolds were further incubated overnight to allow for cell attachment.

Scaffolds for continuous growth were placed on a needle (3 scaffolds/needle) in a 5 ml spinner flask. 3 ml complete MEM Alpha medium supplemented with 10 mM β-glycerophosphate was added and the cell/scaffold constructs were cultured for 21 days on a bench shaker at a constant rate of 55 rpm (5% CO_2/95% air, 37°C) [Declercq H *et al.*, 2008].

Osteocalcin

48 hours prior to assaying (21 days), each scaffold was placed into 3 ml serum-free osteogenic medium containing 10^{-9} M 1,25-dihydroxyvitamin D_3 in 12 well tissue culture dishes. After 48 hours, the medium was harvested and frozen (-20 °C). At the time of analysis, the medium was thawed and osteocalcin was measured at 450 nm by an enzyme immuno assay according to the company's instructions (EIA kit BT 490, Biomedical Technologies, Inc (Sanbio, Uden, The Netherlands)).

Histology

Scaffolds were rinsed with Ringer solution, fixed with 4 % phosphate (10 mM) buffered formaldehyde (pH 6,9) (4°C, 24 hours), dehydrated in a graded alcohol series (UltraClear is used instead of toluene because toluene dissolves the polymer) and embedded in paraffin. 5-7 μm sections were made and stained with hematoxylin & eosin, Masson's Trichrome, von Kossa and mounted with Canada balsam.

Transmission electron microscopy

Scaffolds were fixed (4°C, 1 hour) in 1 % glutaraldehyde buffered with 0.1 M sodium cacodylate (pH 7.2), washed in the same buffer, postfixed with osmium tetroxide, dehydrated in graded concentration of alcohol and embedded in epoxy-resin. Thin sections (60 nm) were cut with a diamand knife, mounted on cupper grids, stained with uranyl acetate and lead citrate and examined using a JEOL 1200 EX II transmission electron microscope operating at 80 KeV.

2.3. Microcarrier Seeding and Expansion

Macroporous gelatin-based Cultispher S microcarrier beads (diameter 130-380 μm) (Percell Biolytica AB, Astorp, Sweden), were prepared according to the manufacturer's instructions. Briefly, 0,09 g microcarriers/vial were hydrated prior to use in PBS and steam-sterilized (15 minutes, 121 °C). They were rinsed in PBS and twice with MEM Alpha medium (37°C). The microcarriers were transferred and divided over five wells of a 12 well tissue culture dish (for suspension culture). 2 million UMR-106, MC3T3-E1 or bone marrow cells were seeded/0,09 g CultiSpher in MEM Alpha medium respectively MEM Alpha medium with L-ascorbic acid 2-phosphate and dexamethasone (for bone marrow cells) [Declercq

H *et al.*, 2005; Tielens S *et al.*, 2007; Vertenten G *et al.*, 2009]. After 24 hours, the microcarriers were transferred to a 50 ml spinner flask (Schott Duran) (working volume 10 ml) and the stirring speed was set at 70 rpm. The cultures were maintained at 37°C, 5 % CO_2 for 28 days.

2.3.1. Microcarrier colonization

The colonization of the cells on the microcarriers was assessed on day 1, 7, 14, 21 and 28 using propidium iodide staining and histology.

Propidium iodide staining. On day 1, 7, 14, 21 and 28, samples were taken from the microcarrier spinner cultures and transferred to Eppendorf tubes. After rinsing with Ringer solution, they were fixed with 4 % phosphate (10 mM) buffered formaldehyde at 4°C (10 minutes). The microcarriers were rinsed with Ringer solution and 0,5 ml Ringer containing propidium iodide (10 µl PI/10 ml Ringer) was added for 5 minutes. After rinsing with Ringer solution, the microcarriers were transferred to slides and mounted with DABCO containing glycerin and visualized using a fluorescence microscope (excitation 536 nm, emission 617 nm).

Histology. Samples were taken from the microcarrier spinner cultures. After rinsing with Ringer solution, they were fixed with 4 % phosphate (10 mM) buffered formaldehyde (pH 6,9) (4°C, 24 hours). They were dehydrated in a graded alcohol series and embedded in paraffin. 5-7 µm sections were made, stained (hematoxylin & eosin and Masson's Trichrome) and mounted (with Canada balsam).

2.3.2. Cell viability

MTT analysis was used to monitor cell viability. On the specific time points, 4 mg samples from spinner microcarrier cultures were taken, transferred to Eppendorf tubes and the medium was replaced by MTT (Sigma-Aldrich NV/SA, Bornem, Belgium) containing medium (0,5 mg/ml). After 4 hours incubation, the medium was withdrawn, the microcarriers were rinsed twice with Ringer solution and 0,5 ml of 1 % Triton X-100 in isopropanol/0,4 N HCl was added. The formazan was dissolved by shaking at 37°C for 30 minutes. 300 µl was pipetted in a 96 well tissue culture dish and the absorbance was measured at 580 nm (Universal Microplate Reader EL 800, BIO-TEK instruments Inc.).

2.3.3. Cell differentiation

48 hours prior to assaying (21 days), microcarriers seeded with rat bone marrow cells were placed into 5 ml serum-free osteogenic medium containing 10^{-9} M 1,25-dihydroxyvitamin D_3 (Sigma-Aldrich NV/SA, Bornem, Belgium). After 48 hours, the medium was harvested and osteocalcin was measured at 450 nm by an enzyme immuno assay according to the company's instructions (EIA kit BT 490, Biomedical Technologies, Inc (Sanbio, Uden, The Netherlands)).

2.4. Encapsulation of Cell-Seeded Microcarriers onto The Polymer Scaffold

2.4.1. Encapsulation process

Methacrylate-endcapped oligomers (LA_{50}-CL_{50}-$HXD_{20/1}$-bismethacrylate), 15 wt % 2-hydroxyethylmethacrylate (HEMA), katalysator (0,6 wt % D,L-camphorquinone and 0,7 wt % ethyl 4-diaminobenzoate) and gelatin (particle size 250-355 μm) were sterilized by ethylene oxide (12 hours, 60°C, 48 hours air). HEMA was added to the katalysator until dissolved. The HEMA/katalysator solution followed by gelatin is added to the viscous prepolymer and mixed thoroughly.

Before encapsulation, the medium of the microcarriers is removed. The microcarriers are further dried by pipetting them on filter paper. The "dry" microcarriers are added to the katalysator/HEMA/polymer/gelatin paste and carefully mixed. The viscous paste is then brought into cylindrical Teflon molds (diameter 5 mm, height 3 mm) and the prepolymer is photopolymerized for 20 s at one side with the dental lamp curing unit. The visible light curing unit: 3M Unitek™ (500 mW/cm²), Ortholux™ was from 3M Dental products. After polymerization, the scaffolds are removed from the mold and sectioned to 1,5 mm height by means of a scalpel. The scaffolds are placed in MEM Alpha medium and incubated at 37 °C. The medium is replaced after 4 to 7 hours and the following days to remove excess gelatin.

In an additional experiment, triacetin in stead of HEMA was used to dissolve the katalysator. Also the porogen gelatin was omitted.

A typical experimental procedure was as follows. Scaffolds with apparent porosity of 68 were prepared with respectively 20 % and 50 % carriers versus polymer. Scaffolds with 20 % carriers initially contained 0,473 g polymer, 0,8681 g gelatin and 0,09 g carriers. Scaffolds with 50 % carriers initially contained 0,355 g polymer, 0,64833 g gelatin and 0,18 g carriers.

2.4.2. Toxicity of dental curing unit

UMR-106 cells were seeded in a 96-well tissue culture plate at a density of 25000 cells/well and 100000 cells/well in 200 µl of complete DMEM medium. After 24 hours the regular medium was withdrawn and the cells were irradiated for 0, 10, 20, 40, 80, 160 and 240 seconds (in triplicate) with the dental lamp curing unit (500 mW/cm² blue light). 200 µl culture medium was added and the cells were incubated at 37°C. The number of viable cells was determined by MTT-assay immediately or after 2 days incubation. 0,5 mg/ml MTT containing medium was added to the cells and incubated. After 4 hours, the MTT containing solution was withdrawn and 200 µl of 1 % Triton X-100 in isopropanol/0,4 N HCl was added. The formazan was dissolved by shaking at 37°C for 30 minutes. The absorbance was measured at 580 nm (Universal Microplate Reader EL 800, BIO-TEK instruments Inc.) and the viability was calculated as percentage of the control.

2.4.3. Light and electron microscopy analysis

A 7 day culture period was chosen to detect if the encapsulated microcarriers were still present in the scaffold.

Scaffolds were fixed with 4 % formaldehyde buffered with 10 mM phosphate (pH 6,9) (4°C, 24 h). They were dehydrated in a graded alcohol series (UltraClear is used instead of toluene because toluene dissolves the polymer) and embedded in paraffin. 5-7 µm sections were made, stained with hematoxylin & eosin and mounted with MountingClear.

Transmission electron microscopy (TEM) was carried out to analyze the viability of the cells within the microcarriers and encapsulated in the polymer. The microcarriers and the polymer were fixed with 2% glutaraldehyde in 0.1 M cacodylate buffer (pH 7.4) for 1 h and were post-fixed with 2% OsO4 for 1 h. After dehydratation, the samples were embedded in Spurr resin and were sectioned (60 nm). The sections were routinely stained with uranyl acetate and lead citrate. The samples were analyzed using a JEOL 1200 EX II TEM operating at 80 KeV.

2.5. In Vivo Evaluation of Cell Seeded Microcarriers Encapsulated in Injectable Scaffolds

2.5.1. Surgical procedure

Gelatin based microcarriers combined with different preparations of the scaffolds were used in this study and have been described in depth earlier [Vertenten G *et al.*, 2009].

The composite consisted of pure methacrylate-endcapped polymer (P1) with a triacetin solution as plasticizer containing the photoinitiators. This mixture was supplemented with microcarriers loaded with autologous mesenchymal stem cells differentiated in the osteogenic lineage. In some composites, 30% w/w α-TCP was further added to the polymer-triacetin-microcarrier mixture.

The surgery was performed as described earlier [Vertenten G *et al.*, 2008]. Briefly, noncritical-sized defects (6.0 mm diameter) were drilled in the medial diaphyseal cortex of each tibia using a trephine burr (3I; Implant Innovations, Copenhagen, Denmark). In each leg, the composites were randomly assigned to a hole, whereas one hole was left empty to serve as a control. Each composite was first placed on the borders and bottom of the defect, and photopolymerized for 40s (500 mW/cm^2 blue light, Visible Light Curing UnitTM; 3M Unitek, Diegem, Belgium). Afterward, the remaining defect was further filled with a second layer of additional polymer before the standard wound closure.

At respectively 2, 4, 8 and 12 weeks after surgery, goats were euthanized and tibial defects were harvested for histological and immunohistochemical analysis of bone healing.

2.5.2. Histological and immunohistochemical analysis

After euthanasia, the different bone defect sites were harvested from the tibias. Undemineralized samples were embedded in destabilized Technovit 9100 New (Heraeus Kulzer, Wehrheim, Germany) as described earlier [Vertenten G *et al.*, 2008]. Four-micrometer sections were cut with a heavy duty microscope (SM2500; Leica Microsystems, Wetzlar, Germany), stretched with 70% ethanol on a slide, and dried for 12h at 60°C.

The sections were stained with hematoxylin and eosin (H&E). Sections were stained immunohistochemically for vimentin and osteocalcin. Before immunohistochemical staining, sections were deacrylized. Antigen retrieval was performed using proteinase K (Dako, Heverlee, Belgium). The sections were rinsed with distilled water and phosphate-buffered saline (PBS). Endogeneous peroxidise was quenched by incubating the sections in a 3% hydrogen peroxide solution in methanol. This was followed by rinsing in distilled water and PBS. The sections were preincubated in bovine serum albumin and rinsed with PBS. Consequently, immunohistochemical staining was performed using the primary antibody. The sections were rinsed with PBS and incubated with biotinylated secondary antibody. Afterward, the sections were rinsed and incubated with avidin-biotin complex-horseradish peroxidise (Dako). The samples were rinsed with PBS and incubated in 3,3'-diaminobenzedine (Sigma, Bornem, Belgium) and finally rinsed with distilled water. Counterstaining was done by immersing the sections in hematoxylin, tap water, and distilled water. Finally, the sections were dehydrated and covered [Vertenten G *et al.*, 2008; Vertenten G *et al*, 2009].

3. Results

3.1. Migration and Differentiation on Preformed Scaffolds

All scaffolds seeded with rat bone marrow cells were cultured dynamically for 21 days in osteogenic culture medium.

Rat bone marrow derived cells were able to attach and differentiate on these 3D scaffolds. After three weeks, the cells present on the scaffolds expressed high ALP activity and high osteocalcin secretion, independent of the scaffold type [Declercq H *et al.*, 2008].

Cross-sections showed bone formation following the contours of the polymer walls for all the three scaffolds and is shown in Figure 4. Masson's Trichrome (Figure 4 (a, a', b, b')) and H&E (Figure 4 (b, c)) staining showed a typical morphology of cuboidal osteoblasts which secreted an extracellular matrix. Calcium phosphate deposits are present in the ECM as is demonstrated by von Kossa staining (Figure 4 (d, d', e, e', f, f'). Hence, the cells were able to mineralize the extracellular matrix.

Transmission electron micrographs at the polymer/bone interface for the three types of scaffolds showed osteoblast-like cells which secreted an abundant collagen I containing extracellular matrix (P1, P2 and P3 is shown respectively in Figure 5 a, b and c). Among the collagen I fibers, matrix vesicles loaded with bone apatite crystals can be observed (Figure 5 a', c'). The cells cultured for 21 days on the three different scaffolds had a good viability, as seen by the mitochondrial structure (crystae) (Figure 5 b'). Dying cells (seen as cells beginning to lose their shape and break down) were rarely observed.

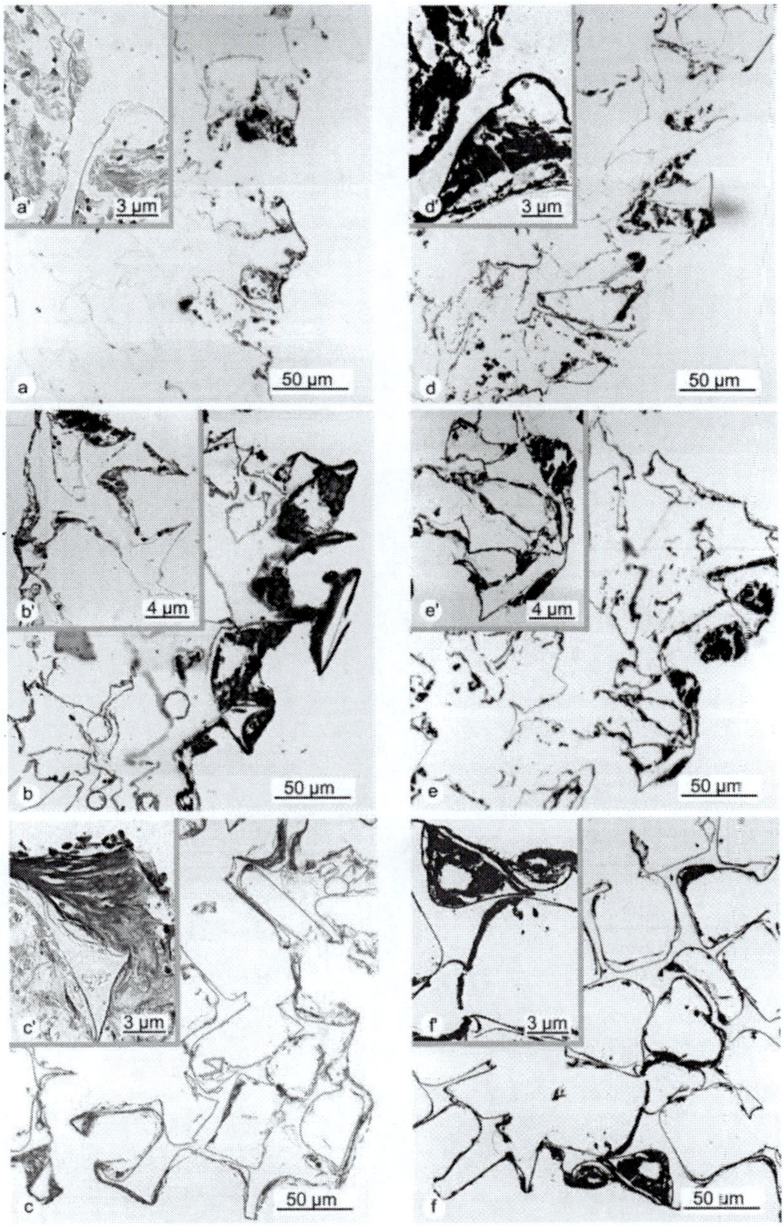

Figure 4. Cross-sections of scaffolds prepared from respectively LA/CL-HXD (P1) (a, a', d, d'), LA/CL-DPENT (P2) (b, b', e, e') and LA/TMC-HXD (P3) (c, c', f, f'). TM stain (a, a', b', c'). Von Kossa stain (d, e, f, d', e', f'). H&E stain (b, c).

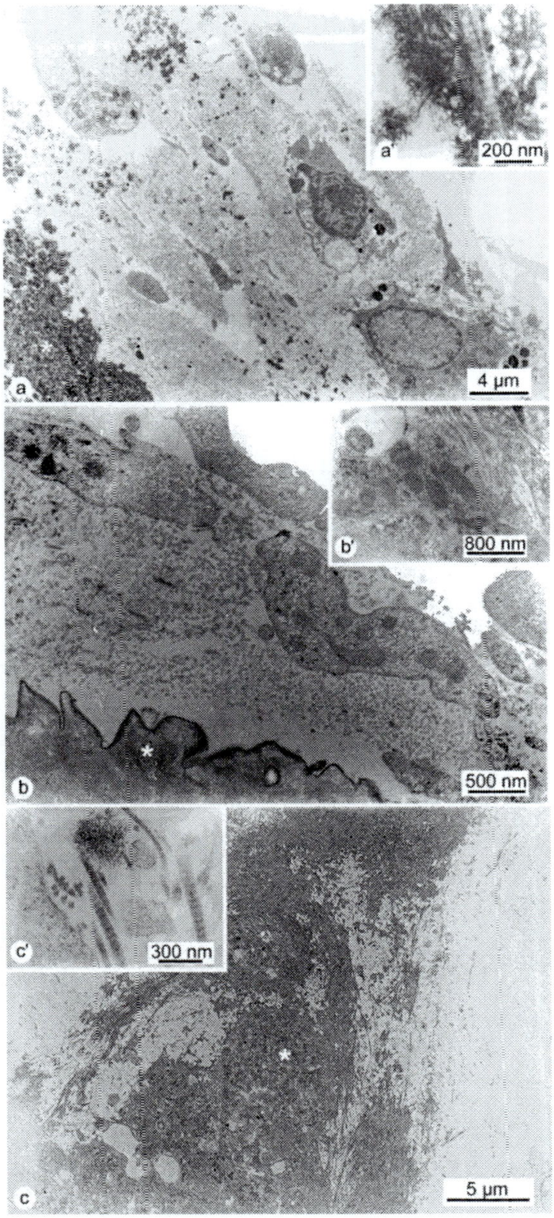

Figure 5. TEM scaffold prepared from respectively LA/CL-HXD (a, a'), LA/CL-DPENT (b, b') and LA/TMC-HXD (c, c'). a, b, c: Bone/polymer interface. A', c': Extracellular matrix formation (collagen I fibers) and apatite crystals, b': mitochondria (crystae). Asterisk: polymer scaffold.

Conclusion

These results demonstrate bone formation at the polymer interface. But cellular infiltration is limited to the edges of the polymer scaffold. In large critical size defects, the infiltration of bone forming cells will be restricted to the edge. In the center of the polymer scaffold, a delay of bone regeneration will occur due to the lack of migration of cells in the scaffold.

Moreover, in large critical size bone defects, it is most likely that the amount of osteogenic cells available at the site of injury will be too limited to repair the defect. Hence, an appropriate source of ostegenic cells should be added at the place of injury. For in vivo applications, it would be beneficial to mix the in situ photopolymerizable scaffolds with osteoprogenitor cells or stem cells to induce bone formation in the center of the scaffold.

A major problem is to deliver cells at the site of injury. Cells are often encapsulated in hydrogels but their mechanical properties are too weak to withstand the molding of the injectable polymer paste. Moreover, as osteoblastic cells are polarized, an "open" structure would be more appropriate. In this hypothesis, the use of porous gelatin based microcarriers, which have an "open" structure, were evaluated.

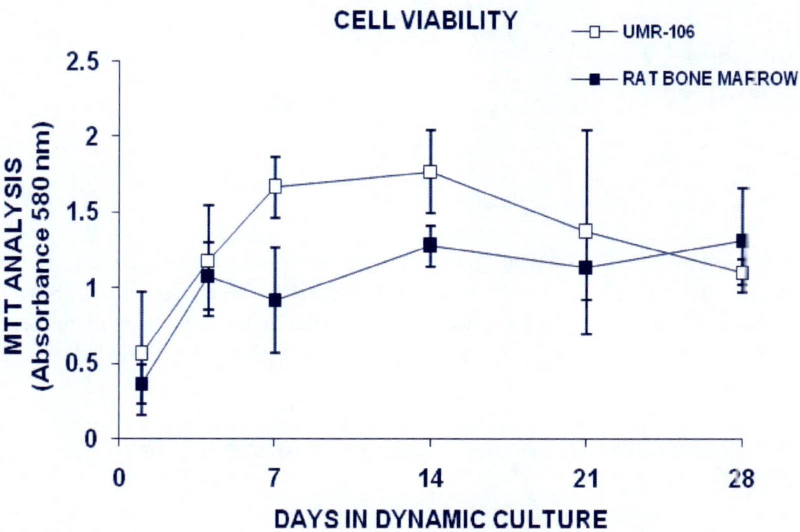

Figure 6. Cell viability of UMR-106 and bone marrow cells (rat) grown on CultiSpher S microcarriers. On specific time points, the amount of viable cells on 0,004 g osteoblast-seeded microcarriers was quantified by MTT assay. Mean and SD of 2 independent experiments in duplicate.

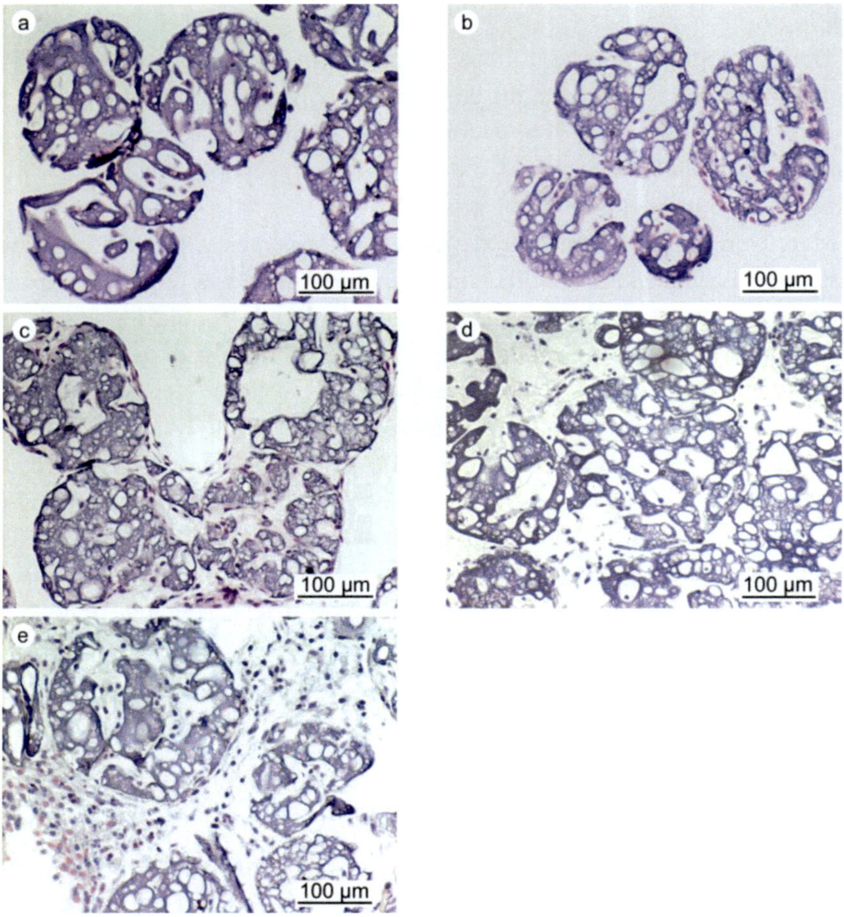

Figure 7. LM pictures of cell-loaded microcarriers (5-µm sections, HE staining) of MC3T3 cells at day 2 (a), 7 (b), 14 (c), 21 (d), and 35 (e). A period of 7-14 days is needed to colonize the microcarriers. A longer culture period results in the formation of cellular bridges between the microcarriers. Some of the microcarriers remain unloaded.

3.2. Expansion and Differentiation of Osteogenic Cells on Microcarriers

The 24 hours static seeding regime allowed the cells to attach to the microcarriers and yielded a relatively homogeneous cell distribution between the carriers (seeded with UMR-106, MC3T3-E1 and bone marrow cells) within the culture as assessed by MTT staining, PI staining and histology. Static seeding was shown to improve the seeding efficiency in comparison to dynamic seeding (data not shown).

For both the osteogenic cell lines and bone marrow cells, cells remained viable and cell number increased over time, as demonstrated by MTT assay (Figure 1). The amount of cells on the carriers reached a plateau after 14 days (Figure 1). On day 21, the amount of cells on the carriers was equal or declined as shown by MTT.

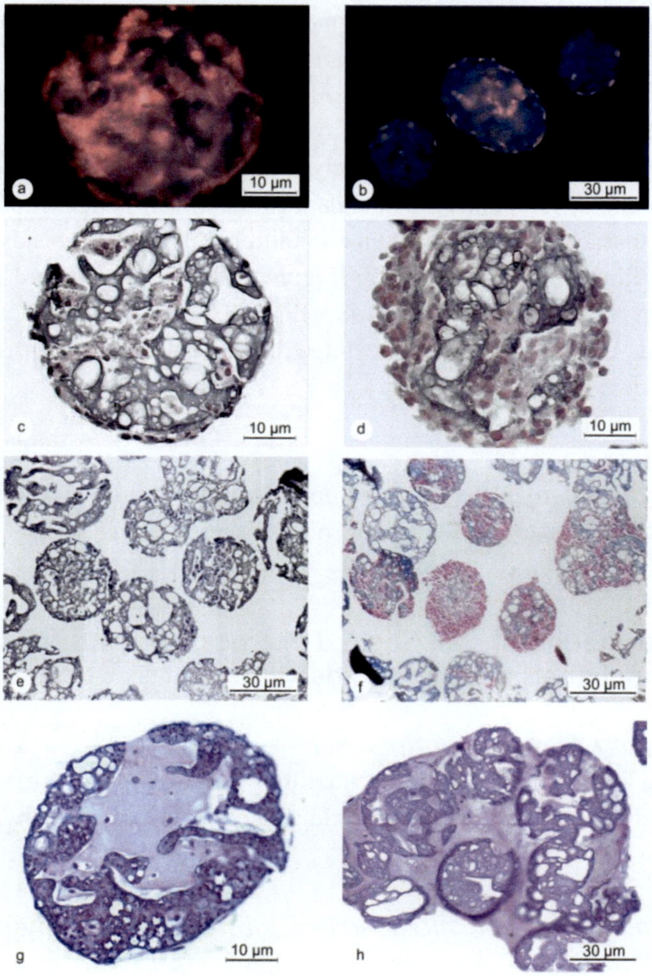

Figure 8. Propidium iodide (a, b), H&E (c, e, g, h) and Trichrome Masson (d, f) staining of UMR-106 (a, c, e) and rat bone marrow cells (b, d, f, g, h) grown on Cultispher S microcarriers after 14 days (a, b, c, e) and 28 days (d, f, g, h).

Light microscopic analyses were performed of MC3T3-E1 seeded microcarriers after 2, 7, 14, 21 and 35 days of culture (Figure 7).

A culture period of 1-2 weeks resulted in a good colonization of the microcarriers. The cells are not only present around the microcarrier (Figure 7 a, b, c, d), but the microcarrier is also fully colonized in the interior (Figure 7 c, d).

The microcarriers have a relatively homogeneous cell distribution between the carriers. Although most carriers are completely colonized by cells, some carriers only have a limited amount of cells (Figure 7, Figure 8 e, f).

Microcarriers seeded with cells formed cellular bridges between the carriers and large clusters were made as shown in Figure 8 e, f, h. The microcarrier aggregates ranged from two cell-seeded carriers to more than a dozen microcarriers and started only after 14 days in culture.

Bone marrow derived cells not only proliferated on the microcarriers, but were also able to differentiate. An abundant extracellular matrix is formed (Figure 8 g, h). The extracellular matrix spanned multiple microcarrier beads and formed bridges resulting in large aggregates (Figure 2 h). Osteocytes are clearly seen embedded in the extracellular matrix (Figure 8 g). The cells were highly differentiated as osteocalcin was secreted in the culture medium (4,2 +/- 0,3 ng/ml) [Declercq *et al.*, 2005].

Conclusion

Bone marrow derived cells seeded on gelatin based microcarriers are able to expand and differentiate into the osteogenic lineage.

3.3. Encapsulation of Cell Seeded Microcarriers into Injectable Photopolymerizable Scaffolds

3.3.1. Toxicity of dental curing unit

UMR-106 cells showed a 100% viability even after 240s of irradiation with visible light for both high and low seeding density as evaluated by MTT-assay (data not shown).

3.3.2. Encapsulation of microcarriers into the three-dimensional scaffold

From the above shown MTT-curves (Figure 6) and previous results concerning the osteoconductivity of the scaffolds, we decided to incorporate cell seeded microcarriers cultured for 14 days in a dynamic system into injectable scaffolds with apparent porosity of 70 obtained by gelatin leaching.

The viscous polymer allowed us to mix the microcarriers properly with the polymer, the katalysator/initiator and the porogen gelatin.

Immediately after crosslinking, the microcarriers and gelatin are evenly distributed throughout the polymer scaffold (Figure 9 a, d).

After 4 to 7 hours incubation in culture medium, the gelatin is leached out completely and a porous osteoconductive 3-D scaffold is formed. Nevertheless, the microcarriers are encapsulated and remain in the 3-D scaffold (Figure 9 b, e). Even after a culture period of 7 days, the microcarriers are still present in the scaffold (Figure 9 c, f). This demonstrates that the microcarriers are well entrapped in the crosslinked polymer scaffold.

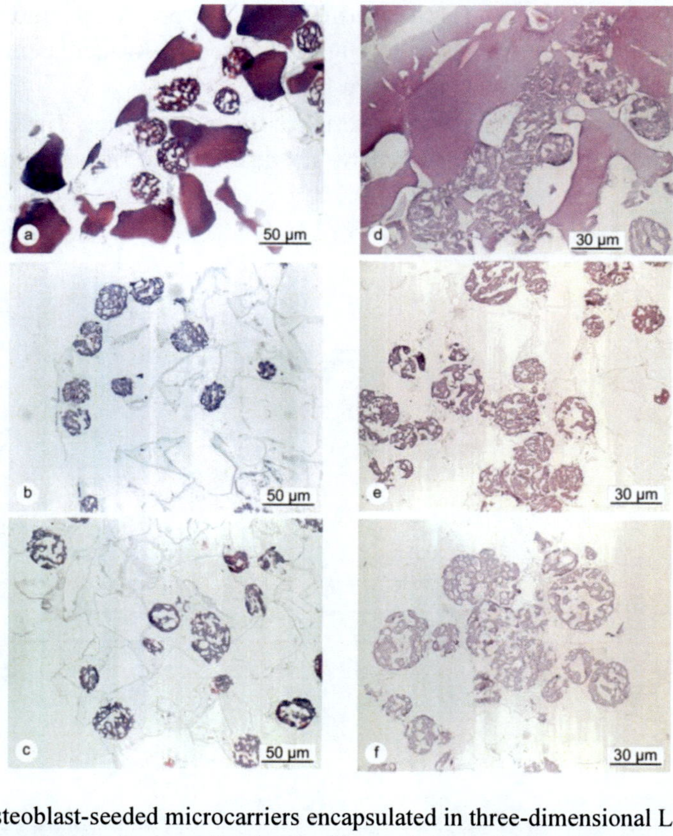

Figure 9. Osteoblast-seeded microcarriers encapsulated in three-dimensional LA/CL scaffolds immediately after crosslinking (a, d), after 7 hours (b, e) and 7 days (c, f) incubation in culture medium. 20 % (a, b, c) versus 50 % (d, e, f) carrier/polymer ratio. H&E stain.

The photopolymerization environment and the molding/mixing of the microcarriers into the polymer/gelatin paste is not detrimental to the mechanical stability of the microcarriers.

When the amount of carriers to polymer is 20 %, the scaffold is well crosslinked, as described above. Increasing the carrier/polymer ratio to 50 % leads to inferior crosslinked scaffolds (Figure 9 d, e, f). The microcarriers are also evenly distributed in the polymer matrix, but are more packed together in the scaffold. Nevertheless, some microcarriers leach out as a result of the minor crosslinking of the scaffold.

3.3.3. Cell viability of encapsulated cells

Histology

Osteoblasts seeded onto microcarriers and encapsulated into the three-dimensional scaffold were analyzed by light microscopy. Inside the microcarriers, cells are clearly present as seen in Figure 10.

The cells are still attached to the interior of the microcarrier. The cell viability of encapsulated cells is studied more in detail with transmission electron microscopy.

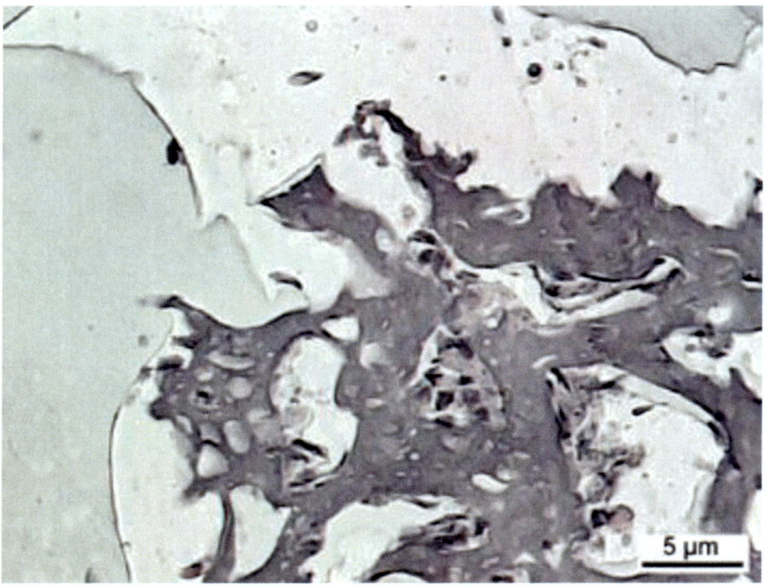

Figure 10. Osteoblast-seeded microcarriers encapsulated in three-dimensional LA/CL scaffolds. H&E stain.

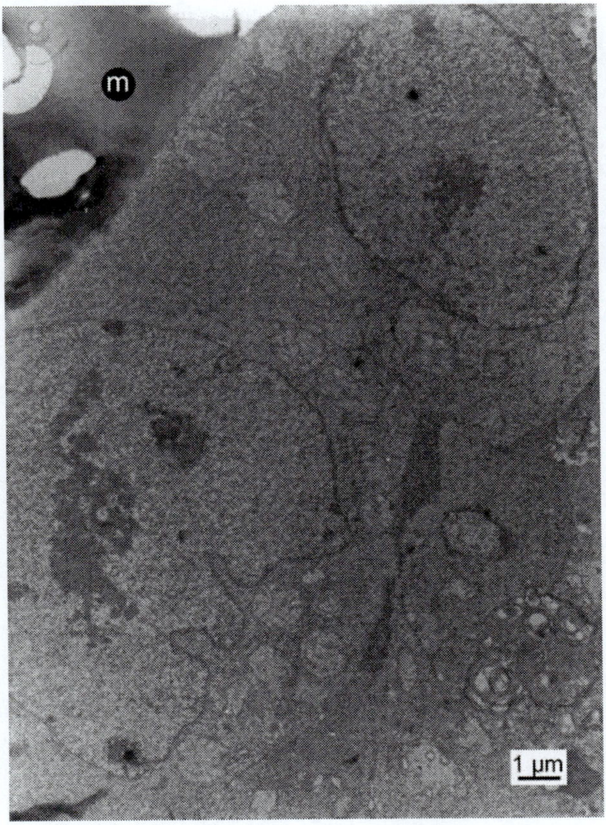

Figure 11. TEM picture of a cell-loaded microcarrier before encapsulation. Viable cells with euchromatic nuclei well attached to the internal surface of the microcarrier are present.

Transmission electron microscopy

Ultrastructural analysis was performed to evaluate the cell viability of the encapsulated cells and to compare the different polymerization conditions. Consequently, cell-seeded microcarriers were evaluated by TEM before and after crosslinking into the polymer with and without addition of gelatin.

Before encapsulation, viable cells showing euchromatic nuclei often containing a prominent nucleolus were observed. No swelling of the mitochondria, indicative for loss of viability, was observed. Cells were attached to the inner surface of the microcarriers (Figure 11).

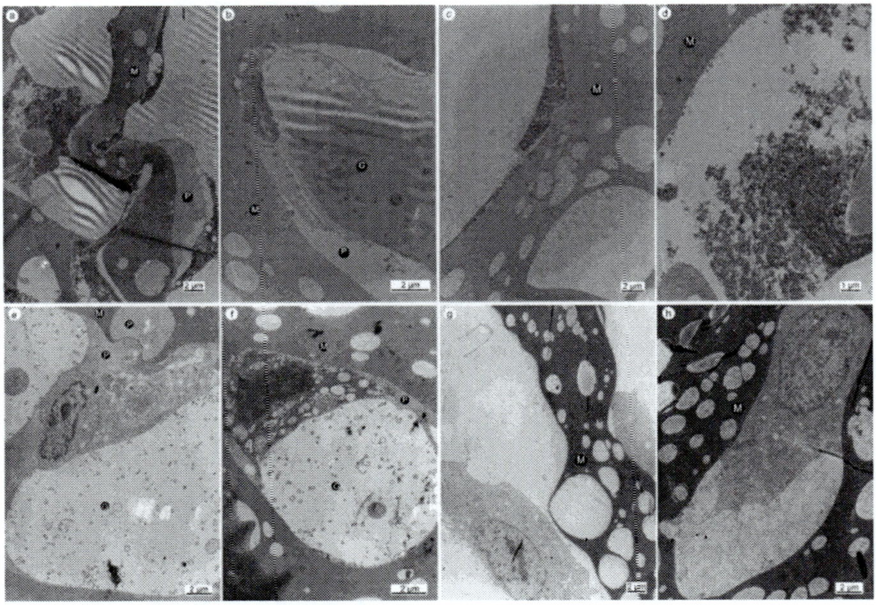

Figure 12. TEM pictures of a cell-loaded microcarrier after encapsulation. (a, b): Encapsulated in polymer combined with HEMA/gelatin; (c,d) HEMA without gelatine; (e, f): TA/gelatine. Cellular debris or disintegrating cells are present. A cell with a pyknotic nucleus and disintegrating cytoplasm is shown. The microcarrier is indicated by (M). Between the cell and the microcarrier surface, material, probably representing the polymer (P), is present. The rest of the pore is filled with gelatine (G). After encapsulation using triacetin without gelatine (g, h), cells remained viable. They are firmly attached to the inner surface of the microcarrier. Euchromatic nuclei and an abundant network of RER are noticed.

After encapsulation, differences correlated with the type of *in situ* forming material were observed. When HEMA (with or without gelatin) or triacetin with gelatin was used to encapsulate the microcarriers, already immediately after the crosslinking procedure, no viable cells could be found in the carriers. Cellular debris and cells with clumped chromatin were present in the pores of the carriers. Often, noncellular material was formed between the cells and the CultiSpher surface, confirming the impression obtained by LM of material infiltrating the microcarrier (Figure 12 a-f). It seemed that the polymer as well as the gelatin infiltrated into the microcarriers.

In contrast, when using triacetin without gelatin during the polymerization, cells remained viable after crosslinking: euchromatic nuclei were present, and no blebbing or dilatation of the nuclear or plasma membrane were seen. In the cytoplasm, an abundant network of RER indicative for a high protein synthesis was present. Cells seemed to be firmly attached to the inner surface of the microcarrier (Figure 12 g, h).

Conclusion

The viability of cells encapsulated in *in situ* photopolymerizable methacrylate-endcapped polyesters is optimal when the porogen gelatine was omitted and triacetin was used in stead of HEMA as a plasticizer.

3.4. In vivo Evaluation of Cell Seeded Microcarriers Encapsulated in an Injectable Photopolymerizable Scaffold

New bone formation was mostly pronounced in the control defects. Control defects were filled with mesenchymal tissue at the center until complete bridging by bone tissue was achieved. Complete healing was not observed in the defects filled with the different composites during the study period. At 8 and 12 weeks post-surgery, composite-filled defects were characterized by the presence of fibrous tissue surrounding the centrally located polymer. New bone formation was present in the periphery of the defects. Polymer degradation was limited over time. Defects filled with composites containing microcarriers were seen to contain several cellular tissue islands consisting of microcarriers, connective tissue, and viable cells (Figure 13).

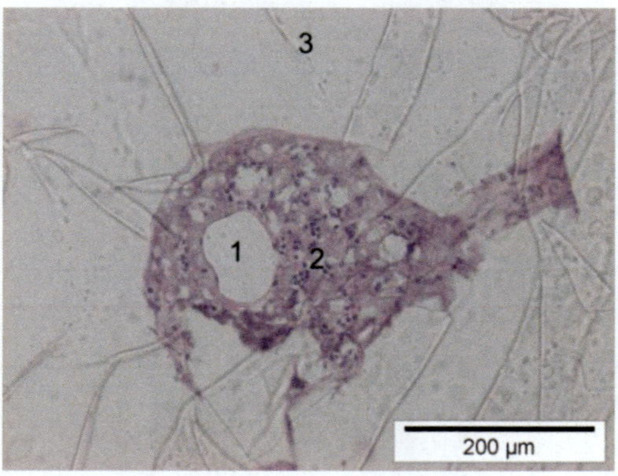

Figure 13. Photomicrograph of a tibial defect treated with composite no. 2, 8 weeks after surgery (magnification 20x, H&E staining). Several centers of tissue consisting of microcarriers (1), colonized and overgrowth by extracellular matrix, and viable cells (2) are present in the polymer (3).

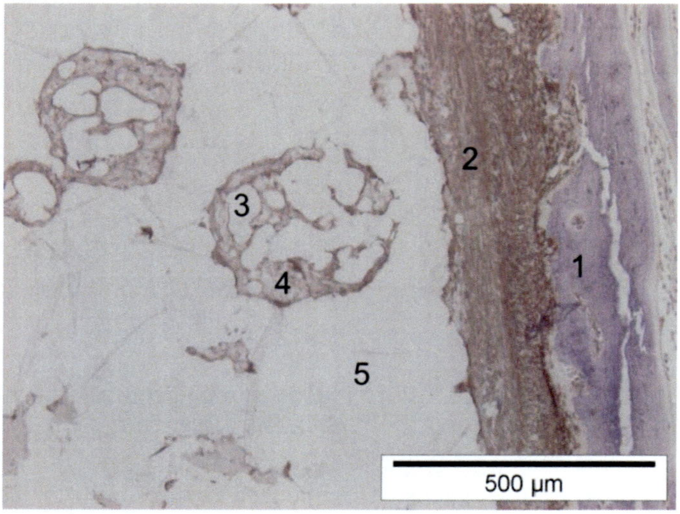

Figure 14. Photomicrograph of a tibial defect treated with composite no. 2, which stained immunohistochemically for vimentin 12 weeks after surgery. The capsule of tissue around the polymer and the tissue centers in the polymer are positively stained (magnification x10). 1, Newly formed bone (metachromatic staining); 2, capsule of tissue around composite (vimentin positive); 3, microcarrier; 4, tissue and cells around the microcarriers (vimentin positive); 5, polymer.

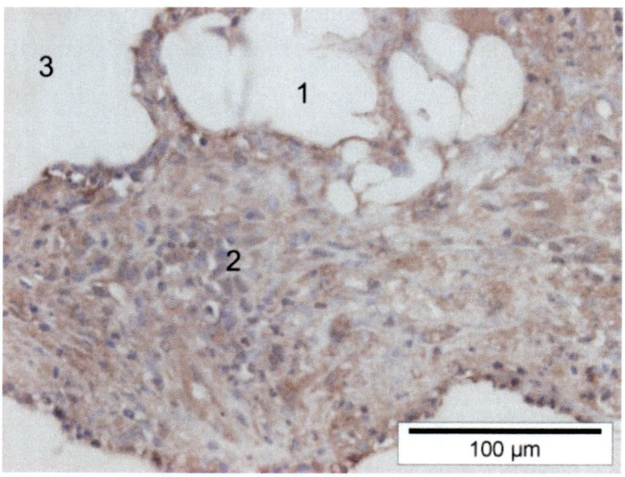

Figure 15. Photomicrograph of a center of tissue in tibial defect treated with composite no. 2, which stained immunohistochemically for osteocalcin 12 weeks after surgery. There is a slightly positive staining of the tissue, and the nuclei are staining negative (magnification x40). 1, Microcarrier; 2, tissue and cells around the microcarrier (slightly positive); 3, polymer.

Intense positive staining for vimentin (mesenchymal tissue) was observed in the tissue concentrations in the composites, the connective tissue surrounding the composites, and the fibrous tissue inside the cortical bone tissue (Figure 14).

Immunohistochemical staining for osteocalcin revealed a slightly positive staining of tissues in and around the microcarriers, the connective tissue surrounding the different composites, and the tissue inside the newly formed trabecular bone (Figure 15).

Conclusion

Although bone marrow stromal cells survived and showed signs of new bone formation, bone healing was still in its initial stages in the polymer composites. The present study demonstrated the implanted cells' ability to proliferate and produce a calcified matrix (osteocalcin positive) inside the polymer as cell proliferation and matrix production were absent in the polymer composite without bone marrow stromal cells.

4. Conclusion

Extra sources of bone forming cells (bone marrow derived mesenchymal stem cells), can be additionally mixed with injectable materials to increase the osteoinductive properties of implant materials. Bone marrow stromal cells which have an osteogenic potential, are anchorage dependent and need a substrate for optimal cell activity. Bone marrow stromal cells can be predifferentiated *in vitro* after seeding on a crosslinked gelatin microcarrier. The cells cultured on the carriers survived the mixing with and curing of the polymer. Unfortunately the cell survival after implantation into the bone defect for a prolonged period of time was limited. A limited amount of new bone formation could be observed. One of the major problems of the crosslinked methacrylate-endcapped poly (D,L-lactide-co-ε-caprolactone) polymer was the low porosity, restricting the essential oxygen and nutrient flow needed for an acceptable bone repair of the defect. Even more, the relative slow degradation of the polymer might also explain the limited bone regeneration.

In this scope, the LA-CL polymer was replaced by a chemically modified hydrogel. The obtained results are promising (publication submitted)

The encapsulation of cell seeded microcarriers in injectable materials is a promising approach and opens new frontiers for bone tissue engineering.

References

Declercq, HA; Gorski, TL; Schacht, EH; Cornelissen, MJ. Osteoblast behaviour on in situ photopolymerizable three-dimensional scaffolds base don D,L-lactide and ε-caprolactone: influence of pore volume, pore size and pore shape. *J Mater Sci-Mater Med*, 2008, 19(9), 3105-3114.

Declercq, HA; Gorski, TL; Schacht, EH; Cornelissen, MJ. Osteoblast behavior on in situ photopolymerizable three-dimensional scaffolds based on D,L-lactide, ε-caprolactone and trimethylene carbonate. *J Mater Sci-Mat Med*, 2006, 17(2), 113-122.

Declercq, HA; Gorski, TL; Tielens, SP; Schacht, EH; Cornelissen, MJ. Encapsulation of osteoblast seeded microcarriers into injectable, photopolymerizable three-dimensional scaffolds based on D,L-lactide and epsilon-caprolactone. *Biomacromolecules*, 2005, 6, 1608-1614.

Gorski, T. Preparation and evaluation of biodegradable composite scaffolds for bone tissue engineering. *PhD thesis*, 2006. Proefschrift voorgelegd tot het behalen van de graad: Doctor in de Wetenschappen, Scheikunde. Promotor: E. Schacht. Polymer Material Research Group, Krijgslaan 281 (S4), Ghent University, B-9000 Ghent, Belgium.

Jackers, G. Ontwikkeling van biodegradeerbare implantaten voor botweefselregeneratie. *PhD thesis*, 2002. Proefschrift voorgelegd tot het behalen van de graad: Doctor in de Wetenschappen, Scheikunde. Promotor: E. Schacht. Polymer Material Research Group, Krijgslaan 281 (S4), Ghent University, B-9000 Ghent, Belgium.

Schacht, E; Gorski, T; Jackers, G; Cornelissen, M; Declercq, H; Verbeeck, R; Van den Vreken, N; Gasthuys, F; Vlaminck, L. Abstract. 40th IUPAC Symposium on Macromolecules, July 4-9th, 2004, Paris.

Tielens, S; Declercq, H; Gorski, T; Lippens, E; Schacht, E; Cornelissen, M. Gelatin-based microcarriers as embryonic stem cell delivery system in bone tissue engineering: an in vitro study. *Biomacromolecules*, 2007, 8, 825-832.

Vertenten, G; Lippens, E; Gironès, J; Gorski, T; Declercq, H; Saunders, J; Van den Broeck, W; Chiers, K; Duchateau, L; Schacht, E; Cornelissen, M; Gasthuys, F; Vlaminck, L. Evaluation of an injectable, photopolymerizable, and three-dimensional scaffold based on methacrylate-endcapped poly (D,L-lactide-co-ε-caprolactone) combined with autologous mesenchymal stem cells in a goat tibial unicortical defect model. *Tissue Engineering Part A*, 2009, 15(00), 1-11.

Vertenten, G; Vlaminck, L; Gorski, T; Schreurs, E; Van den Broeck, W; Duchateau, L; Schacht, E; Gasthuys, F. Evaluation of an injectable, photopolymerizable three-dimensional scaffold based on D,L-lactide and ε-caprolactone in a tibial goat model. *J Mater Sci Mater Med*, 2008, 19, 2761-2771.

In: Bone Regeneration: Growth Factors... ISBN: 978-1-60876-621-5
Editors: V. Legard et al., pp. 233-261

Chapter VIII

The Induction of Bone Formation from Non-Human to Human Primates: Bone Tissue Engineering from Laboratory to Patients

Ugo Ripamonti*, June Teare and Carlo Ferretti
Bone Research Unit, Medical School, Parktown, South Africa.

Abstract

Bone regeneration in pre-clinical and clinical contexts requires three key components: an osteogenic soluble signal; an insoluble signal or substratum which delivers the signal and acts as a scaffold for the initiation of bone formation; and host cells capable of differentiation into bone cells in response to the osteogenic molecular signal. The '*bone induction principle*' has identified important rules for tissue engineering and regenerative medicine at large, i.e. soluble osteogenic molecular signals, insoluble signals or substrata including biomimetic bioactive matrices and responding stem cells. The osteogenic soluble molecular signals of the transforming growth

* Corresponding author: Bone Research Unit, Medical School, 7 York Road, 2193 Parktown South Africa, Tel: + 27 11 717 2144, Fax: + 27 11 717 2300, E-mail: ugo.ripamonti@wits.ac.za

factor-β (TGF-β) superfamily, i.e. the bone morphogenetic/ osteogenic proteins (BMPs/OPs), induce endochondral bone formation in non-human and human primates. Tissue induction in the non-human primate *Papio ursinus* develops as a mosaic structure in which the osteogenic proteins of the TGF-β superfamily singly, synergistically and synchronously initiate and maintain tissue induction and morphogenesis. Tissue engineering starts by erecting scaffolds of biomimetic biomaterial matrices that mimic the supramolecular assembly of the extracellular matrix of bone. Naturally-derived BMPs/OPs and gamma-irradiated recombinant human osteogenic protein-1 (hOP-1) delivered by allogeneic and xenogeneic insoluble collagenous bone matrices initiate the induction of bone formation in heterotopic extraskeletal and orthotopic skeletal sites of the non-human primate *Papio ursinus*. This process culminates in complete calvarial regeneration by day 90 and maintains the regenerated constructs by day 365. Uniquely in the non-human primate *Papio ursinus* and thus by extension to the primate *Homo sapiens*, the three mammalian TGF-β isoforms induce rapid and substantial endochondral bone formation in heterotopic intramuscular sites. In marked contrast, the TGF-β proteins induce only fibro-angiogenesis with potential scarring when implanted in rodents, lagomorphs and canine models. It is proposed that the translation in clinical contexts of the phenomenon of '*Bone: Formation by autoinduction*' may be achieved in humans by the synergistic binary application of relatively low doses of TGF-β proteins with a recombinant human osteogenic protein.

Keywords: Bone induction, bone morphogenetic proteins, transforming growth factor-β, synergistic induction of bone formation, tissue engineering, primates.

Tissue Induction and Morphogenesis

The emergence in postnatal life of complex tissue morphologies rests on a simple and fascinating concept: morphogens exploited in embryonic development can be re-exploited and re-deployed to engineer postnatal tissue regeneration [Reddi, 2000a, 2000b; Ripamonti, 2003; Ripamonti, 2006]. Both embryonic development and postnatal tissue regeneration are equally regulated by selected few and highly conserved family of proteins, members of the transforming growth factor-β (TGF-β) supergene family [Ripamonti, 2006].

Morphogenesis is the developmental cascade of pattern formation, body-plan establishment and interpretation and differentiation of the pattern [Reddi, 1984;

Reddi, 2000a, 2000b]. Morphogenesis is induced by morphogens [Reddi, 2000a, 2000b]. As a prelude to morphogenesis, the generation of cellular diversity, or *differentiation*, must first occur [Ripamonti and Duneas, 1998; Reddi, 2000a, 2000b]. As a corollary to differentiation, there must exist signaling molecules or *morphogens* (first defined by Turing as a '*form generating substances*') [Turing, 1952] that are expressed by specific cells as soluble molecular signals and that are capable of imparting specific differentiation pathways to responding cells [Ripamonti and Duneas, 1998], initiating the ripple like cascade of pattern formation and the attainment of tissue form and function or morphogenesis [Reddi, 1984; Ripamonti and Duneas, 1998; Reddi, 2000a, 2000b].

The analysis of molecular signals in solution, which have been found in several extracellular matrices from the fruit fly *Drosophila melanogaster* to the frog *Xenopus levis* to fractures has dramatically heralded the mechanistic understanding of the biochemical mechanisms of cellular growth, differentiation, development and morphogenesis of vertebrate tissues and organs, above all however, has mechanistically explained the assembly of transformed cellular phenotype translating into the *de novo* induction of bone formation [Ripamonti and Duneas, 1998; Reddi, 2000a, 2000b; Ripamonti, 2006].

Several important studies on "*Bone: Formation by autoinduction*" [Urist, 1965; Reddi and Huggins, 1972] have set the rules of contemporary tissue engineering and regenerative medicine at large, i.e. the *de novo in vivo* and *ex vivo* fabrication of tissues and organs for the mammalian body [Reddi, 2000a, 2000b]. The three critical requirements for successful tissue engineering are a suitable extracellular matrix substratum, soluble inductive molecular signals and responding stem cells [Reddi 2000a, 2000b; Ripamonti *et al.,* 2000a]. Tissue induction and morphogenesis by combining extracellular matrix substrata and soluble morphogenetic signals is epitomized by the cellular cascade of the fascinating phenomenon of the induction of bone formation, the "*bone induction principle*" [Urist, 1965; Urist *et al.,* 1967; Urist *et al.,* 1968; Reddi and Huggins, 1972]. The molecular dissection of bone formation by induction has contributed significantly to our understanding of pattern formation, tissue engineering and regenerative medicine of several tissues and organs of the mammalian body [Reddi, 2000a, 2000b; Ripamonti *et al.,* 2000a; Ripamonti, 2006]. Indeed, the induction of tissue formation or morphogenesis is the essence of the tissue engineering paradigm engineered by combinatorial molecular protocols whereby soluble molecular signals are combined and reconstituted with insoluble signals or substrata [Reddi, 2000a; Ripamonti *et al.,* 2001a; Ripamonti 2004]. The reconstitution or recombination of osteogenic soluble molecular signals with different insoluble signals or substrata when implanted in heterotopic and orthotopic sites of animal models triggers the ripple-like cascade of bone differentiation by induction [Reddi, 2000a].

The challenging problem of the induction of bone formation in non-human and human primates [Urist, 1968] has stimulated the Bone Research Laboratory to develop experimental animal models using adult non-human primate *Papio ursinus* species that share similar bone physiology and remodeling with man [Schnitzler *et al.,* 1993; Ripamonti, 1996a]. Indeed, *Papio ursinus* is also a suitable model for the study of post-menopausal bone loss consistent with and resembling human post-menopausal osteoporosis [Dal Mas *et al.*, 2007] further highlighting similar osteonic bone remodeling with man (Schnitzler *et al.,* 1993]. This article describes the induction of bone formation as evaluated in the non-human primate *Papio ursinus* as a prerequisite for potential clinical application in man [Ripamonti *et al.*, 1992a; Ripamonti *et al.,* 1993; Ripamonti, 1996a; Ripamonti *et al.*, 2001a; Ripamonti *et al.,* 2004a; Ripamonti *et al.*, 2006; Ripamonti, 2006; Ripamonti *et al.*, 2007].

We also discuss the discovery of the induction of bone formation as initiated by the three mammalian TGF-β isoforms when singly implanted in heterotopic intramuscular sites of the *rectus abdominis* muscle of the non-human primate *Papio ursinus* [Ripamonti *et al.*, 1997; Ripamonti *et al.*, 2000b; Ripamonti *et al.,* 2008]. As recently highlighted [Ripamonti *et al.*, 2009a], we are of the opinion that the induction of endochondral bone formation as initiated by the three mammalian TGF-β isoforms in *Papio ursinus* has not been assigned nor understood [Ripamonti *et al.*, 2008; Ripamonti *et al.*, 2009a].

Bone Morphogenetic/Osteogenic Proteins

The induction of bone formation during embryonic development and postnatal osteogenesis including bone regeneration involves a complex cascade of molecular and morphogenetic processes that ultimately lead to the sculpture of precisely organized multicellular structures with mineralized bone surfaced by newly deposited as yet to be mineralized osteoid matrix; elucidating the nature and interaction of the signaling molecules that direct the initiation of bone formation is a major challenge of contemporary molecular, cellular, developmental and tissue engineering biology [Ripamonti, 2002]. Which are the molecular signals that initiate the ripple-like cascade of bone differentiation by induction?

Intramuscular and subcutaneous implantation of demineralized bone matrix results in the induction of endochondral bone formation [Urist, 1965; Reddi and Huggins, 1972; Reddi, 1981]. The '*bone matrix in the solid state*' [Reddi, 1997] thus contains several morphogenetic proteins endowed with the striking prerogative of initiating the induction of bone formation [Urist, 1965; Reddi and Huggins, 1972; Reddi, 1981; Reddi, 1997]. The biochemical and molecular problem of the '*bone matrix in the solid state*' [Reddi, 1997] has been unlocked by the dissociative extraction and reconstitution of the matrix components including osteogenic proteins tightly bound to the extracellular matrix of bone [Sampath and Reddi, 1981; Sampath and Reddi, 1983; Reddi, 1997].

Using highly purified protein fractions dissociatively extracted from baboon bone matrices (Figures 1, 2), we have provided the first morphological evidence of bone formation by induction in heterotopic intramuscular and orthotopic calvarial sites of non-human primates *Papio ursinus* (Figure 2) [Ripamonti *et al.*, 1992a; Ripamonti *et al.*, 1993; Ripamonti, 2006]. In a joint SA MRC and USA NIH research effort in collaboration with AH Reddi of the National Institutes of Health together with Noreen Cunningham, Shu-Shan Ma and Laura Yates, the Bone Research Laboratory purified osteogenin, also known as BMP-3, to homogeneity from baboon bone matrices; final purification to homogeneity was obtained after electroendosmotic elution by means of sodium dodecyl sulphate (SDS) polytacrylamide gel electrophoresis (PAGE) [Ripamonti *et al.*, 1992a]. This resulted in a single band on SDS-PAGE with a corresponding apparent molecular mass of 32kDa with biological activity in the subcutaneous space of the rat [Ripamonti *et al.*, 1992a] (Figure 1 A, B, C and D).

The bone induction potential of naturally-derived highly purified BMPs/OPs fractions was tested in non-healing calvarial defects of *Papio ursinus* (Figure 2) [Ripamonti *et al.*, 1992a; Ripamonti *et al.*, 1993; Ripamonti, 2003; Ripamonti, 2006; Ripamonti *et al.*, 2006]. The '*bone induction principle*' was also tested in human patients under the aegis of the University of the Witwatersrand, Johannesburg [Ripamonti and Ferretti, 2002; Ferretti and Ripamonti, 2002]. Highly purified bovine osteogenic fractions after additional heparin-Sepharose chromatography were used to treat mandibular defects in human patients (Figure 2E) [Ripamonti and Ferretti, 2002; Ferretti and Ripamonti, 2002].

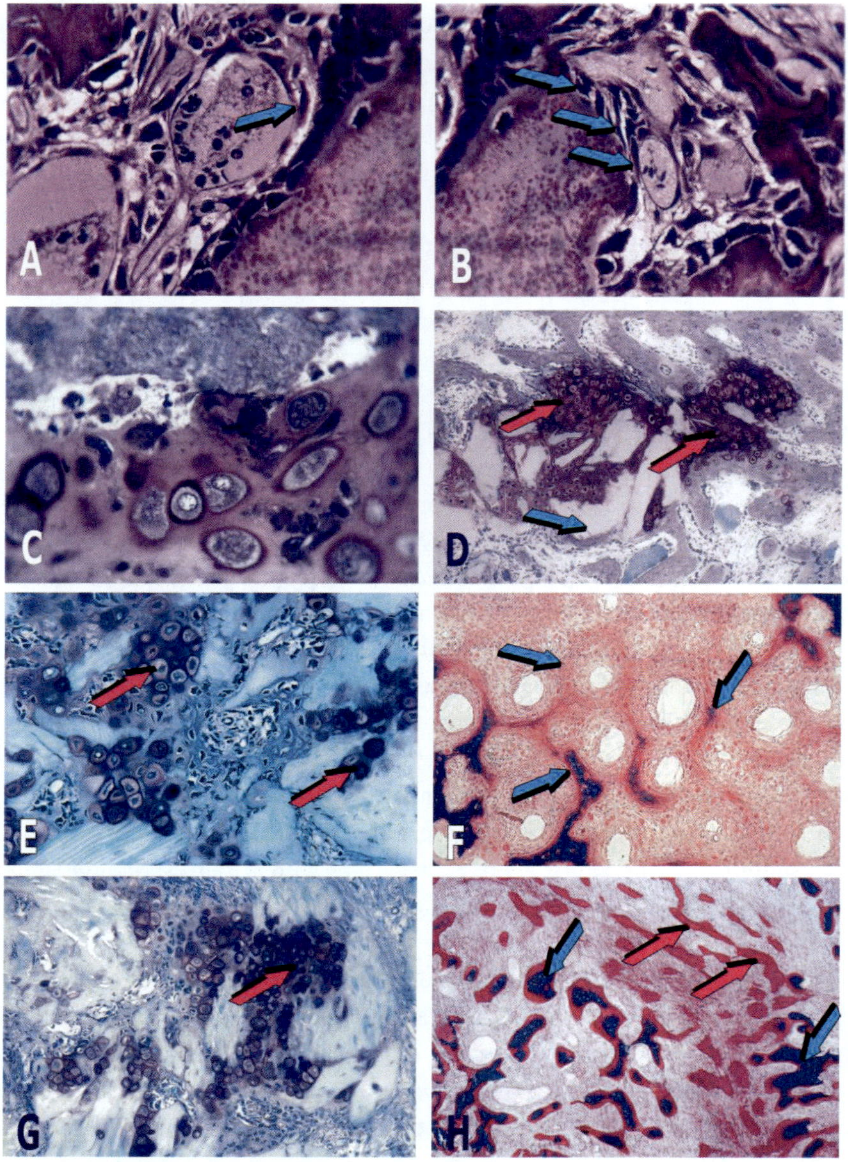

Figure 1. (*Continued*)

(A-D): Angiogenesis, tissue transformation and induction of endochondral bone formation by 0.1-0.5 μg naturally-derived osteogenin purified to apparent homogeneity from baboon bone matrices after electroendosmotic elution 11 days after heterotopic subcutaneous implantation in Long-Evans rats (Ripamonti *et al.* 1992); (A,B): Vascular invasion and capillary sprouting with differentiating endothelial/perycitic cells moving from the endothelial/vascular compartment (***blue*** arrows) to the osteoblastic compartment surfacing the matrix carrier as delivery of osteogenin. (B) ***blue*** arrows indicate the steady-state differentiation of progenitor stem cells, with migration and attachment from the vascular compartment to the osteoblastic microenvironment (C, D): Endochondral bone formation as recapitulation of embryonic development on day 11 after subcutaneous implantation of osteogenin in Long-Evans rats (Ripamonti et al 1992). Chondrogenesis (*magenta* arrows) has formed between the implanted matrix carrier (***blue*** arrow); (E): Endochondral bone formation by 0.1-0.5 μg osteogenin in the subcutaneous space of Long-Evans rats: islands of chondrogenesis (*magenta* arrows) with vascular invasion and differentiation of osteoblastic-like cells secreting bone matrix surrounding invading capillaries; (F): capillary sprouting, vascular invasion with the induction of mesenchymal condensations (***blue*** arrows) with foci of mineralization 30 days after implantation of osteogenin in the non-human primate *Papio ursinus* constructing the osteonic cortico-cancellous structure of the primate osteonic bone; (G,H): Induction of a cartilage anlage (*magenta* arrow) as a recapitulation of embryonic development in the rodent subcutaneous assay by 0.2 μg recombinant human osteogenic protein-1 (hOP-1) and bone differentiation by induction attached to the matrix carrier; (H): Heterotopic induction of prominent osteoid seams(*magenta* arrows) surfacing islands of newly formed and mineralized bone (***blue*** arrows) 30 days after implantation of 100 μg hOP-1/g of gamma-irradiated bovine insoluble collagenous bone matrix into the *rectus abdominis* muscle of *Papio ursinus* (Ripamonti 2005). (A, B, C, D, E, G): Undecalcified sections cut at 2μm stained with toluidine blue; (F, H): Undecalcified sections cut at 6 μm stained free floating **with Goldner's trichrome**.

Figure 1. Tissue induction and morphogenesis by the osteogenic proteins of the transforming growth factor-β (TGF-β) **supergene family** in rodents (A, B, C, D, E and G) and the non-human primate *Papio ursinus* (F, H).

In species ranging from the fruit fly to humans, bone morphogenetic proteins/osteogenic proteins (BMPs/OPs) play a critical role as signaling molecules during embryogenesis [Raftery *et al*., 1995; Sykaras and Opperman, 2003]. They are a unique group of morphogens within the transforming growth factor-β superfamily, which play a key role in the induction, maintenance and repair of bone [Reddi, 2000a; Ripamonti 2006; Sykaras and Opperman, 2003]. BMPs/OPs, originally isolated from the extracellular matrix of bone [Urist and Strates, 1971], are capable of inducing heterotopic cartilage and bone when implanted in muscle [Reddi, 2000a; Ripamonti, 2006; Ripamonti *et al*, 2006a]. Over the years, several BMPs/OPs have been identified [Celeste *et al.,* 1990; Reddi, 1997a; 1997b], with BMPs -2 through -7 known to induce heterotopic bone formation [Wang *et al.,* 1990].

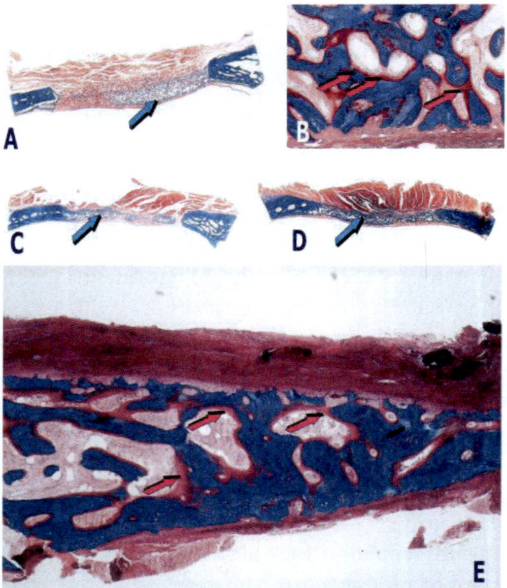

(A, B): Induction of bone formation (***blue*** arrow) surfaced by osteoid seams (*magenta* arrows in B) 30 days after calvarial implantation of highly purified naturally-derived BMPs/OPs extracted and purified greater than 50.000 fold with respect to crude extract after gel filtration chromatography of baboon bone matrix extracts (Ripamonti t al. 1992); (C, D): Induction of newly formed and mineralized bone (***blue*** arrows) 90 days after calvarial implantation of highly purified osteogenic fractions of baboon bone matrices showing complete calvarial regeneration (D); E: Undecalcified section of bioptic material harvested 90 days after implantation of highly purified bovine osteogenic fractions purified greater than 70.000 fold after gel filtration chromatography additionally chromatographed on Heparin-Sepharose and recombined with human collagenous bone matrix and implanted in a mandibular defect of *Homo sapiens* (Ripamonti and Ferretti 2002; Ferretti and Ripamonti 2002). *Magenta* arrows indicate osteoid seams as yet to be mineralized surfacing newly formed mineralized bone (in ***blue***). C c c Undecalcified sections cut at 6 μm stained free floating with **Goldner's trichrome.**

Figure 2. Bone formation by induction in orthotopic craniofacial sites of the non-human primate *Papio ursinus* and the human primate *Homo sapiens*.

The induction of bone formation by BMPs/OPs is one of the most significant discoveries in the field of bone physiology and bone surgery [Urist, 1968; Urist and Strates, 1971]. Bone: formation by autoinduction [Urist, 1965], initiates by invocation of soluble molecular signals that, when combined with insoluble signal or substrata, triggers the ripple-like cascade of bone differentiation by induction [Urist and Mikulski, 1979; Sampath and Reddi, 1981; Ripamonti, 2006;

Ripamonti *et al*, 2006a]. The osteogenic soluble molecular signals of the transforming growth factor-β (TGF-β) superfamily, the BMPs/OPs and, uniquely in primates, the TGF-β isoforms, induce endochondral bone formation as a recapitulation of embryonic development [Ripamonti, 2006; Ripamonti *et al*, 2008].

Factors that govern the bone-inducing capabilities of BMPs/OPs are storage [Raval, 1996], delivery system [Ripamonti *et al.,* 2001a; Li & Wozney, 2001;Salgado *et al.*, 2004; Rose *et al.*, 2004; Cook *et al.,* 2005], dose [Laffargue, 1999], the presence of inhibitors and antagonists [Raval, 1996; ten Dyke, 2006; Heliotis and Tsiridis 2008], tissue, site and animal species, [Ripamonti *et al.,* 1997, 2000b, 2001a, 2001b, 2004a, 2004b; Ripamonti, 2003, 2004a, 2004b; Duneas *et al.*, 1998; Obert *et al.,* 2005], age of recipient [Quarto *et al.,* 1995; Fleet *et al.,* 1996], condition of recipient bed, *i.e.* vascularization [Obert *et al.,* 2005], the presence or absence of responding mesenchymal cells [Puleo, 1997; Obert *et al.,* 2005], the presence of other morphogens, and the time period allocated for BMPs/OPs activity [Puleo, 1997; Obert *et al.,* 2005].

Significant discoveries have been made in the use of both naturally-derived [Ripamonti *et al.,* 2001a; Ferretti and Ripamonti, 2002; Ripamonti *et al.,* 2006; 2007] and recombinant BMPs [Wang *et al.,* 1990; Luu *et al.,* 2007]. Osteogenin, isolated by heparin chromatography from bovine and baboon bone matrices [Luyten *et al.,*1989; Reddi & Cunningham,1990; Ripamonti *et al.,* 1992a;] has a high affinity for native and denatured type IV collagen, binding equally well with both, and to a lesser degree with types I and IX collagen [Paralkar, 1990]. This irresistible binding of osteogenin to type IV collagen was an early indication that other differentiation factors would interact with a variety of extracellular matrices [Paralkar, 1990]. Successful bone induction by osteogenin has been reported in heterotopic [Ripamonti, *et al.,* 1992a, 1992b], calvarial [Hollinger *et al.,* 1989; Ripamonti and Reddi, 1992; Ripamonti, *et al.,* 1992a, 1992c; Ripamonti *et al.,* 1993], and periodontal sites [Ripamonti and Reddi 1992] in non-human [Ripamonti *et al.*, 2009b] and human primates [Bowers *et al.*, 1991].

BMP-2 and BMP-7 [also known as osteogenic protein-1 (OP-1)] are available as recombinant proteins and as such have enjoyed a vast amount of interest for their potential therapeutic worth. BMP-2 is a bone-inducing factor that stimulates new bone formation within 14 days [Toriumi, *et al.,* 1991]. Experimental animal studies have shown that hBMP-2 can be utilized to achieve total osseous repair of large mandibular defects. In a canine mandibular reconstruction study, Toriumi and co-workers [1991] evaluated the functional, histologic and biomechanical attributes of 3-cm full-thickness defects treated with BMP-2 with inactive dog bone matrix as carrier. Six months postoperative histomorphometric results showed complete healing in 68 % of the BMP-2 treated defects compared with 4 % of control defects. Biomechanical strength of the defects reconstructed with BMP-2 increased significantly from 3 to 6 months and was directly associated with the amount of mineralization and thickness of bone bridging the defect. In another canine mandibular reconstruction investigation by Seto *et al.* [2002], BMP-2, combined with gelatin sponge as carrier, induced enough new bone by 8 weeks to allow for the implantation of prosthetic fixtures to restore occlusal function. Clinical veterinary studies have identified hBMP-2 as an effective means of successful and immediate mandibular reconstruction whether due to tumour [Spector *et al.,* 2007] or trauma [Lewis *et al.,* 2008].

In an experimental non-human primate study, Boyne *et al.,* [1999], used 10-year old *Macaque fascicularis* monkeys to evaluate hBMP-2-induced mandibular reconstruction in a surgically-created discontinuity hemimandibulectomy model. After 5 months, core samples of the hBMP-2- implanted defects showed excellent bone density with total regeneration at 16 months when the animals were euthanased. Marukawa *et al.* [2002], achieved similar results after implanting mandibular defects of young *Macaca mulatta* for a period of 30 weeks. At 20 weeks Bränemark screw implants were placed in the regenerated mandible to restore masticatory function. Tibial osteotomy sites reconstructed with hBMP-2 in collagen or gelatin sponge-based carriers, showed radiographic union by 8 weeks indicating that hBMP-2 is useful in long bone defect reconstruction in that it shortens the recovery period and hastens the return of full limb function [Kokubo *et al.,* 2004; Faria *et al.,* 2007].

BMP-7/OP-1 has performed well in the study of long bone defects: Cook *et al.* [1998], in a canine study, tested hOP-1 in large segmental defects of ulnae. After 12 weeks, results showed that 89 % of the hOP-1-treated defects had healed completely, the hOP-1 defects were bridged with lamellar bone which was continuous with the host bone, and the mechanical strength of the hOP-1-treated defects was 65 % that of intact ulnae. Salkeld *et al.* [2001], determined the effect of implanting hBMP-7/OP-1 in various combinations (1:2 or 2:1) with both autogenous and allograft bone in canine ulnae. Twelve weeks postoperative results showed that any combination of OP-1 with autogenous or allograft bone or OP-1 alone yielded improved healing radiographically, biomechanically and histologically when compared with results from implanting autogenous or allograft bone alone. BMP-7/OP-1 has performed well in cartilage repair studies as well [Cook *et al.,* 2003; Chubinskaya *et al.,* 2007]. In a study by [Cook *et al.* 2003] the ability of hOP-1 to repair osteochondral defects in dogs was examined.

Osteochondral defects 5mm diameter x 6mm deep were surgically created in femoral condyles. RhOP-1 plus bovine derived collagen type I were used. Bovine collagen solo and defects left untreated were used as controls. Specimens were evaluated both grossly and histologically. RhOP-1-treated defects showed intact articular cartilage which appeared translucent and continuous.

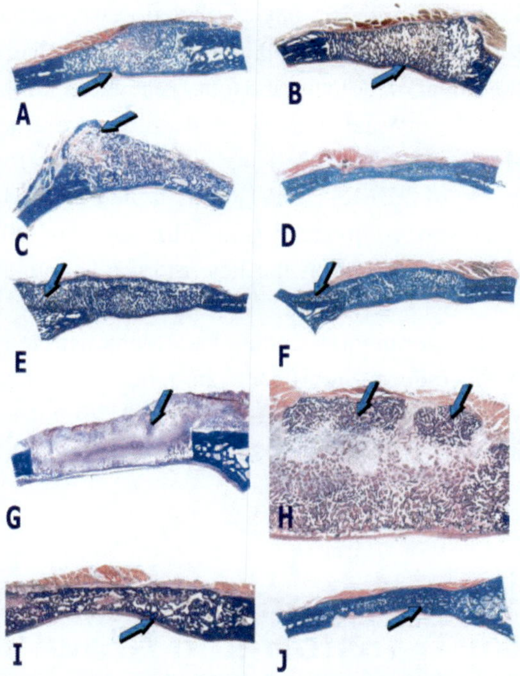

(A, B, C, D, E and F): Complete calvarial regeneration by 0.1 (A), 2.5 (B, C), 0.1 (D) and 2.5 mg recombinant human osteogenic protein-1 (hOP-1) delivered by bovine insoluble collagenous bone matrix as carrier and harvested on day 90 (A, B, C, D) and 365 (D, E, F) after implantation. Exuberant induction of bone formation (B, C) after implantation of 2.5 mg hOP-1 osteogenic devices resulting in substantial induction of bone formation up to one year after implantation (E, F). (G, H, I and J): Morphology of calvarial regeneration by 0.1 (G), 0.5 (H, I) and 2.5 mg of gamma-irradiated hOP-1 reconstituted with gamma-irradiated bovine collagenous bone matrix and harvested on day 15 (G), 30 (H), 90 (I) and 365 (J) after implantation with remodeling and regeneration up to one year after surgical implantation (J). Undecalcified sections cut at 4 to 6 μm stained free floating **with Goldner's trichrome**.

Figure 3. Morphology of calvarial regeneration by recombinant human OP-1 (hOP-1) osteogenic devices implanted in non-healing calvarial defects of the non-human primate *Papio ursinus* ad harvested at different time periods.

Pre-clinical studies using BMP-7/OP-1 in non-human primates [Ripamonti *et al.* 1996b; 2000a; 2001c], have been equally encouraging (Figure 3): Ripamonti *et al.* [1996b], studied the effect of a single application of rhBMP-7/OP-1 when used to regenerate large critical size calvarial defects in the adult baboon, *Papio ursinus.* Baboon insoluble collagenous matrix was used as carrier and combined with ascending doses (0.1, 0.5, and 2.5 mg) of hBMP-7/OP-1. Specimens harvested at 15, 30 and 90 and 365 days revealed newly-formed bone at day 15, even with the 0.1 mg dose. By day 90, the lower doses had induced complete regeneration of the calvarial defects and one year after implantation there was complete restoration of internal and external cortices of the calvaria. Further studies by Ripamonti *et al.,* [2001c], using sintered porous hydroxyapatite discs pretreated with doses of hBMP-7/OP-1 and implanted in both calvarial and heterotopic sites, showed complete regeneration by day 90 for the calvarial specimens, and evidence of bone in the heterotopically implanted specimens as early as day 15.

Although animal experimental studies have demonstrated by overwhelming evidence that BMPs induce significant amounts of bone, clinical studies, however, have been perceived often as equivocal [Friedlaender *et al.,* 2001; Govender *et al.,* 2002; Boyne *et al.,* 2005; Ripamonti *et al*, 2006a; Ripamonti *et al.*, 2007].

Endochondral Bone Induction by the Mammalian Transforming Growth Factor- β Isoforms and Synergistic Induction of Bone Formation

We have shown in a number of systematic studies in different microenvironments of heterotopic intramuscular and orthotopic craniofacial sites including the *rectus abdominis* muscle, the calvarium and the mandible, respectively, that primate tissues and microenvironments respond remarkably differently when compared to rodents, lagomorph and canine tissues at identical doses of osteogenic proteins of the TGF-β superfamily [Ripamonti *et al.*, 1997; Ripamonti *et al.,* 2001a; Ripamonti 2003; Ripamonti 2006; Ripamonti *et al.*, 2009a, 2009b]. The three mammalian TGF-β isoforms, the TGF-β_1, -β_2 and -β_3 proteins, to induce rapid and substantial endochondral bone formation in heterotopic sites of the *rectus abdominis* muscle of the non-human primate *Papio ursinus* (Figure 4) [Ripamonti *et al.*, 1997; Ripamonti *et al.*, 2000b; Ripamonti *et al.,* 2008]. The induction of substantial endochondral bone formation in heterotopic sites of *Papio ursinus* is remarkable and raises several questions related to mechanistic pathways profoundly different amongst mammalian species [Ripamonti *et al.,* 2009b]. Importantly, the biological evidence of apparent redundancy in non-human primates and thus by extension to *Homo sapiens* raises several questions with major significance for our deeper understanding of the '*bone induction principle*' [Urist *et al.*, 1967; Urist *et al.*, 1968; Urist, 1968; Urist and Strates, 1972] as well as the optimal utilization of this knowledge for tissue engineering in clinical context [Ripamonti *et al.*, 2009a]. Still we do not know why the TGF-β proteins induce endochondral bone formation in non-human primates *vs*. the lack of bone differentiation in rodents, lagomorphs and canine species [Roberts *et al.,* 1986; Ripamonti *et al.*, 1997; Ripamonti *et al.*, 2001; Ripamonti *et al*., 2006; Ripamonti, 2006; Ripamonti *et al.*, 2009b]. Which are the molecular signals that control the biological significance of apparent redundancy initiating the induction of endochondral bone formation? Why the mammalian TGF-β isoforms have such a different and critically important effect in non-human primates as compared to rodents and/or lagomorphs? Perhaps the most important question is: can we extrapolate results from animal models including *Papio ursinus* to *Homo sapiens*? [Ripamonti *et al.*, 2009].

The induction of endochondral bone formation by the mammalian TGF-β isoforms in non-human primates *Papio ursinus* is truly remarkable also considering the extent of tissue induction and morphogenesis with comparatively limited μg amount of the recombinant protein (Figure 4) [Ripamonti *et al.,* 1997; Ripamonti *et al.*, 2000b; Ripamonti *et al.*, 2008]. In particular, we have shown that the hTGF-β_3 isoform at the μg doses tested and delivered by the insoluble collagenous bone matrix as well as macroporous sintered calcium phosphate based bioceramic as carriers is the most powerful osteogenic protein of the TGF-β supergene family so far tested in the *rectus abdominis* muscle of *Papio ursinus* [Ripamonti *et al.*, 2008]. In the *bona fide* heterotopic bioassay for bone induction in rodents [Reddi, 2000a], the mammalian TGF-β isoforms do not initiate endochondral bone formation [Roberts *et al.*, 1986; Sampath *et al.*, 1987; Ripamonti, 2003; Ripamonti, 2006]. The presence of several related but molecularly different isoforms with osteogenic activity poses important questions as to the biological significance of this apparent redundancy, additionally indicating multiple interactions during both embryonic development and bone regeneration in postnatal life [Ripamonti, 2003; Ripamonti, 2006].

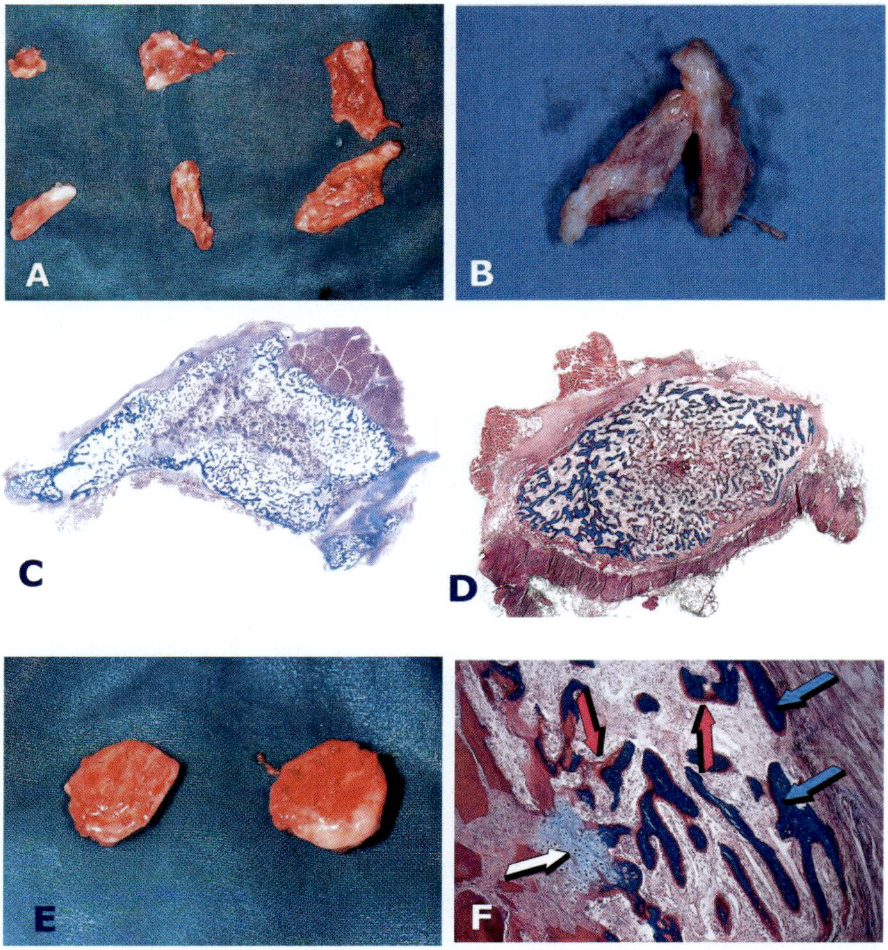

(A, B, C): Induction of large heterotopic ossicles by 125 μg hTGF-β_3 with corticalization of the newly formed and mineralized bone in blue (C) 90 days after implantation; (D): Large corticalized mineralized ossicle induced by 5 μg hTGF-β_1 on day 90 after implantation in the *rectus abdominis* of *Papio ursinus*. (E, F): Heterotopic ossicles (E) induced by 25 μg hTGF-β_2 with chondrogenesis (F) (white arrow) and newly formed mineralized bone (***blue*** arrows) surfaced by osteoid seams (*magenta* arrows). Undecalcified sections cut at 3 μm stained free floating **with Goldner's trichrome.**

Figure 4. Tissue induction and morphogenesis by the three mammalian transforming growth factor-β (TGF-β) **isoforms** when implanted in the *rectus abdominis* muscle of *Papio ursinus*.

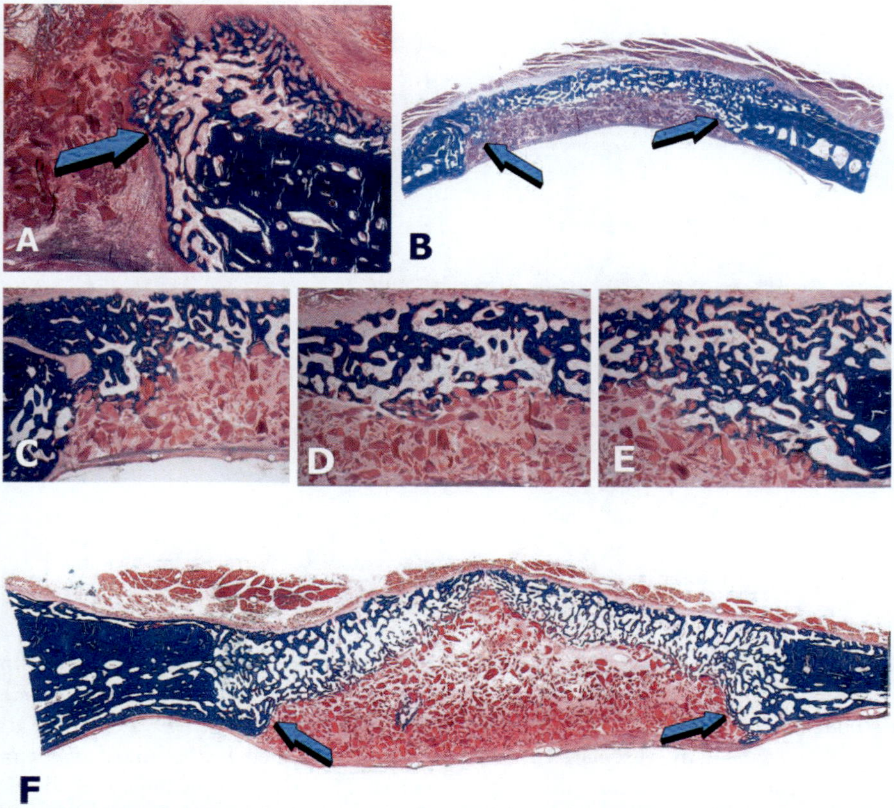

(A): Lack of bone differentiation 30 days after calvarial implantation of 125 μg hTGF-β_3; ***blue*** arrow points to the lack of bone formation and inhibition of osteogenesis within the defect microenvironment; (B, C, D and E): Bone differentiation albeit limited across the defect with lack of bone induction on the endocranial site of the specimens with scattered inactive collagenous matrix combined with 125 μg hTGF-β_3 90 days after calvarial implantation.; ***blue*** arrows point to the inhibition of osteogenesis within the defect microenvironment endocranially; (F): Identical morphology of calvarial regeneration 90 days after implantation of 100 μg hTGF-β_2 reconstituted with baboon insoluble collagenous bone matrix as carrier (Ripamonti et al. 2000). There is newly formed bone pericranially but lack of bone differentiation endocranially above the dural layer with inhibition (***blue*** arrows) of bone differentiation day 90 days after implantation. Undecalcified sections cut at 6 μm stained free floating with Goldner's trichrome.

Figure 5. Limited induction of bone formation by the transforming growth factor-β isoforms when implanted in non-healing calvarial defects of *Papio ursinus*.

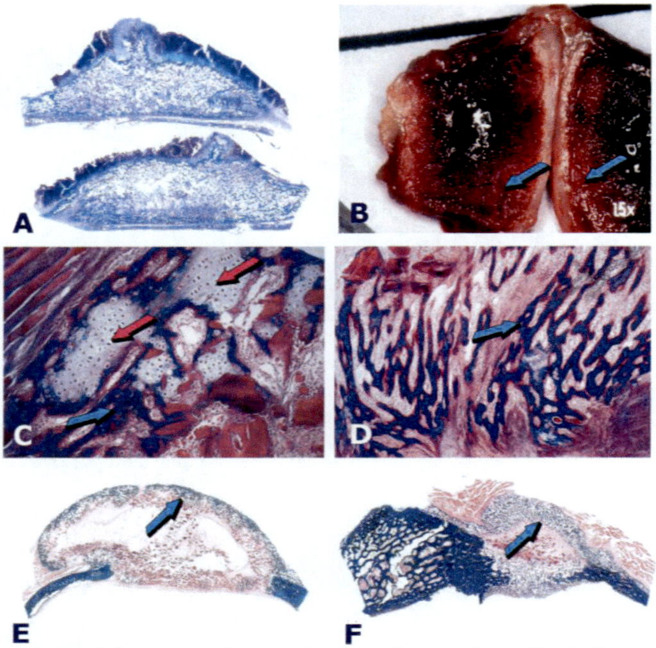

(A): Interposed recombinant morphogens implanted contralaterally in the *rectus abdominis* of *Papio ursinus* synergize to construct massive heterotopic ossicles in the *rectus abdominis* muscle of the recipient animal (Ripamonti *et al.* 1997). (B): Large corticalized highly vascularized ossicle *de novo* generated in the *rectus abdominis* after heterotopic intramuscular implantation of 25 μg hOP-1 in binary application with 1.5 μg hTGF-β_1 15 days after implantation. (C): Islands of cartilage and of endochondral bone formation recapitulating embryonic development with vascularization on day 15 after implantation of synergistic binary applications of 25 μg hOP-1 and 1.5 μg hTGF-β_1; (D): Rapid trabeculation of newly formed mineralized bone (in ***blue***) surfaced by large osteoid seams populated by contiguous osteoblasts 15 days after heterotopic intramuscular implantation of 25 μg hOP-1 with 1.5 μg hTGF-β1; (E): **Extensive pericranial osteogenesis with gross displacement** of the pericranial tissues 30 days after implantation of 100 μg hOP-1 combined with 5 μg platelet-derived porcine TGF-β_1 (Duneas et al. 1998); (F): Rapid induction of mineralized bone formation with displacement of the pericranial *vs.* the endocranial tissues with pronounced fibrovascular invasion between the newly induced and mineralized bone with pronounced pericranial osteogenesis after calvarial implantation of 100 μg hOP-1 combined with 15 μg platelet-derived porcine TGF-β_1.Undecalcified sections cut at 6 μm stained free floating **with Goldner's trichrome.**

Figure 6. Synergistic induction of bone formation in heterotopic intramuscular *rectus abdominis* and orthotopic calvarial sites after implantation of binary applications of recombinant human osteogenic protein-1 (hOP-1) and transforming growth factor-β_1 (TGF-β_1).

To date, more than 20 related proteins with BMP/OP like sequences and activities have been cloned, but little is known about their interaction during the cascade of bone formation by induction, or about the biological and therapeutic significance of this apparent redundancy [Ripamonti *et al.*, 2001a; Ripamonti, 2003; Ripamonti *et al.*, 2004a; Ripamonti, 2006; Ripamonti *et al.*, 2006; Ripamonti *et al.*, 2009b]. It is likely that the endogenous mechanisms of bone repair and regeneration in postnatal life necessitate the deployment and concerted action of several proteins members of the TGF-β supergene family including the three mammalian TGF-β isoforms.

Indeed research data in the non-human primate *Papio ursinus* have shown that the recombinant osteogenic protein-1 device (hOP-1) directly influences the expression levels of OP-1, type IV collagen, BMP-3 and TGF-β_1 mRNAs during the induction of bone formation (Ripamonti 2005). Upon implantation of hOP-1, the temporal and spatial expressions of TGF-β_1 mRNAs are relatively high on day 30 as compared with expression patterns on days 15 and 90 [Ripamonti, 2005]. This has indicated a specific temporal transcriptional window during which TGF-β_1 mRNA expression is mandatory for optimal osteogenesis [Ripamonti, 2005].

The presence of several molecularly different but homologous isoforms within the TGF-β supergene family points to synchronous and synergistic interactions during the cascade of endochondral bone formation; indeed we have shown a potent and accelerated synergistic induction of endochondral bone formation of hOP-1 together with recombinant or naturally-derived TGF-β_1 in both heterotopic *rectus abdominis* and orthotopic calvarial sites of *Papio ursinus* [Ripamonti *et al.*, 1997; Duneas *et al.*, 1998]. By studying the expression of mRNA of gene products of the TGF-β superfamily upon implantation of a recombinant human osteogenic protein [Ripamonti, 2005], we have shown that the induction of bone in *Papio ursinus* develops as a mosaic structure with distinct spatial and temporal patterns of gene expression of members of the TGF-β superfamily that singly, synergistically and synchronously initiate and maintain tissue induction and morphogenesis [Ripamonti, 2005].

The temporal pattern of gene expression in both heterotopic and orthotopic sites of *Papio ursinus* indicates a sequence of steady-state mRNA expression during osteogenic differentiation and the induction of bone formation [Ripamonti, 2005; Ripamonti *et al.,* 2006]. The high expression levels of auto-induced OP-1 mRNA together with expression of type IV collagen mRNA mechanistically explain the continuous and sustained osteogenesis in angiogenesis as evaluated by morphology and histomorphometry [Ripamonti, 2005]. The high and sustained expression of type IV collagen mRNA, a marker for capillary invasion and angiogenesis, mechanistically pre-dates the superior induction of heterotopic bone formation by the synergistic binary application of hOP-1 with relatively low doses of hTGF-β_1 or platelets-derived TGF-β_1 [Ripamonti *et al.*, 1997; Duneas *et al.*, 1998; Ripamonti, 2006].

The spatial and temporal window during which TGF-β_1 mRNA expression is mandatory for optimal osteogenesis has been unambiguously demonstrated by the endochondral osteoinductivity of the three mammalian TGF-β isoforms (Figure 4) [Ripamonti *et al.*, 1997; Ripamonti *et al.*, 2000b; Ripamonti *et al.,* 2008]. Our systematic studies on the synergistic induction of bone formation have shown that maximal response occurred at doses of 25 and 125 μg hOP-1 in combination with 1.5 and 5 μg hTGF-β_1, representing an optimal ration of 20:1 by weight [Ripamonti *et al.*, 1997]. On day 30, binary applications of 25 μg hOP-1 with 1.5 μg hTGF-β_1, and of 25 μg hOP-1 with 5 μg hTGF-β_1 resulted in generated tissue areas greater than the sum of the effects of both morphogens implanted singly [Ripamonti *et al.*, 1997]; such interactions are defined as synergistic [Ripamonti *et al.*, 1997].

Whether the observed endochondral bone formation by the mammalian TGF-β isoforms in the non-human primate *Papio ursinus* is initiated singly by the morphogens or is the result of an as yet unknown combinatorial action of subliminal amounts of locally produced BMPs/OPs [Ripamonti *et al.,* 1997] has not yet been elucidated; implantation of the mammalian TGF-β isoforms in the *rectus abdominis* muscle of *Papio ursinus* results in the expression of mRNA of single bone morphogenetic proteins of the TGF-β superfamily [Duneas *et al.*, 1998; Ripamonti *et al.*, 2000b; Ripamonti *et al.*, 2008]; potentially thus, the induction of bone formation by the TGF-β isoforms in *Papio ursinus* could be the result of BMPs/OPs mRNAs expression after the implantation of the TGF-β isoforms [Ripamonti *et al.*, 2008].

There may be several mechanisms by which TGF-β proteins synergize with hOP-1 in both heterotopic and orthotopic sites; the TGF-β proteins act as a chemotactic and mitogenic factor for responding precursor cells for subsequent induction by hOP-1; the synergistic induction of bone formation is possibly thus the result of superior cell migration and recruitment at the site of surgical implantation, resulting in the observed temporal and spatial enhancement of bone formation [Ripamonti *et al.*, 1997]. Northern blot analyses of tissue samples generated by the synergistic induction of bone formation have shown a three to four fold increase in collagen type IV mRNA in ossicles harvested on day 15 and 30 after heterotopic implantation [Duneas *et al.*, 1998] correlating thus with greater vascular invasion and angiogenesis supported by the binary application of the morphogens [Duneas *et al.*, 1998; Ripamonti, 2003; Ripamonti, 2006].

The rapidity of tissue morphogenesis complete with mineralization of the outer cortex and bone marrow formation bodes well for regeneration of bone in human patients particularly in the elderly where repair phenomena are temporally delayed and healing progresses slower than in pre-clinical animal models even when compared to non-human primate animal species.

Perspective of Tissue Engineering of Bone in Clinical Contexts

Regenerative medicine and bone tissue engineering in clinical contexts have proven to be an elusive endeavor compared to data obtained in pre-clinical trials including non-human primate species [Williams, 2006; Ripamonti, 2006; Ripamonti *et al.*, 2006; Ripamonti *et al.*, 2007]. Bone tissue engineering in several preclinical animal models including the non-human primate *Papio ursinus* has indicated that as single recombinant hBMP/OP would result in tissue induction and morphogenesis in human patients. This theoretical potential has not been translated in acceptable results in human patients. Clinical trials of orthopedic and craniofacial applications such as mandibular reconstructions and sinus-lift operations have indicated that supra physiological doses of single recombinant gene products are required in human patients incurring significant costs without achieving equivalent performance to autogenous bone grafts [Friedlander *et al.*, 2001; Govender *et al.*, 2003; Garrison *et al.*, 2007]; this suggests a flaw in the single morphogen approach to human osteoinduction [Ripamonti *et al.*, 2006; Ripamonti *et al.*, 2007].

In his classic Editorial Comment titled '*The Reality of a Nebulous and Enigmatic Myth*' [1968], Marshall Urist stated that the pre-clinical and clinical research on the bone induction principle '*are bound to dispel the myth and appreciate the reality of bone induction for the benefit of patients with crippling diseases of the bones and joints*'. Forty years later, the Bone Research Laboratory not in Los Angeles but in Johannesburg still feels '*The reality of a Nebulous Enigmatic Myth*' when reading that several tens of milligrams of recombinant hBMPs/OPs are needed to induce an uninspiring bone volume in humans. Human bone regeneration has proven to be an elusive target when compared to results obtained in pre-clinical trials including non-human primates. The induction of bone has dramatically shown that regenerative medicine in clinical contexts is on a different scale altogether when compared to animal models that may not adequately translate and reproduce morphogens-related therapeutic responses in *Homo sapiens*. We have to start to identify systematically the molecular and cellular basis responsible for the significant differences in the healing patterns amongst mammals, Major research efforts should now be devoted to analyze genetically the mammalian wound healing tract controlling the extent of tissue regeneration [McBrearty *et al.*, 1998].

The molecular cloning of the osteonic proteins of the TGF-β supergene family and the results of the several pre-clinical studies in a variety of animal species including non-human primates has convinced molecular biologists, tissue engineers and skeletal recostructionists alike to believe that a single recombinant human osteogenic protein would result in tissue induction and morphogenesis in human patients. Indeed pre-clinical systematic studies in *Papio ursinus* have shown that single applications of recombinant hBMPs/OPs did result in complete regeneration of craniofacial and axial skeletal defects. This theoretical potential has not been translated to acceptable predictable results in human patients. The induction of bone formation by supra-physiological doses of single hBMPs/OPs has been often punctuated by the lack of total and/or convincing regeneration of the missing part of bone so often needed in unfavorable clinical conditions. The extrapolation of data from animal models including non-human primate species is difficult since dosage is tissue-specific. The induction of bone in human patients has dramatically shown that regenerative medicine in clinical contexts is on a different scale altogether when compared to animal models including non-human primates. Only time will tell if the tissue engineering paradigm will be used routinely in clinical contexts and if the fundamental major discoveries in molecular and cellular biology and experimental surgery that have led to the emerging science of regenerative medicine will be rigorously applied to the predictable fabrication and grafting of new tissues and organs for the human body [Williams, 2006]. Finally, animal models including non-human primates may not adequately reproduce morphogens-related therapeutic responses in *Homo sapiens*.

To date bone tissue engineering has been tackled in a rather crude single morphogen approach and has resulted in uninspiring clinical performance at massive and expensive doses. This approach denies the biological evidence that has demonstrated unambiguously, at least in non-human primates *Papio ursinus*, the benefit of binary applications of recombinant morphogens to massively sustain continuous osteogenesis in angiogenesis at sites of surgical applications [Ripamonti *et al.*, 1997; Ripamonti *et al.*, 2006; Ripamonti *et al.*, 2007). The application of a single morphogen incompletely harnesses the biological machinery for bone regeneration requiring mega doses of recombinant morphogens to repair skeletal defects with doses that far exceed the total amount of BMPs/OPs present in the entire human skeleton.

Acknowledgments

We acknowledge the continuous support of the Medical Research Council of South Africa, the University of the Witwatersrand, Johannesburg and the National Research Foundation of South Africa since the opening the Bone Research Laboratory attached to the Medical School of the University. We acknowledge with thank collaborative research and the use of human recombinant morphogens from Creative BioMolecules/Stryker Biotech. USA, Novartis AG Switzerland, Genzyme Corporation, and Genentech Inc. USA. We thank MR Urist and AH Reddi for the introduction to the fascinating phenomenon of '*Bone: Formation by autoinduction*'; a special thank to Barbara van den Heever for establishing unique undecalcified histology techniques for outstanding microphotography on undecalcified sections cut at 3 to 6 μm.

References

Bowers, G; Felton, F; Middleton, C; Glynn, D; Sharp, S; Mellonig, J; Corio, R., Emerson, J., Park, S; Suzuki, J. Histologic comparison of regeneration in human intrabony defects when osteogenin is combined with demineralized freeze-dried bone allograft and with purified bovine collagen. *J Periodntol*, 1991, 62, 690-702.

Boyne, PJ; Lilly, LC; Marx, RE; Moy, PK; Nevins, M; Spagnoli, DB; Triplett, RG. De novo bone induction by recombinant human bone morphogenetic protein-2 (rhBMP-2) in maxillary sinus floor augmentation. *J Oral Maxillofac Surg*, 2005, 63, 1693-1707.

Boyne, PJ; Nakamura, A; Shabahang, S. Evaluation of the long-term effect of function on rhBMP-2 regenerated hemimandibulectomy defects. *British Journal of Oral and Maxillofacial Surgery*, 1999, 37, 344-352.

Celeste, AJ; Iannazzi, JA; Taylor, RC; Hewick, RM; Rosen, V; Wang, EA; Wozney, JM. Identification of transforming growth factor β family members present in bone-inductive protein purified from bovine bone. *Proc Natl Acad Sci*, USA, 1990, 87, 9843-9847.

Chubinskaya, S; Hurtig, M; Rueger, DC. OP-1/BMP-7 in cartilage repair. *Int Orthop*, 2007, 31, 773-781.

Cook, SD; Patron, LP; Salkeld, SL; Rueger, DC. Repair of articular cartilage defects with osteogenic protein-1 (BMP-7) in dogs. *J Bone Joint Surg*, 2003, 85, 116-123.

Cook, SD; Salkeld, SL; Brinker, MR; Wolfe, MW; Rueger, DC. Use of an osteoinductive biomaterial (rhOP-1) in healing large segmental bone defects. *J Orthop Trauma*, 1998, 12, 407-412.

Cook, SD; Salkeld, SL; Patron, LP. Bone defect healing with an osteogenic protein-1 device combiner with carboxymethylcellulose. *Appl Biomater*, 2005, 75, 137-145.

Dal Mas, I; Biscardi, A; Schnitzler, CM; Ripamonti, U. Bone loss in the ovariectomized baboon *Papio ursinus:* densitometry, histomorphometry and biochemistry. *J Cell Mol Med*, 2007, 11, 852-867.

Duneas, N; Crooks, J; Ripamonti, U. Transforming growth factor-β1: induction of bone morphogenetic protein genes expression during endochondral bone formation in the baboon, and synergistic interaction with osteogenic protein-1 (BMP-7). *Growth Factors*, 1998, 15, 259-277.

Faria, MLE; Lu, Y; Heaney, K; Uthamanthil, RK; Muir, P; Markel, MD. Recombinant human bone morphogenetic protein-2 in absorbable collagen sponge enhances bone healing of tibial osteotomies in dogs. *Veterinary Surgery*, 2007, 36, 122-131.

Ferretti, C; Ripamonti, U. Human segmental mandibular defects treated with naturally derived bone morphogenetic proteins. *J Craniofac Surg*, 2002, 13, 434-444.

Fleet, JC; Cashman, K; Cox, K; Rosen, V. The effects of aging on the bone inductive activity of recombinant human bone morphogenetic protein-2. *Endochrinology*, 1996, 137, 4605-4610.

Friedlaender, GE; Perry, CR; Cole, JD Cook, SD; Cierny, G; Muschler, GE; Zych, GA; Calhoun, JH; LaForte, AJ; Yin, S. Osteogenic protein-1 (bone morphogenetic protein-7) in the treatment of tibial nonunions. *J Bone Joint Surg*, 2001, 83, 151-S158.

Garrison, KR; Donnell, S; Ryder, J; Shemilt, I; Mugford, M; Harvey, I; Song, F. Clinical effectiveness and cost-effectiveness of bone morphogenetic proteins in the non-healing of fractures and spinal fusion: a systematic review. *Health Technol Assess*, 2007, 11, 1-168.

Govender, S; Csimma, C; Genant, HK; Valentin-Opran, A. Recombinant human bone morphogenetic protein-2 for treatment of open tibial fractures: a prospective, controlled, randomized study for four hundred and fifty patients. *J Bone Joint Surg*, 2002, 84, 2123-2134.

Heliotis, M; Tsiridis, E. Suppression of bone morphogenetic protein inhibitors promotes osteogenic differentiation: therapeutic implications. *Arthritis Res Ther*, 2008, 10, 115-116.

Hollinger, J; Mark, DE; Bach, DE; Reddi, AH; Seyfer, AE. Calvarial bone regeneration using osteogenin. *J Oral Maxillofac Surg*, 1989, 47, 1182-1186.

Kokubo, S; Mochizuki, M; Fukushima, S; Ito, T; Nozaki, K; Iwai, T; Takahashi, K; Miyata, K; Sasaki, N. Long-term stability of bone tissues induced by an osteoinductive biomaterial, recombinant human bone morphogenetic protein-2 and a biodegradable carrier. *Biomaterials*, 2004, 25, 1795-1803.

Laffargue, PH; Hildebrand, HF; Rtaimate, M; Frayssinet, O; Amoureux, JP; Merchandise, X. Evaluation of human recombinant bone morphogenetic protein-2-loaded tricalcium phosphate implants in rabbits' bone defects. *Bone*, 1999, 25, 55S-58S.

Lewis, JR; Boudrieau, RJ; Reiter, AM; Seeherman, HJ; Gilley, RS. Mandibular reconstruction after gunshot trauma in a dog by use of recombinant human bone morphogenetic protein-2. *J Am Vet Med Assoc*, 2008, 233, 1598-1604.

Li, RH; Wozney, JM. Delivering on the promise of bone morphogenetic proteins. *Trends Biotechnol*, 2001, 19, 255-265.

Luu, HH; Song, WX; Luo, X; Manning D; Luo, J; Deng, ZL; Shariff, KA; Montag, AG; Haydon, RC; He, TC. Distinct roles of bone morphogenetic proteins in osteogenic differentiation of mesenchymal stem cells. *J Orthop Res*, 2007, 25, 665-677.

Luyten, FP; Cunningham, NS; Ma, S; Muthukumaran, N; Hammonds, RG; Nevins, WB; Woods, WI, Reddi AH. Purification and partial amino acid sequence of osteogenin, a protein initiating bone differentiation. *J Biol Chem*, 1989, 264, 13377-13380.

Marukawa, E; Asahina, I; Oda, M; Seto, I; Alam, M; Enomoto, S. Functional reconstruction of the non-human primate mandible using recombinant human bone morphogenetic protein-2. *Int J Oral Maxillofac Surg*, 2002, 31, 287-295.

McBrearty, BA; Clark, LD; Zhang, XM; Blankenhorn, EP; Heber-Katz, E. Genetic analysis of a mammalian wound-healing trait. *Proc Natl Acad Sci*, USA, 1998, 95, 11792-11797.

Obert, L; Deschaseaux, F; Garbuio, P. Critical analysis and efficacy of BMPs in long bones non-union. *Injury*, 2005, 36, 38-42.

Paralkar, VM; Nandedkar, AKN; Pointers, RH; Kleinman, HK; Reddi, AH. Interaction of osteogenin, a heparin binding bone morphogenetic protein, with type IV Collagen. *J Biol Chem*, 1990, 265, 17281-17284.

Puleo, DA. Dependence of mesenchymal cell responses on duration of exposure to bone morphogenetic protein-2 *in vitro*. *J Cell Physiol*, 1997, 173, 93-101.

Quarto, R; Thomas, D; Liang. CT. Bone progenitor cell deficits and the age-associated decline in bone repair capacity. *Calcif Tissue Int*, 1995, 56, 123-129.

Raftery, LA; Twombly, V; Wharton, K; Gelbart, WM. Genetic screens to identify elements of the *decapentaplegic* signaling pathway in *Drosophila Genetics* 1995, 139, 241-254.

Raval, P; Hsu, HHT; Schneider, DJ; Sarras Jr., MP; Masuhara, K; Bonewald, LF; Anderson, HC. Expression of bone morphogenetic proteins by osteoinductive and non-osteoinductive human osteosarcoma cells. *J Dent Res*, 1996, 75, 1518-1523.

Reddi AH. Extracellular matrix and development. Piez KA 1984, A. H. Reddi (Ed.). *Extracellular Matrix Biochemistry*, Elsevier, New York, USA, 375-412.

Reddi, AH. Bone morphogenesis and modeling: soluble signals sculpt osteosomes in the solid state. *Cell*, 1997, 89, 309-319.

Reddi, AH. Bone morphogenetic proteins: An unconventional approach to isolation of first mammalian morphogens. *Cytokine Growth Factor Rev*, 1997, 8, 11-20.

Reddi, AH. Cell biology and biochemistry of endochondral bone development. *Coll Relat Res*, 1981, 1, 209-226.

Reddi, AH. Morphogenesis and tissue engineering of bone and cartilage: inductive signals, stem cells, and biomimetic biomaterials. *Tissue Eng*, 2000a, 6, 351-359.

Reddi, AH. Morphogenetic messages are in the extracellular matrix: biotechnology from bench to bedside. *Biochem Soc Trans*, 2000b, 28, 345-349.

Reddi, AH; Huggins, CB. Biochemical sequences in the transformation of normal fibroblasts in adolescent rats. *Proc Natl Acad Sci*, USA, 1972, 69, 1601-1605.

Ripamonti U. Tissue engineering of bone by novel substrata instructing gene expression during the *de novo* bone formation. *Science in Africa* 2002; http://www.scienceinafrica.co.za/2002/march/bone2.htm

Ripamonti, U. Bone induction by recombinant human osteogenic protein-1 (hOP-1, BMP7) in the primate *Papio ursinus* with expression of mRNA of gene products of the TGF-beta superfamily. *J Cell Mol Med*, 2005, 9, 911-928.

Ripamonti, U. Molecular signals in geometrical cues sculpt bone morphology. *S Afr J Sci*, 2004, 100, 355-367.

Ripamonti, U. Osteogenic proteins of the TGF-β superfamily. In: Henry, HL; Norman, AW, editors. Encyclopedia of Hormones. Austin Academic Press, 2003, 80-86.

Ripamonti, U. Soluble osteogenic molecular signals and the induction of bone formation. Leading opinion paper, *Biomaterials*, 2006, 27, 807-822.

Ripamonti, U. Soluble, insoluble and geometric signals sculpt the architecture of mineralized tissue. *J Cell Mol Med*, 2004, 8, 169-180.

Ripamonti, U. The generation of bone in primates by bone morphogenetic proteins 1996a, T. S. Lindholm (Ed.). *Bone morphogenetic proteins: biology, biochemistry and reconstructive surgery*, Academic Press, Inc., San Diego, California USA, P131-141.

Ripamonti, U; Crooks, J; Matsaba, T; Tasker, J. Induction of endochondral bone formation by recombinant human transforming growth factor-β2 in the baboon (*Papio ursinus*). *Growth Factors*, 2000b, 17, 269-285.

Ripamonti, U; Crooks, J; Rueger, DC. Induction of bone formation by recombinant human osteogenic protein-1 and sintered porous hydroxyapatite in adult primates. *Plast Recon Surg*, 2001b, 107, 977-988.

Ripamonti, U; Duneas, N. Tissue morphogenesis and regeneration by bone morphogenetic proteins. *Plast Reconstr Surg*, 1998, 101, 227-239.

Ripamonti, U; Duneas, N; van den Heever, B; Bosch, C; Crooks, J. Recombinant transforming growth factor-β1 induces endochondral bone in the baboon and synergizes with recombinant osteogenic protein-1 (bone morphogenetic protein-7) to initiate rapid bone formation. *J Bone Miner Res*, 1997, 12, 1584-1595.

Ripamonti, U; Ferretti, C. Mandibular reconstruction using naturally-derived bone morphogenetic proteins: A clinical trial report. In: T. S. Lindholm (Ed.), *Advances in Skeletal Reconstruction Using Bone Morphogenetic Proteins*. World Scientific Publ, Co. Singapore, 2002, 277-289.

Ripamonti, U; Ferretti, C; Heliotis, M. Soluble and insoluble signals and the induction of bone formation: molecular therapeutics recapitulating development. *J Anat*, 2006a, 209, 447-468.

Ripamonti, U; Ferretti, C; Teare, J; Blann, L. The transforming growth factor-β isoforms and the induction of bone formation: implications for reconstructive craniofacial surgery. *J Craniofac Surg*, 2009a, 20, 1544-1555.

Ripamonti, U; Heliotis, M; Ferretti, C. Bone morphogenetic proteins and the induction of bone formation: from laboratory to patients. *Oral Maxillofac Surg Clin North Am*, 2007, 19, 575-589.

Ripamonti, U; Ma, S; Cunningham, NS; Yeates, L, Reddi, AH. Initiation of bone regeneration in adult baboons by osteogenin, a bone morphogenetic protein. *Matrix*, 1992a, 12, 369-380.

Ripamonti, U; Ma, SS; Cunningham, NS; Yeates, L; Reddi, AH. Reconstruction of the bone-bone marrow organ by osteogenin, a bone morphogenetic protein, and demineralized bone matrix in calvarial defects of adult primates. *Plast Reconstr Surg*, 1993, 91, 27-36.

Ripamonti, U; Ma, SS; Reddi, AH. Induction of bone in composites of osteogenin and porous hydroxyapatite in baboons. *Plast Reconstr Surg*, 1992b, 89, 731-739.

Ripamonti, U; Ma, SS; van den Heever, B; Reddi, AH. Osteogenin, a bone morphogenetic protein, adsorbed on porous hydroxyapatite substrata, induces rapid bone differentiation in calvarial defects of adult primates. *Plast Reconstr Surg*, 1992c, 90, 382-393.

Ripamonti, U; Petit, JC; Teare, J. Cementogenesis and the induction of periodontal tissue regeneration by the osteogenic proteins of the transforming growth factor-β superfamily. *J Periodont Res*, 2009b, 44, 141-152.

Ripamonti, U; Ramoshebi, LN; Matsaba, T; Crooks, J; Teare, J. Bone induction by BMPs/OPs and related family members in primates. The critical role of delivery systems. *J Bone Joint Sur*, 2001a, 83, 16-27.

Ripamonti, U; Ramoshebi, LN; Patton, J; Matsaba, T; Teare, J; Renton, L. Soluble signals and insoluble substrata: novel molecular cues instructing the induction of bone. In Massaro, EJ; Rogers, JM, editors. The skeleton. Humana Press, 2004a, p. 217-227 [chapter 15].

Ripamonti, U; Ramoshebi, LN; Patton, J; Matsaba, T; Teare, J; Renton, L. Sculpting the architecture of mineralized tissue: tissue engineering of bone from soluble signals to smart biomimetic matrices. In: Kayser, O; Müller, RH, editors. Pharmaceutical biotechnolby: drug discovery and clinical applications. Wiley-VCH, 2004b, p. 281-297.

Ripamonti, U; Ramoshebi, LN; Teare J; Renton L; Ferretti C. The induction of endochondral bone formation by transforming growth factor-β3: experimental studies in the non-human primate *Papio ursinus*. *J Cell Mol Med*, 2008, 12, 1029-1048.

Ripamonti, U; Reddi AH. Growth and morphogenetic factors in bone induction: role of osteogenin and related bone morphogenetic proteins in craniofacial and periodontal bone repair. *Crit Rev Oral Biol Med*, 1992, 3, 1-14.

Ripamonti, U; van den Heever, B; Crooks, J; Tucker, MM; Sampath, TK; Rueger, DC; Reddi, AH. Long-term evaluation of bone formation by osteogenic protein 1 in the baboon and relative efficacy of bone-derived bone morphogenetic proteins delivered by irradiated xenogeneic collagenous matrices. *J Bone Miner Res*, 2000a, 15, 1798-1809.

Ripamonti, U; van den Heever, B; Sampath, TK; Tucker, MM; Rueger, DC; Reddi, AH. Complete regeneration in the baboon by recombinant human osteogenic protein-1 (hOP-1, bone morphogenetic protein-7). *Growth Factors*, 1996b, 13, 273-289.

Roberts, AB; Sporn, MB; Assoian, RK; Smith, JM; Roche, NS; Wakefield, LM; Fauci, AS. Transforming growth factor type ß: rapid induction of fibrosis and angiogenesis *in vivo* and stimulation of collagen formation *in vitro*. *Proc Natl Acad Sci*, USA, 1986, 83, 4167-4171.

Rose, FRAJ; Hou, Q ; Oreffo, ROC. Delivery systems for bone growth factors – the new players in skeletal regeneration. *J Pharm Pharmacol*, 2004, 56, 415-427.

Salgado, AJ; Coutinho, OP ; Reis, RL. Bone tissue engineering: State of the art future trends. *Macromol Biosci*, 2004, 4, 743-765.

Salkeld, SL; Patron, LP; Barrack, RL; Cook, SD. The effect of osteogenic Protein-1 on the healing of segmental bone defects treated with autograft or allograft bone. *J Bone Joint Surg*, 2001, 83, 803-816.

Sampath, TK; Muthukumaran, N; Reddi, AH. Isolation of osteogenin, an extracellular matrix-associated, bone-inductive protein, by heparin affinity chromatography. *Proc Natl Acad Sci*, USA, 1987, 84, 7109-7113.

Sampath, TK; Reddi, AH. Dissociative extraction and reconstitution of extracellular matrix components involved in local bone differentiation. *Proc Natl Acad Sci*, USA, 1981, 78, 7599-7603.

Sampath, TK; Reddi, AH. Homology of bone-inductive proteins from human, monkey, bovine and rat extracellular matrix. *Proc Natl Acad Sci*, USA, 1983, 80, 6591-6595.

Schnitzler, CM; Ripamonti, U; Mesquita, JM. Histomorphometry of iliac crest trabecular bone in adult male baboons in captivity. *Calcif Tissue Int*, 1993, 52, 447-454.

Seto, I; Tachikawa, N; Mori, Masaji; Hoshino, S; Marukawa, E; Asahina, I; Enomoto, S. Restoration of occusal function using osseointegrated implants in the canine mandible reconstructed by rhBMP-2. *Clin Oral Impl Res*, 2002, 13, 536-541.

Spector, DI; Keating, JH; Boudrieau, RJ. Immediate mandibular reconstruction of a 5 cm defect using rhBMP-2 after partial mandibulectomy in a dog. *Veterinary Surgery*, 2007, 36, 752-759.

Sykaras, N; Opperman, LA. Bone morphogenetic proteins (BMPs): How do they function and what can they offer the clinician? *J Oral Sci*, 2003, 45, 57-73.

ten Dyke, P. Bone morphogenetic protein signal transduction in bone. *Curr Med Res Opin*, 2006, 22, 7-11.

Toriumi, DM; Kotler HS; Luxenberg DP; Holtrop, ME; Wang, EA. Mandibular reconstruction with a recombinant bone-inducing factor. Functional, histologic, and biomechanical evaluation. *Arch Otolaryngol Head Neck Surg*, 1991, 117, 1101-1112.

Turing, AM. The chemical basis of morphogenesis. *Philos Trans R Soc Lond*, 1952, 237, 37-41.

Urist, MR. Bone: Formation by autoinduction. *Science*, 1965, 150, 893-899.

Urist, MR. The reality of a nebulous, enigmatic myth. *Clin Orthop* 1968, 59, 3-5.

Urist, MR; Dowell, TA; Hay, PH; Strates, BS. Inductive substrates for bone formation. *Clin Orthop Related Res*, 1968, 58, 59-96.

Urist, MR; Mikulski, AJ. A soluble bone morphogenetic protein extracted from bone matrix with a mixed aqueous and nonaqueous solvent. *Proc Soc Exp Biol Med*, 1979, 162, 48-53.

Urist, MR; Silverman, BF; Büring, K; Dubuc, FL; Rosenberg, JM. The Bone Induction Principle. *Clin Orthop Related Res*, 1967, 53, 243-283.

Urist, MR; Strates, BS. Bone morphogenetic protein. *J Dent Res Suppl*, 1971, 6, 1392-1406.

Wang, EA; Rosen, V; D'Allessandro, JS; Bauduy, M; Cordes, P; Harada, T; Israel, DI; Hewick, RM; Kerns, KM; LaPan, P; Luxenberg, DP; McQuaid, D; Moutsatsos, IK; Nove, J; Wozney, JM. Recombinant human bone morphogenetic protein induces bone formation. *Proc Natl Acad Sci*, USA, 1990, 87, 2220-2224.

Williams, DF. Tissue engineering: the multidisciplinary epitome of hope and despair, 2006, R. Paton, & L. McNamara (Eds.), *Studies in multidisciplinarity*, Elsevier, Amsterdam, 483-524.

Wozney, JM; Rosen, V; Celeste, AJ; Mitsock, LM; Whitters, MJ; Kriz, RW; Hewick, RM; Wang, EA. Novel regulators of bone formation: molecular clones and activities. *Science*, 1988, 242, 1528-1534.

In: Bone Regeneration: Growth Factors...
Editors: V. Legard et al., pp. 263-273
ISBN: 978-1-60876-621-5

Chapter IX

Mechanisms of Bony Repair at the Injured Growth Plate Cartilage

Cory J. Xian*
Sansom Institute for Health Research, and School of Pharmacy and Medical Sciences, University of South Australia, City East Campus, GPO Box 2471, Adelaide 5001, Australia.

Abstract

Growth plate cartilage is responsible for lengthening of long bones in children. However, its trauma injuries are common and are often repaired undesirably by bony tissue which results in bone growth defects in children. This chapter reviews recent studies using a tibial growth plate injury model in young rats, which have characterised the injury responses and bony repair mechanisms. In this model, bony tissue formation at injured growth plate occurs within 2 weeks and sequentially through inflammatory, fibrogenic, osteogenic, and remodelling responses. The initial inflammatory response is rapid and acute at the injured growth plate, with neutrophil-dominant inflammatory infiltration and up-regulation of inflammatory cytokines and growth factors (TNF-α, IL-1β, TGF-β1, PDGF-B, BMP-3 and -4) and inflammatory mediators (Cox-2 and iNOS) peaking on day 1 and subsiding

* Corresponding author: Fax: (618) 8302 1087; Email: cory.xian@unisa.edu.au

on day 3. After the initial inflammatory response, infiltrated mesenchymal cells undergo osteogenic or chondrogenic events, resulting in both intramembranous and endochondral bone formation and bony repair. Taken together, bony repair at the injured growth plate is preceded sequentially by inflammatory, fibrogenic, and osteogenic responses, and is through both intramembranous and endochondral ossification mechanisms.

Introduction

The growth plate, located between the epiphysis and metaphysis at the ends of a long bone, is responsible for longitudinal growth of children's long bones. Bone lengthening occurs through a strictly regulated process called endochondral ossification, which involves a sequence of chondrocyte proliferation, hypertrophy and apoptosis, extracellular matrix formation and calcification. Due to blood vessel invasion and coordinated action of bone forming cells (osteoblasts) and bone/cartilage resorptive cells (osteoclasts), this mineralized cartilaginous template is remodeled and converted into bony trabeculae in the adjacent metaphyseal bone [1].

Since the growth plate is the most fragile area of the growing long bones in children and it has limited ability to regenerate itself, injuries to the growth plate can represent a significant problem for the developing long bone particularly in a young child who has a significant remaining growth period. While the fractures are limited and contained within the growth plate (Salter type I), healing normally occurs without major disruption of growth plate structure [2]. However, when the defects are large enough and involving the whole width of the growth plate extending from the metaphysis to the epiphysis (Salter types III or IV), bony repair and formation of a bone bridge normally occur at the injured growth plate [2, 3], and cause angular and/or limb length discrepancy deformities [4].

Although growth plate injury is common and is a significant clinical problem, the molecular and cellular processes leading to the undesirable bony repair are relatively unclear. This chapter reviews recent histological, immunohistochemical, cellular and molecular studies which have characterized the injury responses and bony repair mechanisms in a tibial growth plate injury model in young rats which mimics Salter type IV injury.

Growth Plate Injury Responses

Previous experimental investigations using a growth plate drill-hole injury model (mimicking Salter type IV injury) in rabbits, rats, and mice have revealed a rapid time course of bone bridge formation at the growth plate injury site, showing either initial or full osseous bar formation at the injury site by 1 to 3 weeks post surgery [5-9]. In a drill-hole model (Figure 1A) of growth plate injury in young rats in our lab, bony bridge trabeculae normally appear on day 7, increase in number and enlarge in size on day 10, become well-constructed on day 14 with well-formed marrow, and mature on day 25 [10-14] (Figure 1). In these studies, we have identified four different phases of injury responses after an injury is inflicted, namely the inflammatory, fibrogenic, osteogenic and bone bridge maturation remodelling responses, which occur during days 1-3, 3-7, 7-14, 10-25, respectively.

Inflammatory Response: As with bone fractures and soft tissue injuries, the foremost cellular response to trauma at the growth plate is the initial injury-induced inflammatory response, in which inflammatory cells (predominantly neutrophils and to a lesser extent macrophages or monocytes and lymphocytes) were found to infiltrate into the injured growth plate [10-12]. The inflammatory infiltrate was found rapid and transient, with infiltration being apparent at 8 hours, peaking on day 1 (Figure 1B) and subsiding on day 3. Consistent to the neutrophilic predominance and the transient nature of the infiltration, a large induction on day 1 followed by a rapid decline by day 4 in the gene expression of neutrophil chemokine CINC-1 (equivalent to interleukin-8 in human) [15, 16], was revealed in injured growth plate [12].

During the inflammatory response, inflammatory infiltrates (within the injury site) secrete cytokines and growth factors which are thought to be important regulators of subsequent healing responses (see below). Tumour necrosis factor-α (TNF-α) and interleukin-1β (IL-1β), two known important pro-inflammatory cytokines involved with the regulation of the inflammatory response to tissue injury or bone fracture, were shown up-regulated in mRNA expression at the injured growth plate during the inflammatory response, peaking at 8 to 16 hours or day 1 post injury [17]. Consistently, Chung et al (2006) and Arasapam et al (2006) also found a significant increase in the levels of these cytokines on day 1 [11, 12]. Immunohistochemical studies have revealed that inflammatory cells such as neutrophils, macrophage/monocytes and lymphocytes synthesised IL-1β and TNF-α [17].

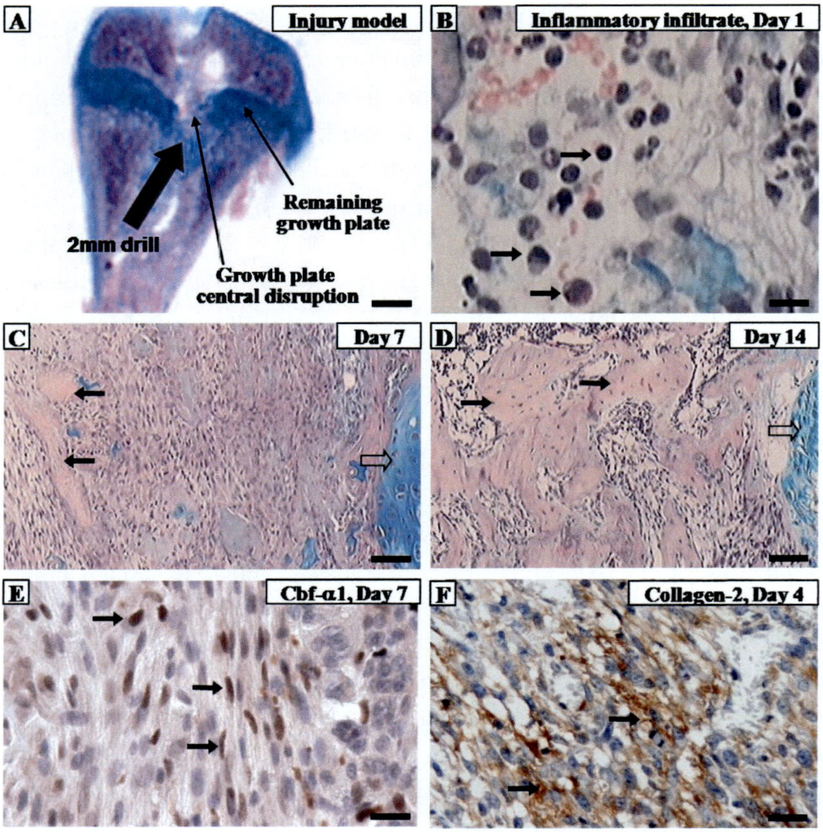

Figure 1. Growth plate injury responses and bone bridge formation at the injury site. (*A*) A drill-hole injury model in the proximal tibial growth plate of young rats with a central disruption inflicted using a 2-mm wide dental burr inserted from metaphyseal bone. (*B*) Day 1 after injury showing inflammatory infiltrate. (*C*) Bony trabeculae (small arrows) appear at the growth plate injury site on day 7. (*D*) Bony trabeculae increase in number and enlarge in size on day 14. (*E*) Immunostaining of osteoblast differentiation factor core binding factor alpha 1 (cbf-α1) (nuclear staining) among some infiltrating mesenchymal cells (day 7). (*F*) Immunostaining of cartilage matrix protein collagen-2 among some infiltrating mesenchymal cells and/or differentiated chondrocyte-like cells at the growth plate injury site (day 4). *A – D*, **Barbara's stain** [14]); block arrows in *C* and *D* indicate adjacent growth plate cartilage. Bar in *A* = 1 mm, bar in *B* = 12.5 μm, bars in *C* and *D* = 200 μm, and bars in *E* and *F* = 25 μm.

Pro-inflammatory cytokines (such as TNF-α and IL-1β) have been reported to activate the p38 mitogen activated protein kinase, a key regulator of the inflammatory response [18], and activation of p38 has also been shown to lead to production and signal transduction of these inflammatory cytokines [19]. In the injured growth plate during the inflammatory event, p38 was shown to have increased activation, with the activated p38 being immuno-localised in inflammatory cells at the injury site and in growth plate chondrocytes adjacent to the injury site [14]. In addition, activity of TNF-α was shown to be required in p38 activation in injured growth plate since inhibition of TNF-α receptor caused a reduction in p38 activation at the injured growth plate [14]. Furthermore, two important inflammatory mediators, cyclo-oxygenase (COX-2) enzyme (responsible for synthesis of prostaglandin-E2) and inducible nitric oxide synthase (iNOS) enzyme (important for the synthesis of nitric oxide), were found up-regulated during the inflammatory response at the injured growth plate [11].

Apart from production of cytokines and inflammatory mediators or regulators discussed above, the inflammatory response is also known to be an important source of growth factors, which, together with the cytokines and regulators, may play important roles in regulating downstream events leading to tissue repair. Transforming growth factor-β1 (TGF-β1), being a potent inhibitor of lymphocyte-activating factor and IL-1 and thus an anti-inflammatory peptide, as well as an important regulator for bone fracture healing [20, 21], was found up-regulated on day 1 [17]. Similarly, the mRNA expression of platelet-derived growth factor-B (PDGF-B) which is important for bone fracture healing [22], was shown to increase by 2- fold on day 1 after growth plate injury [12]. Interestingly, immunohistochemical studies have demonstrated that several bone morphogenetic proteins (BMPs), which are known important growth factors for bone fracture healing, were expressed amongst inflammatory cells at the growth plate injury site during the inflammatory response [13]. Together with the cytokines, these growth factors are believed to be involved in regulating bone fracture healing and bony bridge formation after growth plate injury.

Fibrogenic phase: In our rat growth plate injury model and similar in other animal models, following the initial inflammatory event (up to day 3 after the injury), a fibrogenic response is apparent during days 3-7, involving the infiltration and proliferation of fibrous vimentin-immunopositive mesenchymal cells within the injury site [10]. Some known fibrogenic growth factors have been shown induced during this phase, including fibroblast growth factor-2 (FGF-2) and PDGF-B, suggesting their potential role in regulating the fibrogenic response in growth plate injury repair [17].

During bone fracture healing, FGFs are produced by various cells such as monocytes or macrophages, mesenchymal cells, osteoblasts and chondrocytes [22]. In our rat growth plate injury model, the up-regulated FGF-2 was found to be present in the mesenchymal cells at the injury site, and it is possible that FGF-2 may have a role in the migration and proliferation of mesenchymal cells as well as osteoprogenitor cells and their differentiation into osteoblasts during growth plate injury healing [14].

PDGF-B is an essential factor in bone repair, important in mediating the critical initial events of migration and proliferation of fibroblasts and osteoblasts [23] and inducing new bone formation *in vivo* [22]. In our growth plate injury model, PDGF-B was up-regulated prior to and during the fibrogenic response, found to be detectable on a myriad of cells including inflammatory cells and infiltrated mesenchymal cells during the late inflammatory phase and fibrogenic response [17]. Furthermore, our recent study has recently shown that inhibition of PDGF-B signalling by delivering a PDGF-receptor inhibitor in the animals with injured growth plate significantly suppressed infiltration of mesenchymal cells into the injury site and delayed the subsequent healing responses [24].

In addition, our previous study in a rat growth plate injury model revealed that expression of several BMPs (BMP-2, -3, -4 and -7), which are known to be important for bone fracture healing, was up-regulated during the fibrogenic phase [13], with BMPs and their receptors being localized in infiltrating mesenchymal cells, suggesting that specific BMPs may have a role in regulating mesenchymal cell migration at the injured growth plate prior to and during the fibrogenic phase. Similarly, levels of TGF-β1 increased during the fibrogenic phase at the injured growth plate [17]. During bone fracture healing, up-regulated TGF-β1 at bone fracture site is believed to have an essential role in inducing the migration and proliferation of primitive mesenchymal cells as well as their differentiation into bone cells (osteoblasts) and cartilage cells (chondrocytes) [21].

***Osteogenic and maturation phases*:** Following the fibrogenic event (days 3-7), osteogenic response (days 7 to 14) is dominant at the growth plate injury site (Figure 1C, 1D). During this phase, while some of the cells among the mesenchymal infiltrate may have already been pre-determined as osteoprogenitor cells, some of the cells may undergo bone cell differentiation, as some of them express core-binding factor-alpha1 (cbf-α1) (a master regulator of osteoblast differentiation) (Figure 1E) and alkaline phosphatase (a osteoblast maturation marker) [10-12, 14]. However, a further gene expression study revealed up-regulation of cbf-α1 during this osteogenic phase, suggesting presence of increased osteoblast differentiation during this phase of healing response [11]. Consistently, a higher mRNA level of bone matrix protein osteocalcin was present in the injured growth plate during the osteogenic phase [11, 14], and osteocalcin can be immuno-localised in the bony trabeculae-lining osteoblasts [10, 11], demonstrating the presence of active bone deposition on the new trabecular bone surface. In addition, IGF-I mRNA levels were found up-regulated during the bone bridge formation phase, and IGF-I was immunohistologically found in the cytoplasm of osteoblasts and in the bone matrix within the growth plate injury site [17].

During the bone bridge maturation remodelling phase (days 14-25), most of bone trabeculae are lined by flattened spindle-like osteoblasts in resting phase and producing little osteocalcin and are separated by abundant marrow cells [10]. Interestingly, levels of cytokines and growth factors TNF-α, IGF-I, FGF-2 and TGF-β1 rose again during days 25 to 35 [17], suggesting involvement of these molecules in bone remodelling.

Bony Repair Mechanisms at the Injured Growth Plate

Bone formation during normal bone growth or in fracture repair is achieved by adding newly synthesized bone to existing bone by two mechanisms, which are named by the intermediate structures that are to be converted to form bone. While the endochondral ossification is the process by which bone forms via a cartilaginous intermediate, intramembranous ossification occurs when osteoprogenitor cells differentiate into osteoblasts and produce and add new bone matrix peripherally over a tissue.

Earlier studies have suggested that bony repair or bone bridge formation at the injured growth plate was related to damage to the progenitor cells of the growth plate resting zone [25], or related to physical damage to the epiphysis basement bone plate [26] or to the growth plate-epiphyseal border [2]. It has also been suggested that a physical damage to the growth plate and mixing of the epiphyseal and metaphyseal blood circulations contribute to the bone formation in the growth plate defect [27].

Previously, Lee et al (2000) in a mouse model and Xian et al (2004) in a rat model have demonstrated that bone bridge formation at the injured growth plate did not appear to involve endochondral ossification at the time points examined, but appeared to be a result of direct bone formation by the process of intramembranous ossification [9, 10]. In the rat model, we demonstrated infiltration of mesenchymal cells into the injury site, entering from the metaphysis and epiphysis, many of which expressed cbf-α1 (Figure 1E) and differentiated into osteoblasts that expressed alkaline phosphatase and osteocalcin [10]. Consistent with the direct bone formation response during bone bridge formation at the injured growth plate, Lee et al (2000) did not observed changes in expression of collagen-2, Indian hedgehog and vascular endothelial growth factor, molecules characteristic of endochondral bone formation [9], and Zhou et al (2004) demonstrated a lack of up-regulation of cartilage specific genes, chondrogenic transcription factor Sox-9 and cartilage matrix protein collagen-2a [17].

On the other hand, in some newer studies which examined injury responses at earlier time points after injury, Arasapam et al (2006) found increased expression of cartilage-related genes (collagen-2, collagen-10 and SOX-9) as well as bone-related genes (osteocalcin and cbf-α1) [11]. This suggests an involvement of both endochondral and intramembranous ossification in the bony repair of the injured growth plate cartilage. Similarly, Chung et al (2006) found an increase in the expression of SOX-9 and collagen-2 at the injured growth plate and observed positive immunostaining for collagen-2 (Figure 1F) and -10 among some newly-formed chondrocytes at the injury site, supporting the involvement of the endochondral ossification mechanism in the bony bridge of the injured growth plate [12].

Conclusion

The cellular and molecular events after growth plate injuries often lead to bony repair of the injured growth plate which results in bone growth defects in children. It is now clear that the injury repair responses (sequentially the inflammatory, fibrogenic, osteogenic and remodelling events) for tissue repair and bony bridge formation in growth plate injuries are similar to those seen in bone fracture repair, involving complex cellular and molecular events associated with both endochondral ossification and intramembranous bone formation, although the intramembranous ossification mechanism appears to be more dominant during the bony repair in injured growth plate.

Acknowledgments

The work discussed in this chapter was funded by Bone Growth Foundation and in part by project grants from Channel-7 Children's Research Foundation of and National Health and Medical Research Council (NHMRC) of Australia. The authors thank Fiona Zhou, Rosa Chung, Carmen Macsai, Rosa McCarty, Michaela Scherer, Ganesan Arasapam, Thai Ngo, Jo Cool and Bruce Foster for their contribution in the work discussed.

References

[1] Ianotti JP. Growth plate physiology and pathology. *Orthop Clin N Am*, 1990, 21, 1-17.

[2] Wattenbarger, JM; Gruber, HE; Phieffer, LS. Physeal fractures, part I: Histologic features of bone, cartilage, and bar formation in a small animal model. *J Pediatr Orthop*, 2002, 22, 703-709.

[3] Bailey, RW; Dubow, HI. Evolution of the concept of an extensible nail accommodating to normal longitudinal bone growth. *Clin Orthop*, 1981, 159, 157-170.

[4] Ogden, JA. *Skeletal Injury in the Child*. 3rd ed. New York: Spring-Verlag, 2000.

[5] Makela, EA; Vainionpaa, S; Vihtonen, K; Mero, M; Rokkanen, P. The effect of trauma to the lower femoral epiphyseal plate. An experimental study in rabbits. *J Bone Joint Surg Br*, 1988, 70, 187-91.

[6] Jaramillo, D; Shapiro, F; Hoffer, FA; Winalski, CS; Koskinen, MF; Frasso, R; Johnson, A. Posttraumatic growth-plate abnormalities: MR imaging of bony-bridge formation in rabbits. *Radiology*, 1990, 175, 767-73.

[7] Garces, GL; Mugica Garay, I; Lopez Gonzalez Coviella, N; Guerado, E. Growth-plate modifications after drilling. *J Pediatr Orthop*, 1994, 14, 225-8.

[8] Cottalorda, J; Jouve, JL; Bollini, G; Panuel, M; Guisiano, B; Jimeno, MT. Epiphyseal distraction and centrally located bone bar: an experimental study in the rabbit. *J Pediatr Orthop*, 1996, 16, 664-8.

[9] Lee, MA; Nissen, TP; Otsuka, NY. Ultilization of a murine model to investigate the molecular process of transphyseal bone formation. *J Pedeatr Orthop*, 2000, 20, 802-806.

[10] Xian, CJ; Zhou, FH; McCarty, RC; Foster, BK. Intramembranous ossification mechanism for bone bridge formation at the growth plate cartilage injury site. *J Orthop Res*, 2004, 22, 417-26.

[11] Arasapam, G; Scherer, M; Cool, JC; Foster, BK; Xian, CJ. Roles of COX-2 and iNOS in the bony repair of the injured growth plate cartilage. *J Cell Biochem*, 2006, 99, 450-461.

[12] Chung, R; Cool, JC; Scherer, MA; Foster, BK; Xian, CJ. Roles of neutrophil-mediated inflammatory response in the bony repair of injured growth plate cartilage in young rats. *J Leukoc Biol*, 2006, 80, 1272-1280.

[13] Ngo, TQ; Scherer, MA; Zhou, FH; Foster, BK; Xian, CJ. Expression of bone morphogenic proteins and receptors at the injured growth plate cartilage in young rats. *J Histochem Cytochem*, 2006, 54, 945-54.

[14] Zhou, FH; Foster, BK; Zhou, XF; Cowin, AJ; Xian, CJ. TNF-alpha mediates p38 MAP kinase activation and negatively regulates bone formation at the injured growth plate in rats. *J Bone Miner Res*, *2006*, 21, 1075-88.

[15] Mukaida, N; Harada, A; Matsushima, K. Interleukin-8 (IL-8) and monocyte chemotactic and activating factor (MCAF/MCP-1), chemokines essentially involved in inflammatory and immune reactions. *Cytokine Growth Factor Rev*, 1998, 9, 9-23.

[16] Ramos, CD; Heluy-Neto, NE; Ribeiro, RA; Ferreira, SH; Cunha, FQ. Neutrophil migration induced by IL-8-activated mast cells is mediated by CINC-1. *Cytokine*, 2003, 21, 214-23.

[17] Zhou, FH; Foster, BK; Sander, G; Xian, CJ. Expression of proinflammatory cytokines and growth factors at the injured growth plate cartilage in young rats. *Bone*, 2004, 35, 1307-15.

[18] Obata, T; Brown, GE; Yaffe, MB. MAP kinase pathways activated by stress: the p38 MAPK pathway. *Crit Care Med*, 2000, 28, N67-77.

[19] Lee, JC; Kumar, S; Griswold, DE; Underwood, DC; Votta, BJ; Adams, JL. Inhibition of p38 MAP kinase as a therapeutic strategy. *Immunopharmacology*, 2000, 47, 185-201.

[20] Joyce, ME; Jingushi, S; Scully, SP; Bolander, ME. Role of growth factors in fracture healing. *Prog Clin Biol Res*, 1991, 365, 391-416.

[21] Tatsuyama, K; Maezawa, Y; Baba, H; Imamura, Y; Fukuda, M. Expression of various growth factors for cell proliferation and cytodifferentiation during fracture repair of bone. *Eur J Histochem*, 2000, 44, 269-278.

[22] Bolander, ME. Regulation of fracture repair by growth factors. *Proc Soc Exp Biol Med*, 1992, 200, 165-70.

[23] Fujii, H; Kitazawa, R; Maeda, S; Mizuno, K; Kitazawa, S. Expression of platelet-derived growth factor proteins and their receptor alpha and beta mRNAs during fracture healing in the normal mouse. *Histochem Cell Biol*, 1999, 112, 131-8.

[24] Chung, R; Foster, BK; Zannettino, AC; Xian, CJ. Potential roles of growth factor PDGF-BB in the bony repair of injured growth plate. *Bone*, 2009, 44, 878–885

[25] Salter, RB; Harris, WR. Injuries involving the epiphyseal plate. *J Bone and Joint Surg*, 1963, 45A, 587-622.

[26] Ford, LT; Key, JA. A study of experimental trauma to the distal femoral epiphysis in rabbits. *J Bone Joint Surg [Am]*, 1956, 38, 84-92.

[27] Gomes, LSM; Volpon, JB; Goncalves, RP. Traumatic separation of epiphyses. An experimental sutdy in rats. *Clin Orthop*, 1988, 236, 286-295.

In: Bone Regeneration: Growth Factors... ISBN: 978-1-60876-621-5
Editors: V. Legard et al., pp. 275-284

Chapter X

Application of Carbon Nanotubes on Scaffolds for Bone Tissue Regeneration

Naoto Saito**, Yuki Usui, Kaoru Aoki, Nobuyo Narita, Masayuki Shimizu, Kazuo Hara, Nobuhide Ogiwara, Koichi Nakamura, Norio Ishigaki, Hiroyuki Kato, Hisao Hanyiu, Seiichi Taruta and Morinobu Endo

Shinshu University, Nagano, Japan

Abstract

There has been a great deal of progress in regenerative medicine over the last several years, but scaffolds are still required for tissue construction. Various biomaterials have been studied as scaffold candidates, but no functional scaffolds capable of acting synergistically with the tissue regeneration capacity of the cells and cytokines have been developed. Regenerative medicine for bone tissue is currently in the clinical application phase, but without an optimal scaffold, sufficient results have yet to be

* Address correspondence to this author at the Department of Applied Physical Therapy, Shinshu University School of Health Sciences,3-1-1 Asahi, Matsumoto, Nagano, 390-8621, Japan. Tel, Fax: 81-263-37-2409; E-mail: saitoko@shinshu-u.ac.jp

achieved. We found that carbon nanotubes (CNTs) promote osteogenic capability, and are currently investigating its mechanism. CNTs may be useful as a functional scaffold to markedly increase the bone regeneration capacity of the cells and cytokines.

Introduction

Bone tissue regeneration is often necessary in clinical practice; in the field of orthopedics, it is used in cases of intractable bone fracture, spinal fusion, and in repair of bone defects after tumor excision. Bone tissue regeneration is also used for patients in other medical fields, *e.g.*, in bone defect repair for patients with cleft lip and palate, as well as in repair of periodontal bone defects caused by periodontal disease. In such cases, autogenous bone transplantation or bone allograft transplantation is carried out. However, autogenous bone transplantation has limits on the quantity of bone that can be collected, and is invasive to healthy parts of the body at the donor site [1]. In bone allograft transplantation, bone from another person is used, so there is a risk of disease transfer in addition to issues associated with immunogenicity [2]. The cost of operation of a bone bank is also a significant problem [3]. Compared to these traditional methods, bone tissue regenerative therapy using bone marrow stem cells and bone morphogenetic proteins (BMPs) [4] is better in that it is less invasive to healthy tissues, and does not have a risk of disease transfer [5,6]. Considering the number of patients currently requiring bone transplantation, there is an urgent need for bone tissue regeneration to become available in clinical practice. However, the high cost of BMPs and lack of good scaffold prevent the widespread adoption of this technology. Clinical application of bone tissue regeneration using BMPs and bone marrow stem cells requires the development of a new functional scaffold [7-10]. Carbon nanotubes (CNTs) may be useful as such a new functional scaffold.

CNTs

CNTs consist of graphene sheets rolled into cylinders with diameters in the nanometer range [11,12]. CNTs can be classified into 3 main types—Single-walled (SWCNTs), Multi-walled (MWCNTs), and Cup-stacked CNTs—but various other types of CNT have also been reported [13]. There is a great deal of current research and development in many different fields regarding CNTs as a new structural and electronic material. This is because CNTs have unique and useful characteristics, and by combining CNTs with other materials, the performance and function of the material can be greatly enhanced and modified [14-18]. Therefore, the development of biomaterials using CNTs is anticipated for use in a number of medical fields [19].

Many biomaterials make contact with bone in clinical use, such as artificial joint prostheses, plates, and screws for use in the treatment of fractures, as well as artificial bone. Therefore, it is expected that there will be increasing numbers of studies regarding the application of CNTs in biomaterials used in contact with bone in the near future [20,21]. CNTs can also be applied as scaffold for tissue regeneration, and we have been engaged in research regarding the application of CNTs in tissue regeneration.

Figure 1. SEM image of MWCNTs. Scale bars, 1000 nm. (Reprinted from Ref. 19)

First, we examined bone tissue compatibility *in vivo*, which is necessary for application of CNTs to biomaterials that come into contact with bone. Next, we investigated the effect of CNTs on bone formation, which is the most important factor for their application as scaffold for bone tissue regeneration, and our results indicated that CNTs promote osteogenic capability. To determine the mechanism by which the osteogenic capability is promoted by CNTs, we examined the role of CNTs in formation of hydroxyapatite (HA) which is the main composition of bone matrix [22]. We have also studied the effects of CNTs on bone-related cells at the molecular level [23]. High crystalline MWCNTs with an average diameter of 80 nm and a length from 10 to 20 μm, prepared by catalytic chemical vapour deposition and subsequent thermal treatment above 2800 °C in argon, were used for the experiments [24] (Figure 1).

Bone Tissue Compatibility of MWCNTs

It is important to examine the compatibility of MWCNTs for bone tissue for their application as a scaffold for bone tissue regeneration. First, MWCNTs were buried beneath the periosteum in the cranial bone of mice. Little inflammatory reaction was observed and cortical bone in contact with MWCNTs was not eroded indicating that good bone tissue compatibility. Next, MWCNTs were implanted into a bone defect in the tibia. After 3 weeks, normal bone repair was observed, and the MWCNTs were incorporated into the repaired bone tissue. When graphite was implanted in the bone defect instead of MWCNTs, bone tissue repair was inhibited (Figure 2). When bone repair occurred with implantation of MWCNTs, observation using field-emission scanning electron microscopy (FE-SEM) disclosed that MWCNTs adhered closely to the bone matrix with no intervening space (Figure 3). We found that MWCNTs have high bone tissue compatibility and they do not inhibit bone repair, both of which are very useful characteristics for use as a scaffold for bone tissue regeneration.

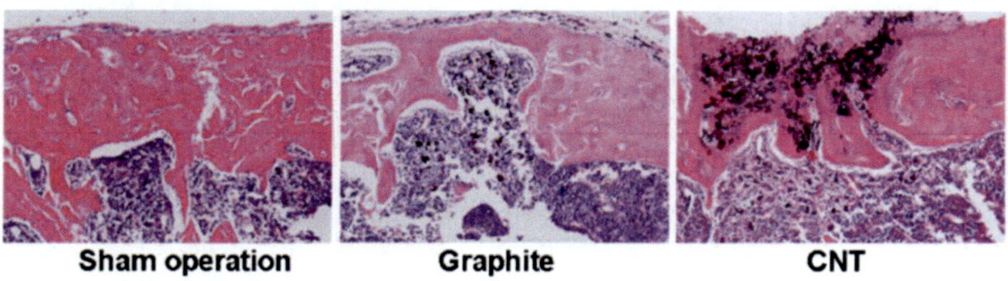

Figure 2. Graphite delayed bone repair, but MWCNTs did not inhibit bone repair. Haematoxylin and eosin staining. (Reprinted from Ref. 22)

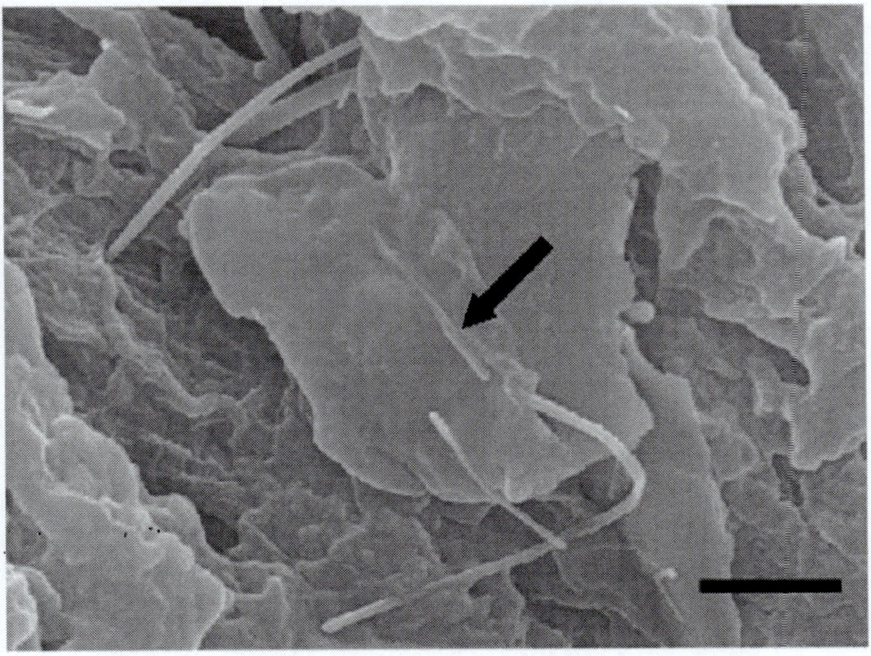

Figure 3. MWCNTs (arrow) adhered closely to the bone matrix with no intervening space, and showed high bone tissue compatibility (FE-SEM). Scale bars, 1000 nm. (Reprinted from Ref. 22)

Promotion of Bone Formation Using MWCNTs

The standard scaffold material for bone regenerative medicine using BMPs is type I collagen. MWCNTs were combined with type I collagen, and collagen-MWCNT composite scaffold was made, mixed with recombinant human (rh) BMP-2, and implanted subfascially in the back muscle of mice (MWCNT group). Collagen without MWCNTs mixed with rhBMP-2 was used as a control. Two weeks after implantation, ectopic bone formation was observed on X-ray images, and bone mineral content (BMC) and histology were evaluated. On X-ray images, shadows of the bone were more prominent in the MWCNT group compared to the controls, and BMC was significantly higher in the MWCNT group (Figure 4). Histologically, the degree of bone maturation was the same, but the MWCNT group had a larger area with bone tissue. MWCNTs promote bone formation, which is an important novel finding for application of MWCNTs as a scaffold material for bone tissue regeneration.

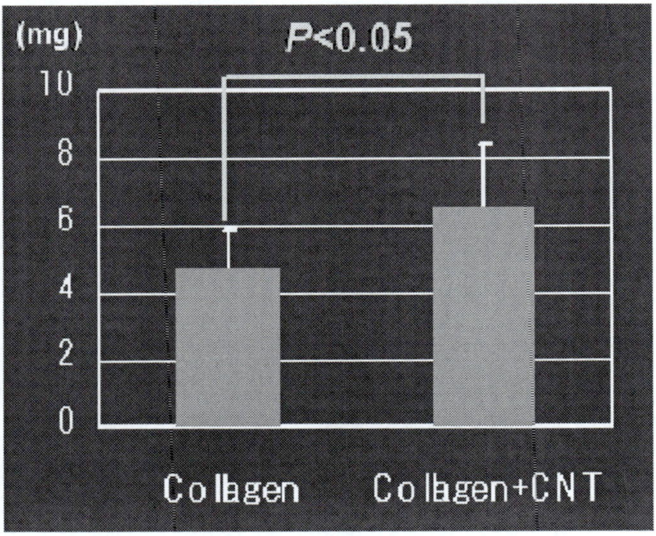

Figure 4. The MWCNT group had a significantly higher bone mineral content than control group, indicating that MWCNTs promote bone formation through BMP. Unpaired Student's t test, two tailed. (Reprinted from Ref. 22)

Effects of MWCNTs on HA Crystallization

MWCNT particles were dispersed in Dulbecco's phosphate-buffered saline (PBS(+)) *in vitro*, and incubated at 37°C. Graphite particles were used as a control. After 6 h, 2 days, and 2 weeks, deposited crystals were evaluated by FE-SEM, and at 2 weeks, the components of the deposited crystals were determined by X-ray diffraction (XRD) analysis. In the MWCNT group, abundant crystals were deposited attached to particles, and the amount of crystals increased with time. In the control group, small amounts of crystals were deposited on some of the particles, but did not increase markedly over time (Figure 5). XRD patterns indicated diffraction peaks of HA were observed in both in both groups. These findings indicated that MWCNTs function as the core for HA crystallization. HA crystals are the primary constituent of the bone matrix. As part of the mechanism by which MWCNTs promote bone formation, it is possible that MWCNTs increase the rate of HA formation.

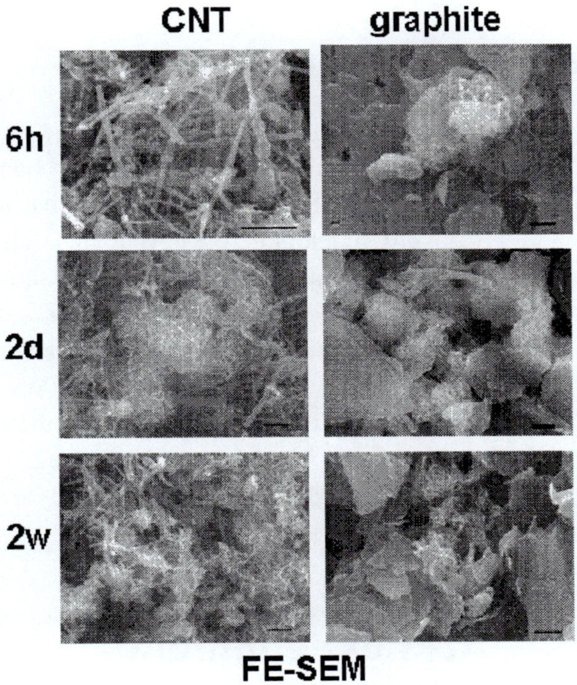

Figure 5. HA crystallization advanced with MWCNTs as a core in PBS(+)) *in vitro* (FE-SEM). Scale bars, 1000 nm. (Reprinted from Ref. 22)

Effects of MWCNTs on Osteoclasts

We examined the effects of MWCNTs on differentiation of precursors into osteoclasts and the functions of osteoclasts, which are cells that resorb bone matrix. When MWCNTs were used at a concentration that did not affect other cells, differentiation of precursor cells into osteoclasts was inhibited. In addition, the function of osteoclasts to resorb bone matrix was specifically inhibited. We found that MWCNTs inhibited the differentiation and function of osteoclasts by blocking translocation of NFATc1, which is the signal transducer, from the cytoplasm to the nucleus within osteoclasts. This result suggested that when

MWCNTs are used as scaffold for bone regeneration, the regenerated bone tissues may be maintained for a long time.

Conclusions

The results described above suggest that CNTs may be applicable in the development of a scaffold that could markedly increase bone regenerative capacity, and thus allow many patients to receive regenerative medicine of the bone safely and more effectively. In addition, the development of scaffolds using CNTs for tissues and organs other than bone will represent a significant advance in the field of regenerative medicine. Research to apply CNTs as scaffolds for bone tissue regeneration has just begun, and many issues remain to be addressed, such as the mechanism by which CNTs promote bone formation, the creation of optimal composite by combining CNTs and existing biomaterials (such as collagen and biodegradable polymers), and biological safety evaluation of CNTs. However, these are very important studies that have the potential to benefit large numbers of patients in future.

Acknowledgments

This research was supported by the CLUSTER program of the Ministry of Education, Culture, Sports, Science and Technology, Japan, by a Grant-in-Aid for Scientific Research from the Ministry of Education, Science, Sports and Culture, Japan, and by the Research and Development of Nanodevices for Practical Utilization of Nanotechnology of the New Energy and Industrial Technology Development Organization, Japan.

References

[1] Einhorn TS. Enhancement of fracture-healing. *J Bone Joint Surg Am* 77: 940-956, 1995.

[2] Butler D, Wadman M, Lehrman S, Schiermeier Q. Last chance to stop and think on risks of xenotransplants. *Nature* 391: 320-324, 1998.

[3] Malinin TI, Martinez OV, Brown MD. Banking of massive osteoarticular and intercalary bone allografts-12 years' experience. Clin Orthop 197: 44-57, 1985.

[4] Urist MR. Bone formation by autoinduction. *Science* 150: 893-899, 1965.

[5] Haynesworth SE, Goshima J, Goldberg VM, Caplan AI. Characterization of cells with osteogenic potential from human marrow. *Bone* 13: 81-88, 1992.

[6] Reddi AH. Symbiosis of Biotechnology and biomaterials: application in tissue engineering of bone and cartilage. *J Cell Biochem* 56: 192-195, 1994.

[7] Saito N, Okada T, Horiuchi H, Murakami N, Takahashi J, Nawata M, Ota H, Nozaki K, Takaoka K. A biodegradable polymer as a cytokine delivery system for inducing bone formation. *Nat Biotechnol* 19: 332-335, 2001.

[8] Saito N, Takaoka K. New synthetic biodegradable polymers as BMP carriers for bone tissue engineering. *Biomaterials* 24: 2287-2293, 2003.

[9] Saito N, Murakami N, Takahashi J, Horiuchi H, Ota H, Kato H, Okada T, Nozaki K, Takaoka K. Synthetic biodegradable polymers as drug delivery systems for bone morphogenetic proteins. *Adv Drug Deliver Rev* 57: 1037-1048, 2005.

[10] Aoki K, Usui Y, Narita N, Ogiwara N, Ishigaki N, Nakamura K, Kato H, Sano K, Ogiwara N, Kametani K, Kim C, Taruta S, Kim YA, Endo M, Saito N. A thin carbon fiber web as a scaffold for bone tissue regeneration. *Small* 5: 1540-1546, 2009.

[11] Oberlin A, Endo M, Koyama T. Filamentous growth of carbon through benzene decomposition. *J Cryst Growth* 32: 335-349, 1976.

[12] Iijima S. Helical microtubules of graphitic carbon. *Nature* 354: 56-58, 1991.

[13] Endo M, Muramatsu H, Hayashi T, Kim YA, Terrones M, Dresselhaus MS. Nanotechnology: 'Buckypaper' from coaxial nanotubes. *Nature* 433: 476, 2005.

[14] Treacy MMJ, Ebbesen TW, Gibson JM. Exceptionally high Young's modulus observed for individual carbon nanotubes. *Nature* 381: 678-680, 1996.

[15] Dai H, Wang EW, Lieber CM. Probing electrical transport in nanomaterials: conductivity of individual carbon nanotubes. *Science* 272: 523-526, 1996.

[16] Ebbesen TW, Lezec HJ, Hiura H, Bennett JW, Ghaemi HF, Thio T. Electrical conductivity of individual carbon nanotubes. *Nature* 382: 54-56 (1996).

[17] Miyagawa H, Misra M, Mohanty AK. Mechanical properties of carbon nanotubes and their polymer nanocomposites. *J Nanosci Nanotechnol* 5: 1593-1615, 2005.

[18] Berber S, Kwon YK, Tomanek D. Unusually high thermal conductivity of carbon nanotubes. *Phys Rev Lett* 84: 4613-4616, 2000.

[19] Saito N, Usui Y, Aoki K, Narita N, Shimizu M, Hara K, Ogiwara N, Nakamura K, Ishigaki N, Kato H, Taruta S, Endo M. Carbon nanotubes: biomaterial applications. *Chem Soc Rev* 38: 1897-1903, 2009.

[20] Zanello LP, Zhao B, Hu H, Haddon RC. Bone cell proliferation on carbon nanotubes. *Nano Lett 6*: 562-567, 2006.

[21] Saito N, Usui Y, Aoki K, Narita N, Shimizu M, Ogiwara N, Nakamura K, Ishigaki N, Kato H, Taruta S, Endo M. Carbon nanotubes for biomaterials in contact with bone. *Curr Med Chem* 15: 523-527, 2008.

[22] Usui Y, Aoki K, Narita N, Murakami N, Nakamura I, Nakamura K, Ishigaki N, Yamazaki H, Horiuchi H, Kato H, Taruta S, Kim YA, Endo M, Saito N. Carbon nanotubes with high bone-tissue compatibility and bone-formation acceleration effect. *Small 4*: 240-246, 2008.

[23] Narita N, Kobayashi Y, Nakamura H, Maeda K, Ishihara A, Mizoguchi T, Usui Y, Aoki K, Simizu M, Kato H, Ozawa H, Udagawa N, Endo M, Takahashi N, Saito N. Multi-walled carbon nanotubes specifically inhibit osteoclast differentiation and function. *Nano Lett* 9: 1406-1413, 2009.

[24] Endo M, Kim YA, Hayashi T, Nishimura K, Matushita T, Miyashita K, Dresselhaus MS. Vapor grown carbon fibers (VGCFs): basic properties and their battery applications. *Carbon* 39: 1287-1297, 2001.

In: Bone Regeneration: Growth Factors...
Editors: V. Legard et al., pp. 285-302 ISBN: 978-1-60876-621-5

Chapter XI

Proteases as Therapeutic Targets in Bone Regeneration

***E.S. Lee*[1] *and T.R. Yoon*[*2]**
[1] Brain Korea 21 Project Biomedical Human Resources Center
[2] Chonnam Natl. University Jeonnam, Republic of Korea

Abstract

Proteases play essential roles in a variety of biological processes such as protein activation and degradation, cell death, development, and maintenance of morphology. Proteolytic activities in bone regeneration are divided into two functions, degradation and proteolysis. Proteolysis is distinct from degradation in that it represents highly specific and limited substrate cleavage that results in a specific change of protein function. In contrast, degradation such as collagen digestion is the breakdown of structural building blocks or the complete destruction of proteins to be recycled or discarded. Bone is a dynamic tissue that undergoes repeated resorption and reformation over its lifetime. Once bone regeneration is initiated, osteoclasts play a role in resorption, and osteoblasts are responsible for bone tissue formation. Matrix metalloproteinases are believed to be specifically implicated in osteoclastic bone resorption. MMP-2, -9, -13, and -14 are involved in endochondral bone growth. MMPs are also associated with

* Correspondence to: TR Yoon, E-mail: tryoon@jnu.ac.kr

estrogen deprivation, which often results in bone disruption. Cathepsin K is a cysteine protease that can degrade key bone matrix proteins such as collagen. While both MMPs and cathepsin K directly degrade skeletal structural proteins, site-1 protease (S1P) functions in the conversion of transcription factors involved in endochondral ossification to their active form. In addition, receptor activator of NF-κB ligand (RANKL), a key regulator of osteoclastogenesis, is converted to a soluble form by proteolysis. Thus, proteases perform essential functions in bone regeneration by either direct degradation of structural proteins or by proteolytic control of regulatory proteins. Knowledge of proteolytic activities in bone regeneration and elucidation of methods to control their activities using inhibitors will facilitate the development of new therapeutic targets and agents for bone diseases.

Introduction

Bone is a mineralized dynamic tissue that undergoes repeated resorption and formation over its lifetime. Bone resorption and formation are tightly coupled. Tissue remodeling is dependent on the balance between osteoclasts and osteoblasts. Almost all osteogenic diseases including osteoporosis are due to a failure of coordination between these two cell types. Agents for the treatment of osteoporosis are classified as either antiresorptive or anabolic. Antiresorptive agents work by inhibiting the activity of osteoclasts, thereby reducing bone resorption. Anabolic agents act by stimulating the formation of new bone. To date, the development of treatments for osteoporosis has been directed toward an excess of antiresorptive agents such as bisphosphonate. However, this type of treatment provides only temporary effects and eventually results in inhibition of bone formation. Therefore, the development of new agents that not only allow bone formation but also inhibit bone resorption has long been considered desirable.

To overcome the limited effects of existing antiresorptive agents, the regulation of proteases and development of protease inhibitors represent alternative goals in pharmaceutical development. Proteases have recently achieved increasing attention as therapeutic targets. Proteases are enzymes that cleave proteins by catalyzing the hydrolysis of peptide bonds [1]. Proteases can regulate the fates and activities of many proteins by controlling their localization. Precise proteolysis controls a wide range of pivotal cellular processes including shedding from cell surfaces, activation of proenzymes and inactivation of mature proteins. Many groups have reported the successful development of potent inhibitors for a variety of proteases [2]. In addition, methods have been developed for screening protease inhibitors [3]. These examples indicate that proteases are attractive therapeutic targets for many diseases.

In the field of bone biology, several proteases have been found to be involved in osteogenesis, or bone development. Accordingly, many researchers are considering these proteases as potential drug targets. Proteolysis is distinct from degradation in that it represents highly specific and limited substrate cleavage, which results in a specific change in protein function [1]. Proteolysis can result in modification of protein structure or function. In contrast, degradation such as collagen digestion is the process of accelerated turnover of structural building blocks or complete destruction of proteins to be recycled or discarded. Most MMPs and members of the cathepsin family are responsible for degradation, which results in structural changes in protein building blocks such as components of the extracellular matrix (ECM). On the other hand, RANKL sheddase and site-1 protease are examples of proteases that induce a functional change in signaling proteins involved in bone regeneration.

This review will discuss the role of proteases in the regulation of bone regeneration and the application of proteolytic events in osteogenesis as therapeutic targets.

Bone Regeneration and Therapeutic Targets

Once bone regeneration is initiated, osteoclasts resorb mineralized bone and osteoblasts form new bone tissue (Fig. 1). Although osteoblasts appear to be involved mainly in induction of bone formation or, at least, to play much more critical roles in bone regeneration, in a strict sense, osteoclasts trigger bone regeneration, or bone remodeling. Without osteoclastic activity, osteoblasts might not be activated for bone regeneration. To date, many drugs for osteogenic diseases such as osteoporosis are intended to promote disruption of osteoclastic activity (Fig. 1). These antiresorptive agents have a significant drawback in that their effects are only temporary, because inhibition of osteoclastic activity eventually results in reducing osteoblastic activity for the reasons described above.

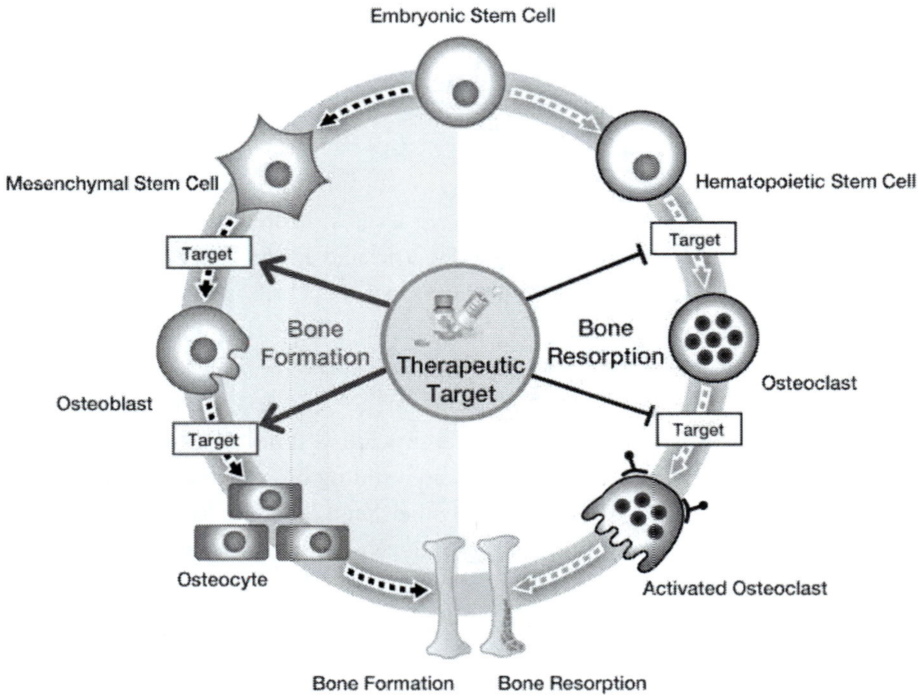

Figure 1. Bone remodeling and therapeutic targets.Osteoclasts play a role in resorption, and osteoblasts are responsible for bone tissue formation. Osteoclasts are differentiated from hematopoietic stem cells and osteoblasts are derived from mesenchymal stem cells. Once embedded in the bone matrix, osteoblasts further differentiate into osteocytes. This process is the target of pharmaceutical agents that control bone resorption and formation.

Bone Regeneration and Proteolytic Events

The new word "degradomics" illustrates the increased attention that is being paid to proteases as therapeutic targets. As noted above, proteolytic activities in bone regeneration are divided into two functions, degradation and proteolysis. Proteolysis is distinct from degradation in that it represents highly specific and limited substrate cleavage that results in a specific change of protein function.

Why are proteolytic activities required for bone regeneration? First, osteoclasts must dissolve bone minerals for resorption of bone in the bone remodeling cycle. To dissolve bone minerals in tissue, osteoclasts provide an optimal environment for proteolytic activity, thereby rendering the ECM more readily degraded and resorbed [4, 5]. Second, proteolysis, along with other protein modifications such as phosphorylation, methylation, and formation of multimer, plays a significant part in signal transduction by activation of proenzymes or inactivation of proteins. Since the early 2000s, it has been known that in response to a specific stimulus such as injury, signals are commonly transmitted from transmembrane proteins to downstream molecules in a series of events that include phosphorylation, glycosylation or methylation [6]. However, many recent reports demonstrate that the transmission of an external signal that triggers the regulation of transcription factors is accomplished through a series of proteolytic cleavage steps. Both osteoclasts and osteoblasts can not help requiring mechanisms of signal transduction such as those requiring proteolysis for induction from their precursor cells and for their continuous differentiation into mature cells that participate in bone regeneration.

Matrix Metalloproteinases (MMPs)

More than 200 known metalloproteinases have been identified, and almost all of them are dependent on zinc at the catalytic triad for their activity [7], including the MMPs. During bone remodeling, connective tissue collagen is rapidly broken down. MMPs have a function in this process. MMPs are a family of endopeptidases that are responsible for the degradation of the ECM proteins, such as collagens, elastin, proteoglycans, and laminins, which indicates that MMPs play a critical role in tissue regeneration and bone remodeling [8]. To date, more than 20 kinds of MMPs have been identified. These are well conserved and divided into four to eight groups according to their domain organization [9-11].

MMPs Expressed in Bone Regeneration

Several MMPs expressed in the skeleton appear to function in endochondral ossification during embryonic development and postnatal remodeling of bone [12]. The membrane-type 1 MMPs (MT-1 MMP, also called MMP-14) and MMP-9 are expressed in preosteoclasts [13]. MT1-MMP null mice exhibit skeletal abnormalities [14]. In addition, MT-1 MMP is one of the candidates proteases implicated in the shedding of RANKL, a key protein in osteoclastogenesis, to convert it into soluble form (we will discuss this in detail in a different section) [15]. All membrane-type MMPs (MMP-14, -15, -16, -17, -24, and -25) are activated by furin-like enzymes [16, 17]. Whereas almost all MMPs are expressed in osteoclasts or preosteoclasts, MMP-13, also called collagenase-3 is expressed in newly recruited osteoblasts [18] and found during postnatal bone remodeling [19]. MMP-13 is a known target of Runx2, an essential transcription factor for bone formation during development [20]. MMP-13 promoter contains a distinct Runx2 binding element [21]. Recent studies on relations between MMP-13 and Runx2 [21] indicate that regulation of MMP-13 via Runx2 is a clue for development of new drugs for osteogenic disorder. On the other hand, treatment with the MMP inhibitor GM6001 results in abnormal ossification [17]. All these reports demonstrate that MMPs are required for normal bone development.

The major collagen composing bone is type I collagen, which is cleaved by MMP-1, -2, -8, -13, and -14. Although MMPs break down collagens, MMPs do not degrade the peptide bonds within the proteins randomly. Collagens are composed of several similar chains that have a regular configuration and bear a specific amino acid sequence that is recognized and cleaved by the proteases. However, degradation of matrix is not their sole function. Recently, several groups reported that MMPs act on cytokines, chemokines, receptors, etc. [22, 23]. Therefore, MMPs may also be considered to be involved in the regulation of signaling events.

Regulation of MMP Activity

How are the catalytic activities of MMPs regulated? Regulation of the activity of MMPs is controlled at several different stages. They are initially controlled at the level of gene expression like almost all proteins, and then the expressed enzymes are spatially localized, which enables MMPs to respond to their specific substrates. Most MMPs are produced in the form of proenzymes that must be activated to have proteolytic activity [12]. The proenzyme contains a so-called cysteine switch, which involves the interaction between a cysteine residue in the prodomain and the zinc ion of the catalytic triad, and maintains the proMMPs in an inactive state. The proenzyme can be activated by controlling the cysteine switch through proteolytic cleavage of the prodomain as well as reduction of the free thiol of the cysteine residue by oxidants and allosteric perturbation of the zymogen [20]. Plasmin is responsible for the proteolysis of

MMPs. Plasmin activates several proMMPs (proMMP-1, -3, -7, -9, -10, and -13), and activated proMMPs can participate in activating other MMPs [24, 25].

Proteolysis is also responsible for the inactivation of MMPs. However, the principal inhibitory regulation of MMP activity is controlled by TIMPs (tissue inhibitors of matrix metalloproteinases). TIMPs are inhibitors of MMPs that interfere with activation by interacting with the intermediate form of an MMP before it is fully activated [25]. Four TIMPs have been identified in vertebrates [26], and each of these inhibits all MMPs, with the one exception that TIMP-1 fails to inhibit MT1-MMP [27].

In conclusion, variations in MMP expression and inhibition by TIMPs affect bone regeneration. Therefore, MMPs are very attractive therapeutic targets for osteogenic diseases such as osteoporosis.

Development of MMP Inhibitors for the Treatments of Osteogenic Diseases

Proteases, including various viral proteases, have been therapeutic targets for the treatment of disease for many years. For example, the 3C protease of human rhinovirus, the single most significant cause of the common cold, was recognized as an attractive therapeutic target, and inhibitors of this protease was successfully developed [28]. The investigators adopted a structure-based design, considering both the structure of the catalytic pocket and the native substrate sequences of the 3C protease. Based on the peptide backbone, peptidomimetic inhibitors were designed with improved binding affinity compared to the protease's native substrates [28], and this approach was successfully applied to development of other protease inhibitors by another group [2]. Requirements for this approach include information regarding native substrate sequences, knowledge of substrate specificities obtained using peptide cleavage assays, and determination of the structure of the catalytic site of the protease, at minimum. In early research, MMPs were reported to hydrolyze synthetic peptide substrates with sequences similar to those near the cleavage site of native type I collagen [29, 30]. The basic domain structures of MMPs are well defined. These indicate that conditions are favorable for designing peptidomimetic compounds as candidate MMP inhibitors.

Attempts have also been made to develop TIMP variants that can selectively cleave MMPs [31], and adenoviral overexpression of TIMP-1 reduced lesions in a mouse model of atherosclerosis [32], indicating that TIMPs and TIMP variants can be utilized for the regulation of osteogenesis, and are potential targets of drugs for the treatment of osteogenic diseases. However, an obstacle must be overcome before TIMPs or TIMP variants can be used as regulators of bone regeneration. As noted above, a single TIMP can inhibit multiple MMPs. Therefore, novel TIMPs having altered specificity must be engineered.

Cathepsins

It has been speculated that cathepsins are involved in intracellular protein turnover and account for non-specific bulk proteolysis within lysosomes [4, 33]. However, the discovery of new cathepsins and subsequent gene-knockout analyses indicate that they participate in a variety of physiological processes, such as bone remodeling, prohormone processing, and wound healing [33-35]. Moreover, cathepsins are involved in several osteogenic pathologies such as osteoporosis and osteoarthritis [33]. Although the cathepsin family includes various members, such as cathepsin B, C, F, H, I, J, K, L, O, S, V, W, and X, here we will focus on the cysteine cathepsins that are involved in bone regeneration and bone pathology.

Although cathepsins B and L were found in bone tissue before other cathepsins, it is now known that cathepsin K is a key enzyme involved in the bone remodeling cycle. Cathepsin K is selectively expressed by osteoclasts for the degradation of bone matrix proteins. Type I and type II collagens, which are components of bone matrix, possess two major structural characteristics, triple helices and single-chain telopeptides. Cathepsin K can cleave the collagens at sites occurring both within helices and in telopeptides, whereas cathepsins L, B, and S cleave only in the telopeptides [36, 37]. In addition, cathepsin K represents the only lysosomal collagenolytic activity [33, 37]. Elimination of cathepsin K in osteoclasts prevents bone resorption [38]. Cathepsin B is ubiquitously expressed, whereas cathepsin K is selectively found in osteoclasts. It is involved in remodeling of the ECM [39]. Recently, proteomic analysis has indicated that cathepsin B is upregulated during osteogenesis (unpublished data). As cathepsin B is important in many pathological processes, numerous inhibitors of cathepsin B have been already described [39]. In addition, many new small molecules as inhibitors of cathepsin B have been designed and synthesized with the use of X-ray crystal structures of cathepsin B [39]. The effects of these compounds during osteogenesis should be investigated. Several cathepsins appear to be involved in bone regeneration, such as cathepsin S, which is implicated in the development of atherosclerotic lesions, and cathepsin L, whose inhibition in mice reduced cortical bone mineral density [40], but the role of these cathepsins in bone regeneration requires further elucidation.

Inhibitors of Cathepsins

Overexpression of cathepsin K induces osteoporosis. How, then, can the expression of cathepsins be balanced to prevent unwanted effects? Cathepsins are synthesized as inactive zymogens, and removal of the prodomain by proteolysis renders them active. It is postulated that prodomain cleavage can be accomplished by autocatalytic activity or other proteases such as cathepsin D [33]. Leupeptin, antipain, and E-64, among others, have been proven to be effective cysteine protease inhibitors [41, 42]. Although cathepsins are cysteine proteases, new inhibitors with high specificity need to be developed to achieve fine and specific control of osteogenic events. Cathepsin K has been targeted for the treatment of osteoporosis. Several inhibitors of cathepsin K have been identified, but development of almost all inhibitors except for odanacatib (MK-0822) has been suspended due to severe side effects [38].

Shedding of Membrane-Bound RANKL by Proteolysis

Transcription factors, proteases, and other proteins involved in signaling pathways are frequently anchored to the cellular membrane, where they cannot participate in biological processes. Proteolysis, along with modifications induced by binding of partner proteins such as ligands, triggers the activation of proteins by releasing a part of the protein containing the catalytic domain from the membrane. Ectodomain shedding is one such process. The shedding of some membrane-bound proteins releases their extracellular domain, thereby resulting in loss of activity or activation for signal transmission to downstream molecules.

The shedding of RANKL is a good example. The RANK/RANKL /osteoprotegerin (OPG) system is a fundamental paradigm in the regulation of osteoclastogenesis. RANKL, a key regulator of osteoclastogenesis, is a type II transmembrane protein of the tumor necrosis factor (TNF) ligand family (Fig. 2)

[15]. Therefore, RANKL is found on the plasma membrane of osteoclasts and on early-stage osteoblasts [15, 43]. However, elevated levels of soluble RANKL have been observed in rheumatoid arthritis (RA) patients. Once RANKL is cleaved, it is released into the local environment [15]. Which protease cleaves RANKL, thereby resulting in the soluble form of RANKL? As noted above, MMP-14 and several metalloproteinases including those of the ADAM family were thought to be candidate RANKL sheddases, but the protease responsible for the shedding has not yet been determined. The effects of RANKL shedding are very controversial. While some groups insist that RANKL shedding is a process that weakens the osteoclastic activity, transgenic mice overexpressing soluble RANKL exhibit osteoporosis [15, 44]. Although the effect of shedding of RANKL is still unclear, the identification of the RANKL sheddase will contribute to the understanding of the mechanism of action of RANK/RANKL/OPG. The subsequent development of inhibitors of the RANKL sheddase may be expected to result in a new drug to treat osteogenic impairment.

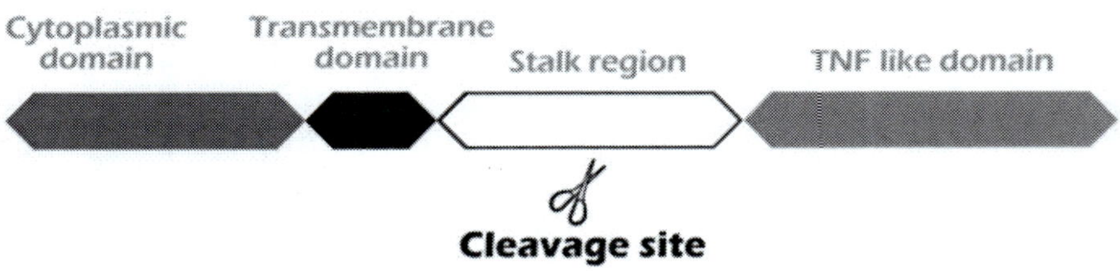

Figure 2. RANKL domain structure.RANKL includes transmembrane domain and stalk region. RANKL is cleaved at stalk region, which results in the soluble form of RANKL.

Proteolytic Processing of Bone Morphogenetic Proteins (BMPs)

Bone morphogenetic proteins (BMPs) belong to transforming growth factor-β (TGF-β) super family [45]. TGF-β is the abundant growth factors in the bone [46] and regulates osteoblast and osteoclast functions [47]. Then, how is TGF-β regulated for osteogenic differentiation? Growth factors are often bound to the ECM, and they readily acquire activity on release from the ECM. TGF-β is produced as inactive dimer with its propeptide. This latent complex is attached to one of the latent TGF-β-binding proteins (LTBPs) that are component of the

ECM. Therefore, the complex is also stored in the ECM like other growth factors. LTBP-1, -3 and -4 can form the complex with TGF- β [48]. Gene knockout studies indicate that LTBPs act as regulators of TGF- β secretion and activation [47]. Especially, LTBP-3 null mice exhibit skeletal defects such as osteosclerosis and osteoarthritis, which is thought to be due to the decreased levels of TGF- β [49]. Thus, LTBPs mediate the sequestration of TGF- β. How (or by what) is the sequestered TGF- β released? Osteoclasts exhibit LTBP1 cleavage, resulting in release of TGF-β. Latent TGF-β –binding protein-1 (LTBP1) cleavage is followed by release of the matrix-bound TGF-β. In contrast, osteoblasts have no cleavage activity for LTBP1, which prevents the release of TGF-β from ECM. To date, plasmin, elastase, MMP-2, and MMP-9 have been found to take part in the cleavage of LTBP1 [50]. Through this proteolytic event, TGF-β plays critical roles in the coupling of bone resorption to bone formation [50]. TGF-β stimulates the development of early osteoblasts, but inhibits their maturation and mineralization. Therefore, proteolytic process of LTBPs can be an attractive target as anabolic agents, but not antiresorptive agents.

Site-1 Protease (S1P)

Endochondral ossification is one of the processes by which most of the long bones of the mammalian skeletal system are formed [51]. S1P (also known as SKI-1, subtilisin kexin isozyme-1) is a proprotein convertase and a key member of the proteolysis pathway involved in the unfolded protein response and cholesterol homeostasis [52]. S1P frees several membrane-bound transcription factors from the membrane. In addition, it was recently revealed that proteolysis of glycoproteins requires S1P [53]. Thus, S1P is closely involved in activation of membrane-bound proteins. Evidence of a role for S1P in bone formation has been reported since 2007. Cartilage-specific S1P knock-out mice revealed that S1P participates in regulating endochondral ossification [54]. Although deletion of various MMPs did not abolish endochondral ossification [55], S1P knockout mice have a lethal defect in endochondral bone formation [54]. Most recently, studies have indicated a potential role for S1P in the mineralization of primary bone [56]. Taken together, this information suggests that S1P may also be an attractive target for regulation of bone regeneration. Sterol regulatory element-binding protein (SREBP) is one of the known substrates targeted by S1P [57]. However, substrates of S1P in bone regeneration have not yet been reported. Fortunately, it is known that S1P is a serine protease and is capable of being inhibited by a serine protease inhibitor, 4-(2-aminoethyl) benzenesulfonyl fluoride hydrochloride (AEBSF) [56]. Therefore, elucidation of its substrates will provide another therapeutic target for the treatment of osteogenic diseases.

Conclusion

We have discussed the roles of proteases in bone regeneration from the point of view of identifying potential therapeutic targets. In this review, we noted several candidate therapeutic targets with possible applications for osteogenic diseases, including MMPs and TIMPs, the cysteine switch, cathepsins, S1P, RANKL sheddase, and LTBPs cleavage resulting in BMP activation.

A large number of reports have demonstrated that MMPs and cathepsins play critical roles in osteogenesis. However, both types of proteases are families with a large number of members: the MMP family contains more than 20 members, and the cathepsin family has more than 10. In addition, the members of these families have similar amino acid sequences and domain structures, but show different substrate specificities. Until now, several protease inhibitors for bone diseases have been developed that have powerful inhibitory activities, but almost all of them show temporary or adverse effects. Low specificity (one inhibitor inhibits multiple proteases) is a major limitation of these drugs.

TIMPs may represent drug targets for osteogenic diseases via control of MMPs. However, a single TIMP can inhibit multiple MMPs. To overcome this challenge, attempts have been made to engineer new TIMP variants. Development of engineered TIMPs having altered specificity is expected. Fortunately, diverse methods are being used to engineer these proteins. The GASP (Genetic Assay for Site-specific Proteolysis) system [58] can be considered a useful tool for screening of TIMP variants with altered specificity for a single MMP.

Compared to the proteolytic activity of MMPs and cathepsins, S1P, RANKL sheddase, and LTBPs cleavage are thought to be highly specific processes. Considering the adverse effects seen with inhibitors of MMPs due to low specificity, these are very attractive targets. However, their mechanisms remain relatively poorly understood. At a minimum, the substrate specificity of these proteases should be elucidated.

There are two predominant methods used to develop inhibitors of proteases. One is the screening of chemical compounds libraries. The other is engineering of preexisting native cellular inhibitors using structure-based design and comparative modeling. This strategy can be combined to the use of peptidomimetic inhibitors, which are small compounds that imitate the amino acid sequences of specific native substrates [2, 28]. What are the requirements for a potent inhibitor to become a drug? The specificity of the inhibitor is a top priority, and it is also advantageous if the inhibitors act non-competitively and irreversibly.

Although the functions of osteoclasts are tightly coupled to those of osteoblasts for osteogenesis and bone regeneration, in a strict sense, osteoclast activity can be understood to be required earlier than osteoblast activity. This is one of the reasons why antiresorptive agents, which have been developed as inhibitors of osteoclastic activity, eventually result in a decrease in osteoblastic activity. Therefore, proteases that play roles in osteoblastogenesis may be more attractive targets for drug discovery in bone diseases in some respects. Regardless, much greater elucidation of the mechanisms of action of all the proteases involved in osteogenesis is needed if potent and successful inhibitors are to be developed as drugs.

References

[1] Carlos LO; Christopher MO. Protease degradomics: A new challenge for proteomics. *Nat Rev Mol Cell Biol,* 2002. 3, 509-519.

[2] Lee ES; Lee WG; Yun SH; Rho SH; Im I; Yang ST; Sellamuthu S; Lee YJ; Kwon SJ; Park OK; Jeon ES; Park WJ; Kim YC. Development of potent inhibitors of the coxsackievirus 3C protease. *Biochem Biophys Res Commun,* 2007. 358, 7-11.

[3] Kim SY; Valencia M; Lee ES; Park D; Oh M; Xue D; Park WJ. Identification of CED-3 substrates by a yeast-based screening method. *Mol Biotechnol*, 2004. 27, 1-6.

[4] Barrett AJ. Cellular proteolysis. An overview. *An N Y Acad Sci,* 1992. 674, 1-15.

[5] Erlebacher A; Filvaroff EH; Gitelman SE; Derynck R. Toward a molecular understanding of skeletal development. *Cell*, 1995. 80, 371-378

[6] Fabret, C; Feher, VA; Hoch, JA. Two-component signal transduction in *Bacillus subtilis*: How one organism sees its world. *J Bacteriol*, 1999. 181, 1975-1983.

[7] Woessner JF; Nagase H. Matrix metalloproteinases and TIMPs. First edition. England Oxford university press: 2000.

[8] Reuben PM; Cheung HS. Regulation of matrix metalloproteinase gene expression by protein kinases. *Front Biosci*, 2006. 11, 1199-1215.

[9] Massova I; Kotra LP; Fridman R; Mobashery S. Matrix metalloproteinases: structures, evolution, and diversification. *Faseb J*, 1998. 12, 1075-1095.

[10] Nagase H; Visse R; Murphy G. Structure and function of matrix metalloproteinases and TIMPs. *Cardiovasc Res*, 2006, 562-573.

[11] Vu TH; Werb Z. Matrix metalloproteinases: effectors of development and normal physiology. *Genes Dev*, 2000. 14, 2123-2133.

[12] Krane SM; Inada M. Matrix metalloproteinases and bone. *Bone,* 2008. 43, 7-18

[13] Ortega N; Behonick D; Stickens D; Werb Z. How proteases regulate bone morphogenesis. *Ann NY Acad Sci*, 2003. 995, 109-116.

[14] Holmbeck K; Bianco P; Caterina J; Yamada S; Kromer M; Kuznetsov SA; Mankani M; Robey PG; Poole AR; Pidoux I; Ward JM; Birkedal-Hansen H. MT1-MMP-deficient mice develop dwarfism, osteopenia, arthritis, and connective tissue disease due to inadequate collagen turnover. *Cell,* 1999. 99, 81-92.

[15] Hikita A; Tanaka S. Ectodomain shedding of receptor activator of NF-kappaB ligand. *Adv Exp Med Biol*, 2007. 602, 15-21.

[16] Sato H; Kinoshita T; Takino T; Nakayama k; Seiki M. Activation of a recombinant membrane type 1-matrix metalloproteinase (MT1-MMP) by furin and its interaction with tissue inhibitor of metalloproteinases (TIMP)-2. *FEBS Lett*, 1996. 393, 101-104.

[17] Kang T; Nagase H; Pei D. Activation of membrane-type matrix metalloproteinase 3 zymogen by the proprotein convertase furin in the trans-Golgi network. *Cancer Res*, 2002. 62, 675-681.

[18] Vu TH; Shipley JM; Bergers G; Berger JE; Helms JA; Hanahan D; Shapiro SD; Senior RM; Werb Z. MMP-9/gelatinase B is a key regulator of growth plate angiogenesis and apoptosis of hypertrophic chondrocytes. *Cell*, 1998. 93, 411-422.

[19] Ståhle-Bäckdahl M; Sandstedt B; Bruce K; Lindahl A; Jiménez MG; Vega JA; López-Otín C. Collagenase-3 (MMP-13) is expressed during human

fetal ossification and re-expressed in postnatal bone remodeling and in rheumatoid arthritis. *Lab Invest*, 1997. 76, 717-728.

[20] Geoffroy V; Kneissel M; Fournier B; Boyde A; Matthias P. High bone resorption in adult aging transgenic mice overexpressing cbfa1/runx2 in cells of the osteoblastic lineage. *Mol Cell Biol*, 2002. 22, 6222-6233.

[21] Mak IWY; Cowan RW; Popovic S; Colterjohn N; Singh G; Ghert M. Upregulation of MMP-13 via Runx2 in the stromal cell of giant cell tumor of bone. *Bone*, 2009. 45, 377-386.

[22] Parks WC, Wilson CL, Lopez-Boado YS. Matrix metalloproteinases as modulators of inflammation and innate immunity. *Nat Rev Immunol*, 2004. 4, 617-629.

[23] Ra HJ; Parks WC. Control of Matrix metalloproteinase catalytic activity. *Matrix Biol*, 2007. 26, 587-596.

[24] Lijnen HR. Plasmin and matrix metalloproteinases in vascular remodeling. *Thromb Haemost*, 2001. 86, 324-333.

[25] Visse R; Nagase H. Matrix metalloproteinases and tissue inhibitors of metalloproteinases; structure, function, and biochemistry. *Circ Res*, 2003. 92, 827-839.

[26] Brew K; Dinakarpandian D; Nagase H. Tissue inhibitors of metalloproteinases: evolution, structure and function. *Biochim biophys Acta*, 2000. 1477, 267-283.

[27] Will H; Atkinson SJ; butler GS; Smith B; Murphy G. The soluble catalytic domain of membrane type 1 matrix metalloproteinase cleaves the propeptide of progelatinase A and initiates autoproteolytic activation; regulation by TIMP-2 and TIMP-3. *J Biol Chem*, 1996. 271, 17119-17123.

[28] Dragovich PS; Prins TJ; Zhou R; Johnson TO; Hua Y; Luu HT; Sakata SK; Brown EL; Maldonado FC; Tuntland T; Lee CA; Fuhrman SA; Zalman LS; Patick AK; Matthews DA; Wu EY; Guo M; Borer BC; Nayyar NK; Moran T; Chen L; Rejto PA; Rose PW; Guzman MC; Dovalsantos EZ; Lee S; McGee K; Mohajeri M; Liese A; Tao J; Kosa MB; Liu B; Batugo MR; Gleeson JP; Wu ZP; Liu J; Meador JW 3rd; Ferre RA. Structure-based design, synthesis, and biological evaluation of irreversible human rhinovirus 3C protease inhibitors. 8. Pharmacological optimization of orally bioavailable 2-pyridone-containing peptidomimetics. *J Med Chem*, 2003. 46, 4572-4585.

[29] Fields GB; Van Wart HE; Birkedal-Hansen H. Sequence specificity of human skin fibroblast collagenase. Evidence for the role of collagen

structure in determining the collagenase cleavage site. *J Biol Chem,* 1987. 262, 6221-6226.

[30] Turk BE; Huang LL; Piro ET; Cantley LC. Determination of protease cleavage site motifs using mixture-based oriented peptide libraries. *Nat Biotech*, 2001. 19, 661-667

[31] Nagase H; Brew K. Designing TIMP variants that are selective metalloproteinase inhibitors. *Biochem Soc Symp,* 2003. 70, 201-212.

[32] Rouis M; Adamy C; Duverger N; Lesnik P; Horellou P; Moreau M; Emmanuel F; Caillaud JM; Laplaud PM; Dachet C; Chapman MJ. Adenovirus-mediated overexpression of tissue inhibitor of metalloproteinase-1 reduces artherosclerotic lesions in apolipoprotein E-deficient mice. *Circulation*, 1999. 100, 533-540.

[33] Olga V; Thomas R; Christoph P; Dusan T; Vito T; Boris T. Emerging roles of cysteine cathepsins in disease and their potential as drug targets. *Curr Pharm Design*, 2007. 13, 387-403.

[34] Screen M; Dean W; Cross JC; Hemberger M. Cathepsin proteases have distinct roles in trophoblast function and vascular remodeling. *Development*, 2008. 135, 3311-3320.

[35] Turk V; Turk B; Turk D. Lysosomal cysteine proteases: facts and opportunities. *EMBO J*, 2001. 20, 4629-4633.

[36] Garnero P; Borel O; Byrjalsen I; Ferreras M; Drake FH; McQueney MS; Foged NT; Delmas PD; Delaissé JM. The collagenolytic activity of cathepsin K is unique among mammalian proteinases. *J Biol Chem*, 1998. 273, 32347-32352.

[37] Kafienah W; Bromme D; Buttle DJ; Croucher LJ; Hollander AP. Human cathepsin K cleaves native type I and II collagens at the N-terminal end of the triple helix. *Biochem J*, 1998. 331, 727-32.

[38] Deal C. Potential new drug targets for osteoporosis. *Nat Clin Pract Rheumatol*, 2009. 5, 20-27.

[39] Frlan R; Gobec S. Inhibitors of cathepsin B. *Curr Med Chem*, 2006. 13, 2309-2327.

[40] Millest AJ; Breen SA; Loveday BE; Clarkson PN; Simpson CA; Waterton JC; Johnstone D. Effects of an inhibitor of cathepsin L on bone resorption in thyroparathyroidectomized and ovariectomized rats. *Bone*, 1997. 20, 465-71.

[41] Delaisse JM; Eeckhout Y; Vaes G. Inhibition of bone resorption in culture by inhibitors of thiol proteinases. *Biochem J*, 1980. 192, 365-368.

[42] Delaisse JM; Boyde A; Maconnachie E; Ali NN; Sear Ch;, Eeckhout Y; Vaes G; Jones SJ. The effects of inhibitors of cysteine-proteinases and collagenase on the resorptive activity of isolated osteoclasts. *Bone*, 1987. 8, 305-313.

[43] Anderson DM; Maraskovsky E; Billingsley WL; Dougall WC; Tometsko ME; Roux ER; Teepe MC; DuBose RF; Cosman D; Galibert L. A homologue of the TNF receptor and its ligand enhance T-cell growth and dendritic-cell function. *Nature*, 1997. 390, 175-179

[44] Mizuno A; Kanno T; Hoshi M; shibata O; Yano K; Fujise N; Kinosaki M; Yamaguchi K; Tsuda E; Murakami A; Yasuda H; Higashio K. Transgenic mice overexpressing soluble osteoclast differentiation factor (sODF) exhibit severe osteoporosis. *J bone Miner Metab*, 2002. 20, 337-344.

[45] Luu HH; Song W-X; Luo X; Manning D; Luo J; Deng Z-L; Sharff KA; Montag AG; Haydon RC; He T-C. Distinct roles of bone morphogenetic proteins in osteogenic differentiation of mesenchymal stem cells. *J Orthop Res*, 2007. 25, 665-677.

[46] Bonewald LF; Mundy GR. Role of transforming growth factor-b in bone remodeling. *Clin Orthop Relat Res*, 1990. 250, 261-276.

[47] Koli K; Ryynanen MJ; Keski-Oja J. Latent TGF- β binding proteins (LTBPs)-1 and -3 coordinate proliferation and osteogenic differentiation of human mesenchymal stem cells. *Bone*, 2008. 43, 679-688.

[48] Chen Y; Ali T; Todorovic V; O'Leary JM; Downing AK; Rifkin DB. Amino acid requirements for formation of the TGF- β -latent TGF- β binding protein complexes. *J Mol Biol*, 2005. 345, 175-186.

[49] Dabovic B; Levasseur R; Zambuto L; Chen Y; Karsenty G; Rifkin DB. Osteopetrosis-like phenotype in latent TGF- β binding protein 3 deficient mice. *Bone*, 2005. 37, 25-31.

[50] Sarah LD; Jennifer LR; Gregory RM; Lynda FB. Proteolysis of latent transforming growth factor- β (TGF-β)-binding protein-1 by osteoclasts. *JBC*, 2002. 21352-21360.

[51] Provot S; Schipani E. Molecular mechanisms of endochondral bone development. *Biochem Biophys Res Commun*, 2005. 328, 658-665.

[52] Brown MS; Ye J; Rawson RB; Goldstein JL. Regulated intramembrane proteolysis: a control mechanism conserved from bacteria to humans. *Cell*, 2000. 100, 391-398.

[53] Rojek JM; Lee AM; Nguyen N; Spiropoulou CF; Kunz S. Site 1 protease is required for proteolytic processing of the glycoproteins of the South

American hemorrhagic fever viruses Junin, Machupo, and Guanarito. *J Virol*. 2008. 82, 6045-6051.

[54] Debabrata P; Xiaoyun X; Sherri D; Jennifer B; Carl F; Ernst BH; Linda JS. Site-1 protease is essential for endochondral bone formation in mice. *J Cell Biol*, 2007. 179, 687-700.

[55] Inada M; Wang Y; Byrne MH; Rahman MU; Miyaura C; Lopez-Otin C; Krane SM. Critical roles for collagenase-3 (Mmp13) in development of growth plate cartilage and in endochondral ossification. *Proc Natl Acad Sci USA*, 2004. 101, 17192-17197.

[56] Gorski JP; Huffman NT; Cui C; Henderson EP; Midura RJ; Seidah NG. Potential role of proprotein convertase SKI-1 in the mineralization of primary bone. *Cells Tissues Organs*, 2009. 189, 25-32.

[57] Bartz R, Seemann J. Mitotic regulation of SREBP and ATF6 by separation of the Golgi and ER. *Cell Cycle*, 2008. 15, 2100-2105.

[58] Kang H; Kim SY; Park WJ. An improved strategy for a genetic assay for site specific proteolysis. *Mol Cells*, 2001. 11, 263-266.

In: Bone Regeneration: Growth Factors... ISBN: 978-1-60876-621-5
Editors: V. Legard et al., pp. 303-324

Chapter XII

Ex Vivo Gene Therapy in Autologous Bone Marrow Stromal Stem Cells for Tissue-engineered Craniomaxillofacial Bone Regeneration

***Long-Bin Jeng*[1] *and Sophia Chia Ning Chang*[2,*]**
[1]Department of Surgery, [2]Department of Plastic Surgery, China Medical University & Hospital, School of Medicine, China Medical University, Taichung, Taiwan.

Abstract

In therapeutic bone repairs, autologous bone grafts conventional or vascularized, allografts, and biocompatible artificial bone substitutes all have their shortcomings. The bone formed from peptides (rhBMPs), demineralized bone powder or combinations of both is small in size. The tissue engineering may be an alternative for cranial bone repair.

*Corresponding author: Sophia Department of Plastic Surgery, China Medical University & Hospital, Taiwan, R.O.C. Tel: 886-4-22052121 ext. 1638; Fax: 886-4-22029083; E-mail: scnchang@www.cmu.edu.tw.

We develop animal models to test the hypothesis that *ex vivo* replication defective Adenovirus mediated human BMP-2 gene transfer to bone marrow mesenchymal stem cells (MSCs) enhances the autologous bone formation for repairing a critical size craniomaxillofacial defect.

The MSC was implanted by using *ex vivo* adenovirus BMP-2 medicated gene transfer to the expanded MSC 7 days before implantation with a concentration was 5×10^7/ml of collagen type I, and the control site using adenovirus β-galactosidase under the same condition.

The BMP-2 cell/construct site showed complete whitish solid bone formation after 3 months and the HE and Von Kossa stains demonstrated that mature woven bone with well mineralization. 3D CT image showed a wholesome repair in both cranium and maxilla. Quantitative analysis demonstrated that there was significant different ($p<0.001$) in bone formation. The biomechanic test did reveal no significant statistical difference in the stiffness of new bone formed by BMP-2 cell construct and the normal maxilla bone. The data evidenced *de novo* bone formation with quality as good as normal craniofacial bone, and thus capable of sustaining gravity load. The results demonstrated that *ex vivo* gene therapy by adenovirus BMP-2 transduction will induce and enhance craniomaxillofacial bone regeneration.

Keywords: rhBMPs, recombinant human Bone Morphogenic protein; MSCs, bone marrow mesenchymal stem cells; βgal, β-galactosidase; 3D CT, Three Dimensional Computer Tomography.

Introduction

There are over one million surgical operations performed each year in the United States involving bone and cartilage repair [1]. Autologous bone grafts are frequently needed in these surgeries. However, harvesting bone for grafting is an invasive procedure with potential morbidity to patients. The use of allogenic bone graft is also limited due to immunological rejection and the risk of transmission of infectious diseases. Additionally, the failure of bone graft to completely resorption, difficulty in shaping the bone graft to fit the osseous defect and lack of sufficient bone grafts to completely fill the defect add the additional limits to use bone grafts in clinical practice. Demineralized bone powder [2] and biocompatible bone substitutes [3] showed unsatisfactory results because of their incapability for repairing critical size bone defects.

In recent years, tissue engineering has made some success in creating functional tissue by employing biocompatible, biodegradable polymers seeded with living cells [1]. Culturing osteoblasts in PGA/PLA fiber meshes have manifested that osteoblasts-PGA/PLA constructs can be fabricated. Furthermore, implantation of this construct into an animal host generated the new bone composition with the final morphology established by the shape of the polymer scaffold. This construct triggered vigorous cellular immunity and failed in the immunocompetent animal model.

It has been proved that osteoblasts are derived from pluripotential mesenchymal stem cells which may develop into various cellular lineage capable of producing bone, muscle, cartilage, fat, and fibrous tissue [4, 5]. The ultimate fate of an undifferentiated stem cell is determined by the local environment to which the mesenchymal stem cell is exposed [6]. The use of inductive proteins to stimulate differentiation has been of particular interest [7-11]. Bone morphogenetic protein (BMP) has been shown to induce the development of osteoblasts from mesenchymal stem cells [12].

Bone marrow mesenchymal stromal stem cells (MSCs) have demonstrated great potential for bone repairs. MSCs are pluripotent progenitor cells, which possess the ability of self-renewal and give rise to multiple mesodermal tissue types [13]. The osteogenic potential of marrow-derived MSC has been well defined as evidenced by bone formation following *in vivo* transplantation of MSCs [14-19]. Bone morphogenetic protein (BMP) has been shown to encourage the development of osteoblasts from MSCs [20-23]. In previous studies we have demonstrated that BMP-2 gene-transfer into MSCs induced stem cells into an osseous lineage and formed bone *in vivo* [21]. In this study, we further asked whether this strategy could use to repair a critical size membranous bone defect. To answer this question, we assessed repair of maxilla bone defect in miniature swine by BMP-2 gene engineered autologous MSCs [24].

Results

Recombinant Adenovirus Construction with Human BMP-2 Expression

These two recombinant adenoviruses have been deleted only in the El region of the genome. We purposely retained the E3 region of the virus genome intact on the vectors because of evidence that the E3 proteins of the virus inhibit the host immune response [25]. Titered in 293 cells, the viral lysate of the vectors is usually 10^8 plaque-forming units per milliliter.

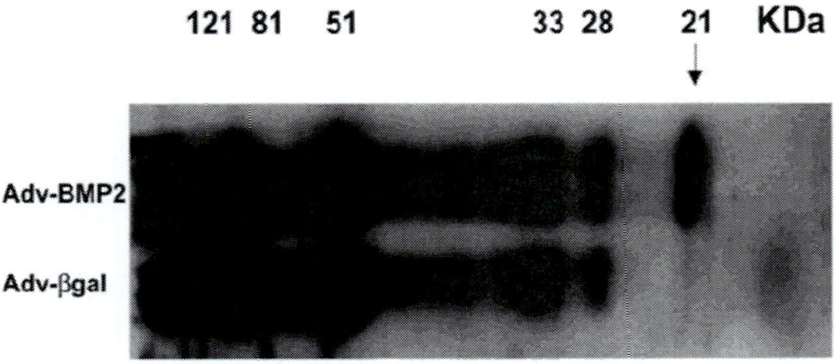

Figure 1. Western blot detection of adenovirus BMP-2 expression in MSCs *in vitro.* MSCs were infected with adenovirus βgalactosidase (*Adv*-βgal) or BMP-2 (*Adv*-hBMP-2) respectively. Upper: MSCs infected with adenovirus BMP-2. Lower: MSCs infected with adenovirus βgalactosidase. Seven days after infection, medium was collected and immunoprecipitation was performed. The noted a band over 21 KDa.

Detecting the coding product BMP-2 would assess the expression of human BMP-2 gene in MSCs, which was mediated by adenovirus BMP-2 gene transfer. A specific monoclonal Ab against BMP-2 was used in assay. Because BMP-2 is a secrete growth factor, the conditioned medium from virus transduced cells were used as a start material. After immune precipitation and western blot specific band of molecular weight 21KDa was detected from conditioned medium from adenovirus-BMP-2-transduced MSCs but not in the conditioned cultured medium from adenovirus-βgal-transduced MSCs. The adenovirus mediated BMP gene seems well expressed and processed in MSCs because the molecular weight of secreted BMP-2 indicated mature status without prepeptide (Figure 1).

The Cranial Defects Repair Model

There was noticeable whitish bone formation at the sites filled with BMP-2 infected cells from 6 weeks on, but irregular islands of bone formation were found at the control site (Figure 2A). Near complete full-thickness cranium defect repair was achieved at 3 months in experimental sites (Figure 2B, C). There is block of bone formation at the experimental sites, not at the control site [26].

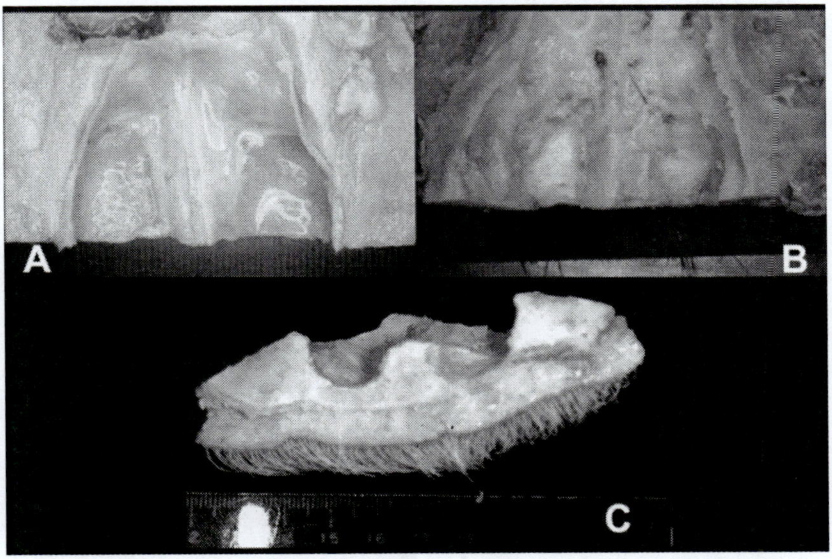

Figure 2. Gross appearance of the cranium after 6 weeks and 3 months implantation *in vivo*. A: Cranium view (6 weeks). B: Cranium view. (3 months). C: Transverse view (3 months) (Left: BMP-2 site, Right: β-Gal site).

Table I. Areas of new bone formation measured by 3D C.T. (cm^2) (n=10/Group).

Time	*Site*		Exp – Cont (95% CI*)
	Experimental	Control	
6 weeks	4.08 ± 0.42	3.56 ± 0.48	0.52 (0.05, 0.99)
3 months	6.23 ± 0.38	4.10 ± 0.40	2.13 (1.72, 25.6)
3 months to 6 weeks	2.15	0.54	

* CI = confidence interval.

The average new bone formation area in the 3D CT image for the two sides at the two studied periods are shown in Table 1. The actual defect area of bilateral cranial defect is near 7.2±0.42 cm^2, owing to its anatomical limitation. While the new bone formation area (cm^2) measured by 3D C.T., the increment from 6 week to 3 month was significant for both the experimental (from 4.08 to 6.23, $p<0.001$) and the control sides (from 3.56 to 4.10, $p=0.014$), the improvement was significantly more in the experimental side than in the control one (2.15 vs. 0.54, $p < 0.001$); as well as at 3 month than at 6 week (2.13 vs. 0.52, $p< 0.001$) [26].

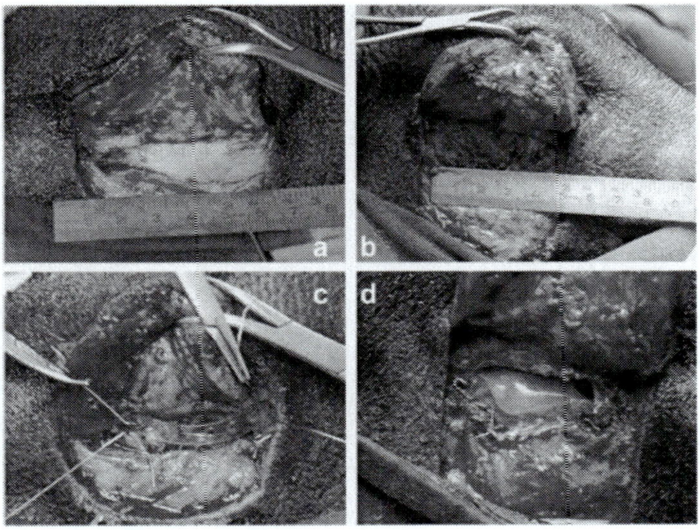

Figure 3. The maxilla defect in miniature swine: Left upper: The infraorbital maxilla was denuded after removal of periosteum. Right upper: The maxilla bone defect was 1.2 x 3 cm^2 bilaterally. Left lower: One PLLA plate was placed to bridge the maxilla defect. Right lower: The MSCs/collagen type I construct was placed on the defect.

The Maxilla Defect Repair Model

Bilateral maxillary defects of 3x1.2 cm^2 each at each side in miniature swine were created by removal of bone and periosteum [24]. No osteoinductive tissues were remained at defects (Figure 3A & B). To bridge the defect, the PLLA plate block the maxilla defect and kept the construct from soft tissue invagination (Figure 3C). 3ml of collagen type I-cell mix construct was implanted to each defect (Figure 3D). The miniature swine were quickly recovered from surgery. No side effects were observed. The operated infraorbital areas were healed well. The indurations increased with the time upon removal of all the soft tissues around the reconstructed areas. 3months post surgery, there was noticeable whitish bone observed at the defect site implanted with BMP-2 cell constructs (Figure 4B), upon removal of all the soft tissue around the reconstructed areas. No bone formation was found at the defect site implanted with βgal cell construct. The infraorbital notching in the control side was obvious (Figure 4A).

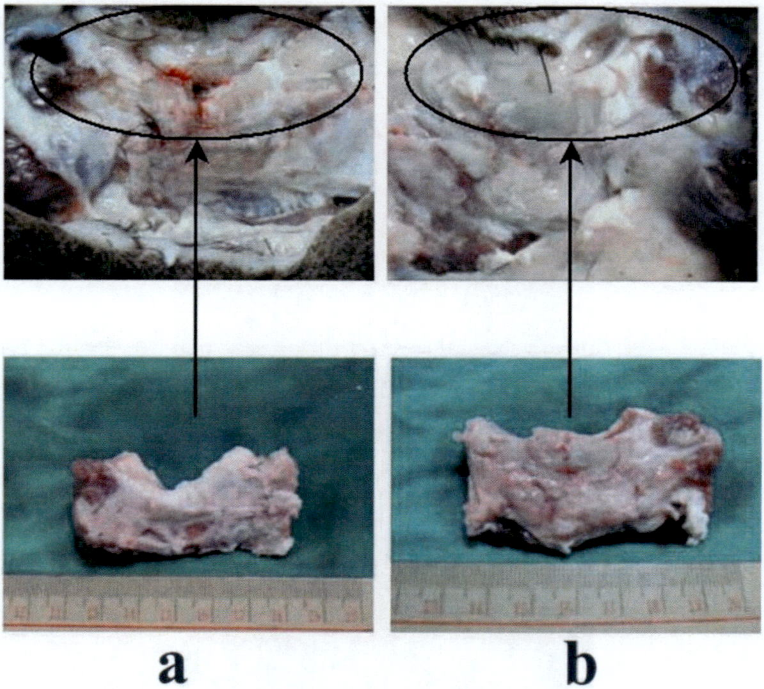

Figure 4. The gross pictures of the regenerated maxilla after 3 months implantation *in vivo*. A: the control site. B: the experimental site.

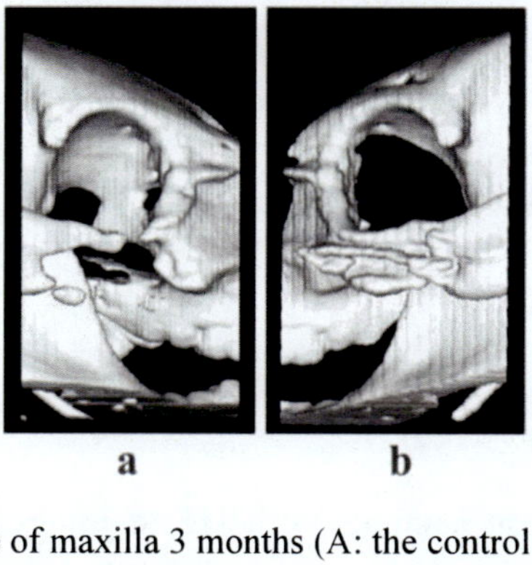

Figure 5. The 3D CT image of maxilla 3 months (A: the control site, B: the experimental site).

3D CT image studies were performed for each sample immediate after harvest. The image demonstrated that the defects implanted with BMP-2 cell construct were completely repaired at 3 month after implantation (Figure 5B). The whole regenerated maxilla at these sites were reserved in normal position. Small islands of bone formation were observed at the defects implanted with βgal cell constructs, however, the defects remained unrepaired (Figure 5A).

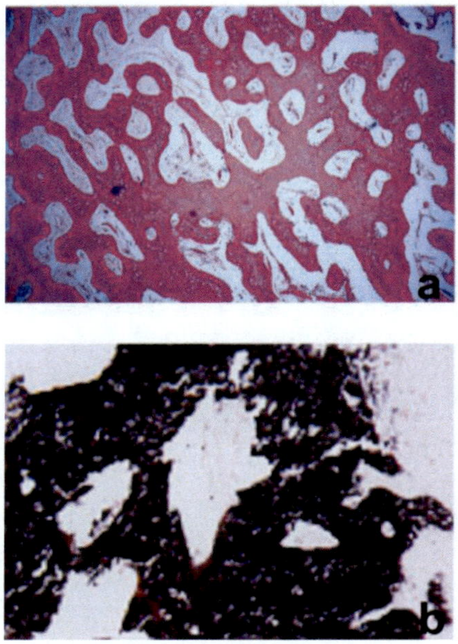

Figure 6. 6A: The H & E staining of the experimental site showed cancellous bone formation (200x). 6B: The higher power field of Von Kossa stain of the BMP-2 site bony formation (200X).

Quantitative analysis with software Analyzer® (Harvard, MA) demonstrated that there was significant difference ($p<0.001$) in bone formation between the defects implanted with BMP-2 cell construct and βgal cell construct (2.63 ± 0.18 cm^2 vs. $1.78\pm0.08cm^2$).

Histology Examination

Histological studies on samples with HE staining demonstrated cancellous bone formation at the defects implanted with BMP-2 cell construct [24]. Mature wove *de novo* bone was observed at the defects (Figure 6A). High magnification of microscopy showed osteoblasts in the peripheral areas of newly formed bone. Stained halt was found in the cytoplasm of cells. The Von Kossa's assay presented strongly positive stain in the repaired defect of BMP-2 cell implantation, indicating well mineralization (Figure 6B).

Biomechanic Test

The biomechanical properties were compared between the bone harvested from repaired defect, which implanted with BMP-2 cell construct 3 months *in vivo* and normal maxilla bone. The results showed that these two originated bones were very similar. The strength of the new bone was similar to that of the normal cranial bone when evaluated by biomechanical test (cranium bone vs. tissue engineered bone: 88.646 $\pm$ 5.121 vs. 80.536 $\pm$ 19.302 MPa; p=0.227) [26]. Specifically, the equilibrium modules revealed that the newly formed bones and normal maxilla bones were alike in stiffness, (80624$\pm$8365 vs. 81644$\pm$5497 kPa, p=0.751) as there was only a 1.2% difference. Moreover, the effect size for this difference was only 0.19, a small effect considered by Cohen [27]. The analysis indicated that newly formed bone in BMP-2 cell repaired defects shared strong biomechanical properties as the original bones.

Immunohistochemistry for Adenovirus Hexon Protein

Specific Ab against adenovirus hexon protein was used in immunohistochemistry study to follow the possible remain of adenovirus in construct in repaired defect site. Results demonstrated that positive staining was found in BMP-2 cell constructs *in vitro* (Figure 7A) [24]. However, there was no hexon protein found in repaired defects three months after implanted with BMP-2 cell constructs (Figure 7B).

It indicated that adenoviral vectors were clean *in vivo* during these 3 months. No risk of continuing expression of BMP-2 protein in implanted sites was remained.

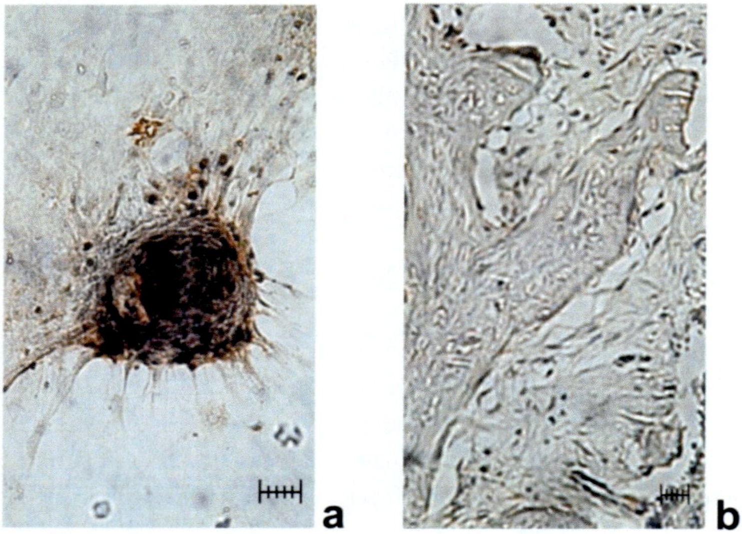

Figure 7. The immunochemical stain for detecting adenovirus: (Scale bar 50um.) 7A: positive brown staining in pre-implantation culture. 7B: negative in 3 months samples.

Discussion

Ever since the genes of the BMP family were successfully cloned, large quantities of individual BMP proteins have been made available by recombinant-DNA technology. Yasko et al. first demonstrated the effects of recombinant human BMP-2 on bone formation in the rat segmental femoral-defect model [28]. Studies of this nature have been performed in the sheep femur [29] and in canine spine and mandible models [30-33]. Although the results are encouraging in studies in which BMP protein is implanted directly to a defective area, a large amount of protein (up to milligram quantities) is often required to stimulate significant new bone formation *in vivo.* This requirement for large quantities of the protein increases the risk of unwanted side effects. Moreover, the lack of a system to deliver proteins in a continuous manner over time [34] may hamper this technique. Recombinant BMP-2 shows great potential in bone repair; however, current approval in FDA approval of recombinant BMP-2 protein in dental small defects repair as a commercial use stresses the importance of developing an alternative procedure for BMP-2 application.

In this study, we utilized an alternative technique to induce bone formation. We used the strategy combining BMP-2 gene transfer and autologous MSCs. The results demonstrated that this strategy lead to completely repair the critical size of weight-bearing maxilla bone defect with BMP-2 gene transferred MSCs. We demonstrated that ex vivo gene therapy by adenovirus BMP-2 transduction will induce and enhance craniomaxillofacial bone regeneration [24].

Near-complete repair of the critical size cranial defect by tissue engineered MSC/collagen type I construct was observed. The actual defect area of bilateral cranial defect is near 7.2±0.42 cm^2, owing its anatomical limitation. While the new bone formation area (cm^2) measured by 3D C.T., the increment from 6 week to 3 month was significant for both the experimental (from 4.08 to 6.23, $p<0.001$) and the control sides (from 3.56 to 4.10, $p=0.014$), the improvement was significantly more in the experimental side than in the control one (2.15 vs. 0.54, $p < 0.001$); as well as at 3 month than at 6 week (2.13 vs. 0.52, $p< 0.001$). Moreover, after Bonferroni adjustment of alpha, the difference between the experimental and control group was not significant at 6 weeks (mean difference = 0.52; $p=0.035$ > adjusted alpha = 0.025), but significant at 3 months (mean difference = 2.13; $p< 0.001$) [26]. The critical sized maxilla defect 3x1.2 cm^2 was created bilaterally. The osteogenic periosteum around maxilla was removed. The BMP-2 cell construct site showed complete whitish solid bone formation, while the βgal cell construct site revealed an infraorbital notch. 3D CT image showed a wholesome infraorbital rim repair in the experimental site. The histology demonstrated that mature woven bone with well mineralization. The biomechanic test did not reveal the significant statistical difference in the new bone formed by BMP-2 cell construct and the original cranium or maxilla bones. The data evidenced *de novo* bone formation with quality as good as normal maxilla bone, and thus capable of sustaining gravity loading and peripheral soft tissues invagination [24].

Periosteum explant culture cells have been proved to be able to be tissue-engineered into bone-like tissue [35-38]. However, the slow expansion rates of periosteum explant culture cells have been noted and then were vulnerable to the enzymatic digestion. The difficult assess to periosteum may limit the clinical application of periosteal cells in tissue-engineered bone. MSCs, on the other hand, were not subjective to enzymatic digestion and were readily expandable in a monolayer culture in vitro [39]. It was easier to access and thus a possible clinical application would be rendered [24].

A recombinant replication deficient adenovirus successfully mediated the human BMP-2 gene transfer and BMP-2 protein expression in the MSCs. Expressed BMP-2 protein was detected in culture medium of the transduced cell. Adenovirus-mediated BMP-2 gene transfer on MSCs directed the cells to proliferation and differentiation. We speculate that the cell response to the gene transfer was through an autocrine or paracrine effect of BMP-2 protein that was expressed from the transferred BMP-2 gene by the cell itself [24].

Recombinant adenovirus-mediated gene transfer has its limitation in the immune-competent host. The host immune response to the adenovirus proteins from leak expression of virus genome will delete the recombinant adenovirus and terminate expression of the transgene. Also, inflammation has been observed. Therefore, the use of a new generation of recombinant adenovirus vectors that has less immunogenecity and longer expression in the immune-competent host will help to reduce these problems and thus be more effective [24].

Developmental biology teaches that undifferentiated stem cells follow a definite sequence of cellular alterations and lineage; this results in a fully differentiated cell with a specific phenotype [40-42]. In previous literature, some gene therapies in osteochodral repair were applied on cell lines [43-45], which are against the principals of tissue engineering and further clinical applications. Optimal biomaterials chosen in this study made good maxilla repair possible [46]. This study demonstrates that BMP-2 gene transfer into MSCs could potentially direct differentiation into bone formation *in vivo.* We demonstrated that the use of MSCs/collagen type I with *adv*-BMP-2 gene transfer would enhance the bony healing of craniofacial defect. The adenoviral vector was cleaned by the host by three months in this study and other reports [47-50]. This is the most encouraging report that autologous tissue-engineered membranous bone repair will be achieved by MSCs with *adv* BMP-2 gene transfer. This may be an alternative treatment for enophthalamus and craniomaxillofacial deformities.

Materials and Methods

Construction of Recombinant Adenovirus

Adenovirus BMP-2, a replication-defective adenovirus vector containing human BMP-2 gene, was constructed as previous description [21]. In this vector, transcription of the BMP-2 gene is controlled by cytomegalovirus early gene promoter/enhancer. The recombinant adenovirus containing *Escherichia coli* lac z gene (adenovirus β-galactosidase) was constructed by the same procedure and used as the control vector in the experiments.

Bone Marrow Mesenchymal Stem Cell Culture and Adenoviral Infection

20 female miniature swine (Mitsae Pig, CGU, Taiwan) were used in this study. Miniature swine weighted around 15 kilograms and aged 4 months old. Miniature swine bone marrow was aspirated under anesthesia with i.m. injection of 1ml/10kg 2% Rompum (Bayer, USA), and 50mg/10kg Ketamine (Yung Shin Pharmaceutical Industrial Co., Taiwan). 20ml bone marrow aspirate was collected from each swine iliac crest and mixed with 2c.c. of heparin Sodium (5000unit/ml, China Chemical & Pharmaceutical Co., Taiwan). Obtained bone marrow was mixed with 20ml DMEM medium. The marrow cells were collected by 1000rpm spin. The pellet was resuspended with 2.5ml DMEM medium. The cell suspension was loaded onto 70% Percoll (Sigma, St. Louis, Missouri) gradients and centrifuged at 460x g for 15min. Three fractions of gradient were formed after centrifuge; the top 25% of the gradients (low density cells), pooled density=1.03g/ml; the middle 50% of the gradients (high density cells), pooled density=1.10g/ml; and the bottom 25% of the gradients, pooled density=1.14g/ml. MSCs were observed in the low-density fraction and were collected and plated. Adherent MSCs were cultured in complete DMEM at 37℃ with 5% CO_2. The medium was changed every 3-4 days.

MSCs were received human BMP-2 gene transfer by infection with recombinant adenovirus BMP-2 at a multiplicity at infection (MOI) of 50. MSCs were infected at 37℃ for overnight and cultured in complete medium for another 5 days. As a control, adenovirus mediated βGal gene transfer was carried out at the same time and for the same condition. The transduced MSCs were trypsinized by 0.4% trypsin and washed three times by DMEM medium. The cells were then mixed with Type I collagen (Pancogene S®, Gattefosse, Saint-Priest, Cedex, France) to achieve the cell concentration at $50x10^6$/ml. After one hour incubation at the 37℃ in CO_2 incubator, Cell/Collagen type I mix become gel-like

Western Blot Detection of Adenovirus BMP-2 Expression in MSC in vitro

MSCs were grown in DMEM with 10% FBS to 90% confluence in roller bottle. The cells were infected with 20ml virus working stock (10^7 PFU/ml) of adenovirus βgal or adenovirus BMP-2 at a multiplicity of infection (MOI) of 50 [24]. The infected cells were maintained in DMEM with 10% FBS. The conditioned medium was collected 7 days later post infection.

10ul mAb against human BMP-2 (R&D systems, Inc. Minneapolis, MN) was added to the 25ml harvested conditioned medium and incubated at 4℃ for 2-3 hrs. 50ul protein A Sepharose (Amersham Pharmacia Biotech. Piscataway, NJ) were then added and rotated overnight. Antigen-Ab-Sephanose complexes were collected by centrifugation at 1500rpm 5min. The pellets were resuspended in 50ul 2x Laemli buffer and boiled for 5min. After brief centrifugation, supernatants were run on a 12.5% SDS-polyacrylamide gel electrophoresis and transferred to a Nitrocellulose membrane (Micron Separations Inc. Westborough, MA). After blocking with 5% BSA solution at room temperature for 1 hr, the transferred membrane was exposed to mAb against human BMP-2 in 1% BSA solution for overnight. Blots were then reacted with goat anti-mouse IgG conjugated to horseradish peroxidase (Santa Cruz Biotechnology, Inc. Santa Cruz, California) in 1% BSA solution for 1hr and developed with enhanced chemiluminesence (Amersham Pharmacia Biotech. Piscataway, NJ).

Open Implantation on the Cranial Defects

All miniature swine were intubated and kept on ventilators. Two cranial defects of 2x5 cm^2 each were created per animal [26]. The so-called critical size defect is the one that will not regenerate spontaneously during a life span of the animal. The 2x5 cm^2 cranial defect would not regenerate within 3 months from our previous study (unpublished data) and other report [51]. The original defect is somewhat oval shape with 2 cm in the transverse widest part and 5 cm of the longest length. Bilateral compartments with defect area (7.2±0.42 cm^2) were separated by a central remaining bar (0.5x5 cm^2). With the thickness of cranial bone near 1.5 to 2 cm and applying the subcutaneous watertight closure for each compartment, this prevents mixing of two different gel-like constructs of two sites. The periosteum from above, the dura from below, and the cranium bone were completely removed to ensure that no osteoinductive tissue was left (Figure

8A, B) [52]. The dura defects were repaired with acellular dermis (Pig Alloderm, LifeCell USA) (Figure 8C).

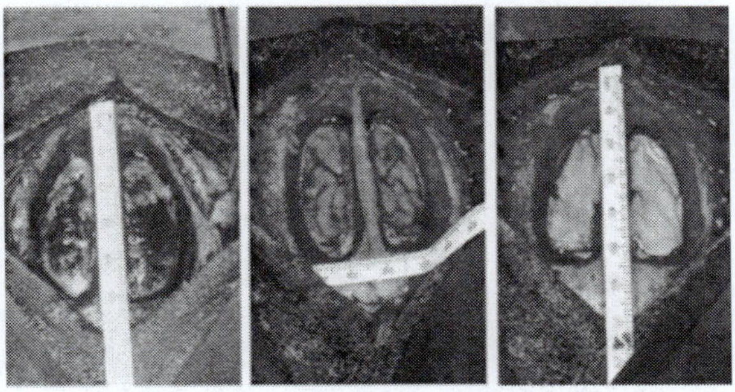

Figure 8. The 2x5 cm^2 sized cranial defect in miniature swine: A: Bilateral bony defect, B: underlying dura removed. C: Alloderm used for dura repair.

Six ml of cell/Collagen type I (50 million cells/ml) constructs were implanted to each site. The adenovirus-mediated BMP-2 gene transferred MSC/Collagen type I construct was used to fill the right side cranial defect, the β-Gal gene transferred MSC/Collagen type I construct was used to fill the left side defect. The scalp wounds were closed in a watertight manner.

Open Implantation after Creating Maxillary Bone Defect

At the secondary stage of surgery, all miniature swine were intubated and on ventilators. Bilateral infraorbital flaps were elevated to expose the maxilla [24]. The maxilla bone was sawed and burred out to create an infraorbital rim defect of $3x1.2cm^2$. There is no osteoinductive periosteum left in bilateral maxillary defects. To bridge the defect, one biodegradeable PLLA (polylactic acid, BioSorb FX™, Bionx Implants Ltd., Tampere, Finland), 6-holes miniplate was used on each maxillary defect as an internal splinting. 3ml of Collagen type I (Pancogen S®, Gattefosse Co., Saint-Priest, France)/Cell mix construct was implanted to each defect. The right side defect (on the animal's right) was covered with BMP-2 gene transferred MSCs construct. Left side defect (on the animal's left) was filled with βgal gene transferred MSCs construct. The facial wounds were closed in a watertight manner. The miniature swine were sacrificed at 3months after tissue engineered constructs implantation.

3D CT Model

Helical scanning technique was applied to the swine heads. Primary data manipulation was performed in the Radiology Department at a 2-mm thickness, and the raw data were transferred to an IBM-compatible personal computer in the Medical Imaging Laboratory [53]. Reformation of the CT data was performed, setting the voxel unit at 0.6x0.6x0.6 mm in size. 3 dimensional images of the swine heads were displayed on the computer. A thresholding technique was used to remove the soft tissue, displaying the osseous structures of the swine heads. The regenerated bone was outlined, and surface area measurement was performed. The CT data reformation, image display, and area measurement were performed using Analyze 4.0 software [54] (Biomedical Imaging Resource, Mayo Foundation, Rochester, Minnesota, USA).

Histologic Examination of Bone Samples Harvested from the Maxilla Defects

Harvested samples were fixed in buffered 10% formalin for 72 hours and then were sawed into one piece of samples was decalcified in Decalcifier I (Surgipath, Northbrook, IL, U.S.A.) solution for 48 hours [24]. The decalcified tissues were underwent procedures for embedding, incising, and staining with hematoxylin and eosin. The nondecalcified other piece of the samples was stained with von Kossa's silver nitrate for demonstration of matrix mineralization.

Immunohistochemistry for Adenovirus Hexon Protein

Paraffin embedded samples were detected adenovirus hexon protein by immunohistochemistry [24]. Embedded tissues were incised after removal from paraffin, slides and then hydrated. The rabbit primary antibody (Ab) against adenovirus hexon protein (Chemicon International, Temecule, CA) was applied on slides at concentration 1:500 dilution. Slides were incubated at 35℃ for 1 hr. A biotinylated goat anti-rabbit secondary Ab was then applied, followed by a streptavidin-horseradish peroxidase conjugate. Diaminobenzidine (DAB) was used as the chemogen, before slides were counterstained with hematoxylin [55].

Biomechanical Analysis

The samples used for biomechanical analysis were stored at –70℃ after harvest. Newly formed bone in repaired defects were burred and finished to 6mm diametric cylinder with 3mm height for compressive testing [24]. Each specimen was then tested to failure in axial compression by using MTS testing machine (Bionix 858, MTS Corp., MN, USA). To ensure perpendicular compression, a 6mm diametric cylindrical rod with self-aligned function was used as plunger, which clamped on the upper side of MTS wedge grip. The specimen was then placed on a flat supporting jig clamped on the lower side of MTS wedge grip connecting to a 2,500N load cell. After the specimens were positioned, the compressive force was applied at a constant loading rate of 20N/sec and the relation between force and displacement was recorded in an increment of 0.02mm by MTS Teststar software. The maximum compression stress for each individual specimen was determined by dividing the magnitude of maximum force to the minimum original cross section area carried by the test specimen [56]. Comparison was performed between newly formed bone with normal maxilla one.

Statistical Analysis

Data are expressed as means ± standard deviations. Paired T test was employed to assess the statistical significance of results for all the 3D CT images of the repaired bone areas. Paired T test was also used to compare the biomechanical property of the new bone and the normal maxilla.

Acknowledgment

The authors would like to thank professor Sing Kai Lo, Institute for International health, University of Sydney for the statistic work in this manuscript. We wish to thank Dr. Lun-Jou Lo and the Medical Imaging Laboratory for providing the CT image processing and measurement, and Dr. Richard A Robb, PhD, Biomedical Imaging Resource, Mayo Foundation, Rochester, Minnesota, USA, for using the Analyze program. This research was supported by NMRPG1074, 2070, 3009, (NSC 93-2314-B-182A-002).

References

[1] Langer, R; Vacanti, JP. *Tissue engineering. Sci*, 1993, 920, 260-266.

[2] Rabie, AB; Chay, SH; Wong, AM. Healing of autogenous intramembranous bone in the presence and absence of homologous demineralized intramembranous bone. *Am. J Orthod Dentofacial Ortho*, 2000, 117, 288-297.

[3] Furukawa, T; Matsusue, Y; Yasunaga, T. Histomorphometric study on high-strength hydroxyapatite/poly (L-lactide) composite rods for internal fixation of bone fractures. *J Biomed Mater Res*, 2000, 50, 410-419.

[4] Metcalf, D. The molecular control of cell division, differentiation commitment and maturation in haemopoietic cells. *Nature*, 1989, 339, 27-30.

[5] Wiles, MV; Keller, G. Multiple hematopoietic lineages develop from embryonic stem (ES) cells in culture. *Development*, 1991, 111(2), 259-267.

[6] Goshima, J; Goldberg, VM; Caplan, Al. The origin of bone formed in composite grafts of porous calcium phosphate ceramic loaded with marrow cells. *Clin. Orthop.*, 1991, 269, 274-283.

[7] Ahrens, M; Ankenbauer, T; Schroder, D; Holinagel, A; Mayer, H; Gross, G. Expression of human bone morphogenetic proteins 2 or 4 in murine mesenchymal progenitor C3H10T1/2 cells induces differentiation into disctinct mesenchymal cell lineage. *DNA Cell Biol.*, 1993, 12(10), 871-880.

[8] Gitelman, SE; Kobrin, MS; Ye, JQ; Lopez, AR; Lee, A; Derynck, R. Recombinant Vgr-1/BMP6-expression tumors induce fibrosis and endochondral bone formation in vivo. *J. Cell Biol.*, 1994, 126(6), 1595-1609.

[9] Katagiri, T; Yamaguchi, A; Komaki, M; et al. Bone morphogenetc protein-2 convers the differentiation pathway of C2C12 myoblasts into the osteoblast lineage. *J. Cell Biol.*, 1994, 127, 1755-1766.

[10] Rosen, V; Nove, J; Song, JJ; Thies, RS; Cox, K; Wozney, JM. Responsiveness of clonal limb bud cell lines to bone morphogenetic protein 2 reveals a sequential relationship between cartilage and bone cell phenotypes. *J. Bone Miner Res.*, 1994, 9(11), 1759-1768.

[11] Yamaguchi, A; Katagiri, T; Ikeda, T; et al. Recombinant human bone morphogenetic protein-2 stimulates osteoblastic maturation and inhibits myogenic differentiation in vitro. *J. Cell Biol.*, 1991, 113(3), 681-689.

[12] Gerhart, TN; Kirker-Head, CA; Kriz, MJ; et al. Healing segmental femoral defects in sheep using recombinant human bone morphogenetic protein. *Clin. Orthop.*, 1993, 293, 317-326.

[13] Pittenger, MF; Mackay, AM; Beck, SC. Multilineage potential of adult human mesenchymal stem cells. *Sci*, 1999, 284, 143-147.

[14] Goshima, J; Goldberg, VM; Caplan, AI. The osteogenic potential of culture-expanded rat marrow mesenchymal cells assayed in vivo in calcium phosphate ceramic blocks. *Clin Orthop*, 1991, 262, 298-311.

[15] Goshima, J; Goldberg, VM; Caplan, AI. The origin of bone formed in composite grafts of porous calcium phosphate ceramic loaded with marrow cells. *Clin Orthop*, 1991, 269, 274-283.

[16] Mardon, HJ; Bee, J; von der Mark, K. Development of osteogenic tissue in diffusion chambers from early precursor cells in bone marrow of adult rats. *Cell Tissue Res*, 1987, 250, 157-165.

[17] Owen, ME; Friedenstein, AJ. in Cell and Molecular Biology of Vertebrate Hard Tissues, Ciba Foundation Symposium (D. Evered, & S. Harnett, Eds.), 1988, 42-60, John Wiley and Sons, Chichester, U.K.

[18] Ohgushi, H; Goldberg, VM; Caplan, AI. Heterotopic osteogenesis in porous ceramics induced by marrow cells. *J Orthop Res*, 1989, 74, 568-578.

[19] Ohgushi, H; Goldberg, VM; Caplan, AI. Repair of bone defects with marrow cells and porous ceramic. Experiments in rats. *Acta Orthop Scand*, 1989, 60, 334-339.

[20] Lou, J; Xu, F; Merkel, K. Gene therapy: adenovirus-mediated human bone morphogenetic protein-2 gene transfer induces mesenchymal progenitor cell proliferation and differentiation in vitro and bone formation in vivo. *J Ortho Res*, 1999, 17, 43-50.

[21] Lou, J; Manske, PR; Aoki, M. Adenovirus-mediated gene transfer into tendon and tendon sheath. *J Ortho Res*, 1996, 14, 513-517.

[22] Hogan, BL; Hogan, BL. Bone morphogenetic proteins: multifunctional regulators of vertebrate development *Genes Dev*, 1996, 10, 1580-1594.

[23] Reddi, AH. Bone and cartilage differentiation. *Curr Opinion Gen Dev*, 1994, 4, 737-744.

[24] Sophia Chia Ning Chang*, Huo Li Chuang, Yu Ray Chen, Jan Kan Chen, Hui-Ying Chung, Yi-Lung Liu, Hui-Yi Lin, Ching-Lung Tai and Jueren Lou. Ex Vivo Gene Therapy in Autologous Bone Marrow Stromal Stem Cells for Tissue-engineered Maxillofacial Bone Regeneration. *Gene Therapy*, 2003, 10(24), 2013-2019.

[25] Horwitz, MS; Thfariello, J; Grunhaus, A; Fejer, G. Model systems for studying the effects of adenovirus E3 genes on virulence in vivo. *Curr. Topics Microbiol. Immunol.*, 1995, 199, 195-211.

[26] Sophia Chia Ning Chang*, Fu-Chan Wei, Huoli Chuang, Yu Ray Chen, Jan Kan Chen, Kuei Ching, Lee, Philip K.T. Chen, Ching-Lung Tai, Jueren Lou. Ex Vivo Gene Therapy in Autologous Critical Size Craniofacial Bone Regeneration. *Plastic and Reconstructive Surgery*, 2003, 112(7), 1841-1850.

[27] Cohen, J. (1977). *Statistical power analysis for the behavioral sciences* (revised edition). New York: Academic Press., 284-288

[28] Yasko, AW; Lane, JM; Fellinger, EJ; Rosen, V; Wozney, JM; Wang, EA. The healing of segmental bone defects, induced by recombinant human bone morphogenetic protein (rhBMP-2). A radiographic, histological, and biomechanical study in rats. *J. Bone Joint Surg.*, 1992, 74, 659-670.

[29] Gerhart, TN; Kirker-Head, CA; Kriz, MJ. Healing segmental femoral defects in sheep using recombinant human bone morphogenetic protein. *Clin Orthop*, 1993, 293, 317-326.

[30] Rosen, V; Nove, J; Song, JJ. Responsiveness of clonal limb bud cell lines to bone morphogenetic protein 2 reveals a sequential relationship between cartilage and bone cell phenotypes. *Bone Miner Res*, 1994, 9, 1759-1768.

[31] Muschler, GF; Hyodo, A; Manning, T. Evaluation of human bone morphogenetic protein 2 in a canine spinal fusion model. *Clin Orthop*, 1994, 308, 229-240.

[32] Toriumi, DM; Kotler, HS; Luxenberg, DP. Mandibular reconstruction with a recombinant bone-inducing factor: functional, histologic, and biomechanical evaluation. *Acr Otolaryngo. Head Neck Sur.*, 1991, 117, 1101-1112.

[33] Cook, SD; Dalton, JE; Tan, EH. In vivo evaluation of recombinant human osteogenic protein implants as a bone graft substitute for spinal fusions. *Spine*, 1994, 19, 1655-1663.

[34] Wang, EA; Rosen, VD; Alessandro, JS. Recombinant human bone morphogenetic protein induces bone formation. *Proc Natl Acad Sci*, USA, 1990, 87, 2220-2224.

[35] Nakahara, H; Bruder, SP; Goldberg, VM. In vivo osteochondrogenic potential of cultured cells derived from the periosteum. *Clin Ortho Res*, 1990, 259, 223-232.

[36] Isogai, N; Landis, WJ; Mori, R. Experimental Use of Fibrin Glue to Induce Site-Directed Osteogenesis from Cultured Periosteal Cells. *Plast Reconstr Surg*, 2000, 105, 953-963.

[37] Breitbart, AS; Grande, DA; Kessler, R. Tissue engineered bone repair of calvarial defects using cultured periosteal cells. *Plast Reconstr Surg*, 1998, 101, 567-574, discussion 575-576.

[38] Vacanti, CA; Kim, W; Upton, J. The efficacy of Periosteal Cells Compared to Chondrocytes in the Tissue Engineered Repair of Bone Defects. *Tissue Eng*., 1995, 1, 301-389.

[39] Lazarus, HM; Haynesworth, SE; Gerson, SL. Ex vivo expansion and subsequent infusion of human bone marrow-derived stromal progenitor cells (mesenchymal progenitor cells): implications for therapeutic use. *Bone Marrow Transplant*, 1995, 16, 557-564.

[40] Beresford, JN. Osteogenic stem cells and the stromal system of bone and marrow. *Clin Orthop*, 1989, 240, 270-280.

[41] Caplan, Al. Biological Mechanisms of Tooth Movement and Craniofacial Adaptation: Proceedings of the International Conference Held at the Great Southern Hotel, Columbus, Ohio, May 8-11, 1991. Davidovitch Z. 1992, 433-438.

[42] Haynesworth, SE; Goshima, J; Goldberg, VM. Characterization of cells with osteogenic potential from human marrow. *Bone*, 1992, 13, 81-88.

[43] Mason, JM; Grande, DA; Barcia, M. Expression of human bone morphogenic protein 7 in primary rabbit periosteal cells: potential utility in gene therapy for osteochondral repair. *Gene Ther*, 1998, 5, 1098-1104.

[44] Mason, JM; Breitbart, AS; Barcia, M. Cartilage and bone regeneration using gene-enhanced tissue engineering. *Clin Orthop*, 2000, 1(Suppl. 379), S171-178.

[45] Lieberman, JR; Le, LQ; Wu, L; Finerman, GA; Berk, A; Witte, ON; Stevenson, S. Regional gene therapy with a BMP-2-producing murine stromal cell line induces heterotopic and orthotopic bone formation in rodents. *J Orthop Res*, 1998, 16, 330-9.

[46] Oakes, DA; Lieberman, JR. Osteoinductive informed consent when patients appear to lack capacity and surrogates. *Clin Orthop*, 2000, 1(Suppl. 379), S101-S112.

[47] Crystal, RG. Transfer of genes to humans: early lessons and obstacles to success. *Sci*, 1995, 270, 404-410.

[48] Roman, M; Axelrod, JH; Dai, Y. Circulating human or canine factor IX from retrovirally transduced primary myoblasts and established myoblast cell lines grafted in murine skeletal muscle. *Somat Cell Mol Gen*, 1992, 18, 247-258.

[49] Resenfeld, MA; Yoshimura, K; Trapnel, BC. In vivo transfer of the human cystic fibrosis transmembrane conductance gene to the airway epithelium. *Cell*, 1992, 68, 143-155.

[50] Wilson, JM. Adenoviruses as gene-delivery vehicles. *N Engl J Med*, 1996, 334, 1185-1187.

[51] Shang, Q; Wang, Z; Liu, Wei; Shi, Y; Cui, L; Cao, Y. Tissue-engineered bone repair of sheep cranial defects with autologous bone marrow stromal cells. *J. Cranio. Surg*., 2001, 12(6), 586-593,

[52] Hopper, RA; Zhang, JR; Fournasier, VL; Protzner, IM; Protzner, KF; Pang, CY. Effect of isolation of periosteum and dura on the healing of rabbit calvarial inlay bone grafts. *Plast. Reconst. Surg*., 2001, 107(2), 454-462,

[53] Lo, LJ; Marsh, JL; Vannier, MW. Craniofacial computer-assisted surgical planning and simulation. *Clin Plast Surg*, 1994, 21, 501-516.

[54] Robb, RA; Hanson, DP. A software system for interactive and quantitative visualization of multidimensional biomedical images. *Australas Phys Eng Sci Med*, 1991, 14, 9-30.

[55] Heise, C; Sampson-Johannes, A; Williams, A; McCormick, F; Von Hoff, DD; Kirn, DH. ONYX-015, an E1B gene-attenuated adenovirus, causes tumor-specific cytolysis and antitumoral efficacy that can be augmented by standard chemotherapeutic agents. *Nat Med*, 1997, 3, 639-645.

[56] Tai, CL; Shih, CH. Mechanical testing of polymethyl methacrylate bone cement impregnated with bone and hydroxyapatite. *Chin. J. Med. Biol. Eng*., 1997, 17, 1-8.

Chapter Sources

Chapter 12 was also published in "Gene Therapy and Cancer Research Progress", edited by Jessica L. Lewis, Nova Science Publishers. It was submitted for appropriate modificationc in an effort to encourage wider dissemination of research.

Index

A

B

C

D

E

F

G

H

I

J

K

L

M

N

O

P

S

T

U